Clean Energy Production Technologies

Series Editors
Neha Srivastava, Department of Chemical Engineering and Technology
IIT (BHU), Varanasi, Uttar Pradesh, India

P. K. Mishra, Department of Chemical Engineering and Technology
IIT (BHU), Varanasi, Uttar Pradesh, India

The consumption of fossil fuels has been continuously increasing around the globe and simultaneously becoming the primary cause of global warming as well as environmental pollution. Due to limited life span of fossil fuels and limited alternate energy options, energy crises is important concern faced by the world. Amidst these complex environmental and economic scenarios, renewable energy alternates such as biodiesel, hydrogen, wind, solar and bioenergy sources, which can produce energy with zero carbon residue are emerging as excellent clean energy source. For maximizing the efficiency and productivity of clean fuels via green & renewable methods, it's crucial to understand the configuration, sustainability and techno-economic feasibility of these promising energy alternates. The book series presents a comprehensive coverage combining the domains of exploring clean sources of energy and ensuring its production in an economical as well as ecologically feasible fashion. Series involves renowned experts and academicians as volume-editors and authors, from all the regions of the world. Series brings forth latest research, approaches and perspectives on clean energy production from both developed and developing parts of world under one umbrella. It is curated and developed by authoritative institutions and experts to serves global readership on this theme.

More information about this series at http://www.springer.com/series/16486

Manish Srivastava • Neha Srivastava
P. K. Mishra • Vijai Kumar Gupta
Editors

Nanomaterials in Biofuels Research

Springer

Editors
Manish Srivastava
Department of Physics and Astrophysics
University of Delhi
Delhi, India

Department of Chemical Engineering
and Technology
IIT (BHU)
Varanasi, Uttar Pradesh, India

P. K. Mishra
Department of Chemical Engineering
and Technology
IIT (BHU)
Varanasi, Uttar Pradesh, India

Neha Srivastava
Department of Chemical Engineering
and Technology
IIT (BHU)
Varanasi, Uttar Pradesh, India

Vijai Kumar Gupta
Department of Chemistry and
Biotechnology
Tallinn University of Technology
Tallinn, Estonia

ERA Chair VALORTECH
Estonian University of Life Sciences
Tartu, Estonia

ISSN 2662-6861 ISSN 2662-687X (electronic)
Clean Energy Production Technologies
ISBN 978-981-13-9332-7 ISBN 978-981-13-9333-4 (eBook)
https://doi.org/10.1007/978-981-13-9333-4

This Springer imprint is published by the registered company Springer Nature Singapore Pte Ltd.
The registered company address is: 152 Beach Road, #21-01/04 Gateway East, Singapore 189721, Singapore

Foreword

Renewable energy is a potential and bright replacement of fossil fuels in the form of bioenergy. Bioenergy options, like biohydrogen, biogas, biomethane, bioethanol, biomethanol, biobutenol, algal biofuels, and biodiesel, are the promising alternative source of renewable energy which can contribute towards green and sustainable environment by vomiting pollution. Numbers of tremendous efforts have been taken worldwide for the development of innovative technologies which can evenly distribute bioenergy options. Nevertheless, in spite of continuous practice towards the development of low-cost technology for easy commercialization of biofuels, practical viability is still far away and unequally distributed. The application of nanotechnology is emerging as a new area in the field of biofuels production via various means such as catalyst, enzyme, and microbial immobilizer. Nanomaterials have enough potential for commercial exploitation, and their industrial market is expected to become more broad in the coming decades for this area. There is a need to merge and explore the potential of the combination of biofuels and nanotechnology area for viable and sustainable economic biofuels production.

Based on the current scenario, publication of the book *Nanomaterials in Biofuels Research* is much required effort in this series, and I am happy to write this positive message with great satisfaction. The book contains 11 essential chapters focusing on nanomaterial synthesis and characterization for biofuels application. Current global scenario of biofuels and role of nanotechnology in biofuels production, different immobilization patterns to improve biofuels production, and synthesis and mechanism of nanomaterial for economic biofuels production with the green approach are presented and discussed in details. The book suggests a new insight with advancement and sustainable solution to overcome existing loopholes in nanotechnology-based biofuels production process. The book predicated that nanotechnology will play a key role in the near future in the area of the biofuels production process, and hence, this book will serve as an asset for the researcher working in the relevant area including scientists, researchers, teachers, and students.

I am taking the opportunity to congratulate Dr. Manish Srivastava (DU, Delhi), Dr. Neha Srivastava (IIT-BHU, Varanasi), Prof. Dr. P.K. Mishra (IIT-BHU, Varanasi), and Dr. Vijai Kumar Gupta (TTU) for their effort to bring out the publication of *Nanomaterials in Biofuels Research* to cater the long-felt need of scientists, teachers, researchers, and students. My sincere thanks are for the editors for their untiring efforts, devotion, and dedication in this endeavor. All the authors and editors of this book deserve sincere appreciation for their commendable achievements.

Enhanced Composites and Structures Center, Dr. Vijay Kumar Thakur
School of Aerospace, Transport and Manufacturing,
Cranfield University,
Bedfordshire, UK
21 March, 2019

Acknowledgments

The editors are thankful to all the academicians and scientists whose contributions have enriched this volume. We also express our deep sense of gratitude to our parents whose blessings have always prompted us to pursue academic activities deeply. It is quite possible that in a work of this nature, some mistakes might have crept in text inadvertently, and for these, we owe undiluted responsibility. We are grateful to all the authors for their contribution to the present book. We are also thankful to Springer Nature for giving this opportunity to the editors and the Department of Chemical Engineering and Technology, IIT (BHU), Varanasi, UP, India, for all their technical support. We thank them from the core of our heart. Editor Manish Srivastava acknowledges the DST, Govt of India for awarding the DST-INSPIRE Faculty Award [IFA13-MS-02] 2014 and also Science and Engineering Research Board for SERB-Research Scientist Award-2019.

Contents

About the Editors

Dr. Manish Srivastava has worked as DST INSPIRE faculty in the Department of Physics and Astrophysics, University of Delhi, India during June 2014 to June 2019. Currently he is working as SERB-Research Scientist in the Department of Chemical Engineering and Technology IIT (BHU), Varanasi, India. He has published 45 research articles in peer-reviewed journals, authored several book chapters, and filed one patent. He worked as a postdoctorate fellow in the Department of BIN Fusion Technology, Chonbuk National University, from August 2012 to August 2013. He was an assistant professor in the Department of Physics, DIT School of Engineering, Greater Noida, from July 2011 to July 2012. He received his PhD in Physics from the Motilal Nehru National Institute of Technology, Allahabad, India, in 2011. Presently, he is working on the synthesis of graphene-based metal oxide hybrids and their applications as catalysts. His areas of interest are synthesis of nanostsructured materials and their applications as catalyst for the development of electrode materials in energy storage, biosensors, and biofuels production. His *fields of expertise* are synthesis of nanomaterials and their application as catalyst for the development of electrode materials in energy storage, biosensors, and biofuels productions.

Dr. Neha Srivastava is currently working as postdoctorate fellow in the Department of Chemical Engineering and Technology, IIT (BHU), Varanasi, India. She has published 25 research articles in peer-reviewed journals with three patents and has one technology transfer. She has completed her PhD from the Department of Molecular and Cellular Engineering, SHIATS, India, in 2016 in the area of bioenergy and has received six Young Scientist Awards. Presently, she is working on biofuels production (cellulase enzymes, production and enhancement, biohydrogen production from waste biomass, bioethanol production). Her *fields of expertise* are biofuels production, microbial bioprocessing, and enzyme technologies.

Professor P. K. Mishra is currently professor and head in the Department of Chemical Engineering and Technology, Indian Institute of Technology, Banaras Hindu University, Varanasi, India, where he obtained his PhD degree in Chemical Engineering in 1995. He has authored and coauthored over 60 technical papers published in reputed national and international journals and supervised more than 20 doctoral students. He has received several awards at national/international levels and has five patents with one technology transfer. He has also made significant contribution towards the development of entrepreneurship ecosystem in Eastern part of the country. He is fellow of the Institution of Engineers, India; coordinator of Technology Business Incubator at the Indian Institute of Technology, Banaras Hindu University; and member of the executive committee of NIESBUD, Ministry of Skill Development, Government of India. His *fields of expertise* are biofuels production, microbial bioprocessing, and enzyme technologies.

Dr. Vijai Kumar Gupta, ERA chair of Green Chemistry, Department of Chemistry and Biotechnology, School of Science, Tallinn University of Technology, Tallinn, Estonia, is one of the leading experts in the area of microbial biology and biotechnology. He is member of the International Sub-commission on Trichoderma and Hypocrea, Austria, and International Society for Fungal Conservation, UK, and secretary of the European Mycological Association. He is the fellow of the prestigious The

Linnean Society of London, UK, of Indian Mycological Association, and of Mycological Society of India. He has been honored with several awards in his career, including Indian Young Scientist Award, for his advanced research achievements in the field of fungal biology and biotechnology. He is the editor of few leading scientific journals of high repute, having many publications in his hands with h-index 21. He has edited many books for publishers of international renown, such as CRC Press, Taylor and Francis, USA; Springer, USA; Elsevier Press, the Netherlands; Nova Science Publisher, USA; Walter de Gruyter, Germany; and CABI, UK.

Biofuel: Types and Process Overview 1

Ajay Kumar Chauhan

Abstract

Excessive use of conventional fossil fuels resulted in a hike in price, exhaustion, and change in climatic conditions. Therefore, a novel route to biofuel generation is another feasible option of sustainable process development. Continuous upgradation of technologies for biofuel generations from the first generation (1G) to the fourth generation (4G) gives new hopes to fulfill energy demands. Biofuel generation from multiple approaches such as physical, biological (includes microbial and enzymatic), chemical, and biochemical catalysis with nanotechnology from multiple feedstocks is the key to biofuel generation. Suitable conversion of cellulosic biorefinery and lignin biorefinery (via lignin valorization) is the key for complete utilization of lignocellulosic biomass. Utilization of the biological methods includes the use of microbial machinery from different domains which might open the door toward an environmentally benign process. Nanotechnology and its potential application with 1G to 4G have future promises for increasing yield and integration of technology. So, this book chapter covers a detailed process overview of biofuel generations and its challenges with the hope of overcoming it.

Keywords

Sustainable · Microbial machinery · Catalysis · 1G · 4G · Lignin valorization · Nanotechnology

A. K. Chauhan (✉)
Department of Biotechnology, Indian Institute of Technology, Roorkee, Uttarakhand, India
e-mail: achauhan@bt.iitr.ac.in

M. Srivastava et al. (eds.), *Nanomaterials in Biofuels Research*,
Clean Energy Production Technologies, https://doi.org/10.1007/978-981-13-9333-4_1

Abbreviations

1G	First generation
2G	Second generation
3G	Third generation
4G	Fourth generation
BGA	Blue-green algae
C	Carbon
CBDH	Cellobiose dehydrogenase
Centimeter	cm
CH4	Methane
CNG	Compressed natural gases
CO	Carbon monoxide
CO_2	Carbon dioxide
DME	Dimethyl ether
DNA	Deoxyribonucleic acid
ED	Entner-Doudorof
EMP	Embden-Meyerhof-Parnas
ETBE	Ethyl butyl ether
FAME	Fatty acid methyl esters
FT	Fischer-Tropsch
GHG	Greenhouse gases
H_2	Hydrogen
HC	Hydrocarbons
HDO	Hydrodeoxygenation
Lignin peroxidase	LiP
LPG	Liquefied petroleum gas
LPMO	Lytic polysaccharide monoxygenase
Manganese peroxidase	MnP
Mn	Manganese
MPa	Mega Pascal
N_2	Nitrogen
NO_x	Nitrogen oxide
NPs	Nanoparticles
PBR	Photobioreactor
PEG	Polyethylene glycol
PM	Particulate matter
SNG	Syngas

1.1 Introduction

Increase in population and economic growth has increased the energy demand, and fossil fuels fulfill this energy demand as per the data revealed from the International Energy Agency (IEA) in 2011. As the consumption of fossil fuels is nonrenewable and emits greenhouse gases (GHG) which severely deplete environmental conditions (Buragohain et al. 2010), therefore, the notable increase in renewable energy sources such as solar, wind, and biomass gives new hope of a sustainable process. Biofuels are the renewable energy sources defined as the energy derived from biological carbon fixation, which plays a pivotal role in new energy sources. Therefore, the production of biofuel requires extensive research and development of technologies. These biomass require suitable conversion of biomass into biobutanol, bioethanol, biomethanol, bio-dimethyl ether, biomethane, Fischer-Tropsch (FT) fuel, biohydrogen, algae, and halophilic lipid-based biofuels.

Plant biomass is an abundant source of renewable energy which comprises of the various components such as carbon, hydrogen, oxygen, and traces of minerals. Availability of plant biomass is more reliable compared to another energy sources such as solar energy, wind energy, and hydropower, and it can easily grow in variable environmental conditions. Moreover, the low density of biomass over fossil fuel makes it more feasible for storing and transportation purposes. In developing countries, utilization of plant biomass for household purposes (burning of wood) underestimates its value as a potential biofuel. The burning of fossil fuels and plant biomass is the main contributor to the increase in CO_2 level and has a direct impact on global warming. GHG emissions and global warming open up the door for the novel processes of biofuel generation.

Lignocellulosic biomass is made up of highly complex biopolymer of cellulose (40–60%), hemicellulose (10–40%), and lignin (15–30%) as major component which gives support to plant, and it is resistant to various microbial degradations as well as biochemical conversions (Himmel et al. 2007; DeMartini et al. 2013; Schutyser et al. 2018). Cellulose and hemicellulose mainly consist monomeric sugar unit, and after pre-treatment, lignocellulosic biomass sugar component via further saccharification is converted into suitable biofuel (Chen 2014). However, the lignin component is made up of the phenylpropane unit, which is more recalcitrant than the cellulose and hemicellulose. During kraft pulping process, lignin is utilized for the co-power generation, which can be separated, and suitably valorized into useful precursors for biofuel generation. However, the hydrogenation process is being used as a suitable method for converting lignin into a valuable product and converting raw material into biofuel (Chen 2014; Abdullah et al. 2017). Lignin valorization strategy utilizes the complete conversion of lignocellulosic biomass into biofuel generation.

Different methodologies are adapted from time to time for treating the plant biomass for the biorefinery route. This methodology has its advantages and limitations. Here, pulp and paper industries are the best examples of lignocellulosic biomass biorefinery, and they are emerging as fermentative production of bioethanol (Ragnar et al. 2000; Hahn-Hägerdal et al. 2006; Limayem and Ricke 2012). Various existing conventional techniques result in low yield of primary products obtained from the lignocellulosic biomass, but in case of lignin, it can be suitably valorized into valuable products (Davis et al. 2013). Further, the optimization process is done by understanding the process. Even biofuels productions have a direct impact on carbon sink; therefore, biofuels produced from oil-based fuels always have a better choice compared to others GHG emissions into the environment critics of biofuels, which ultimately depends on the route of production. Another selection of biofuel has the debate over food versus fuel; land used for cultivation for feedstock always competes with food demand and land uses for productions. Therefore, the governments of the countries take so many initiatives for the continuous growth of the biofuel industry, which includes providing fund for research and development along with mandating laws of Environmental Protection Agency (EPA) for blending of biofuel with conventional fuel. Renewable Fuel Standard (RFS) provides harness power of biofuels and infrastructure, and high productions of biofuels must be produced and widely available for customers with the competitive price of conventional biofuel (Richards 2013).

1.2 Classification of Biofuels

On the basis of origin of raw material biofuels are classified into first-generation (1G), second-generation (2G), third-generation (3G), and fourth-generation (4G) (Demirbas 2011). The first-generation biofuel, sugar, starch, vegetable oils, and fats are converted into bio-alcohol from (n = 1–4) fatty acid methyl esters (FAME). Second-generation biofuel depends on the carbon negative in terms of the carbon dioxide concentration in the environment, which majorly depends on the lignocellulosic biomass (plant material) (Gomez et al. 2008). The third and fourth generation of the biofuel utilize the algae and the blue-green algae (BGA) machinery for converting lipid into biofuel. In third generation, algae is directly used for the production of the biofuel, where in the fourth generation, metabolic engineered algae from the oxygenic photosynthetic microorganism create artificial carbon reservoir (Lü et al. 2011). Further, they are known as chemically synthesized biofuel and biologically synthesized biofuel. In chemical synthesis, biofuel is the generation with the help of a catalyst, and the action of various parameters such as pH, temperature, and pressure has played a significant role. Moreover, chemical methods, not benign environmental methods, in terms of the yield are preferable over biological methods. In the chemical process, the use of the catalyst is preferred over the various physical processes because of the more specific product formation and low cost of separation. Generally, physical processes yield more complex product formations, which are not easily separable and thus require a higher price for downstream processing. In biological methods, lignocellulosic biomass is microbially treated with bacteria, halophilic archaea, fungi, and algae. Moreover, other than these biological processes include various kinds of biocatalysts known as the enzymes used for the production of the biofuels' generation. These enzymes are cellulases, xylanases, and lignolytic enzymes such as lignin peroxidase (LiP), manganese peroxidase (MnP), laccase, lytic polysaccharide monooxygenase (LPMO), multicopper oxidases, cellobiose dehydrogenase (CBDH), and lipases. Moreover, enzymatic use of the various processes of biofuel generations is the most environmentally friendly process which overcomes the numerous barriers of the chemical processes. Benefits of the enzymatic process yield specific product formation at optimum pH, temperature, and pressure. Therefore, enzyme from extremophile enzymes have added extra-advantages of various product formations at broad pH range and different temperature ranges from 4 to 80 °C, and they even act on higher pressure making them suitable for industrial process. To date, all the biological processes require a higher cost of production than the chemical process, and the cost-effectiveness of the biological process can only be reduced by the processes understanding of the microbial pathways, enzymatic reduction, and its substrate utilization strategy. However, new biological modified strains have shown various promises to overcome all barriers. In the last decade, there had been four industrial plants that started for the commercial production of the biofuel listed as "Project Liberty" by the joint venture of POET-DSM, 2G ethanol at Dupont, Abengoa Bioenergy Hybrid Kansas, and Crescentino by Beta Renewables in Europe (Valdivia et al. 2016).

Raw Materials

Availability of raw material is the key to success to achieve better economy and viability of the process. Biodiversity available on the earth is a good sink of energy as in the form of hydrocarbon and the production of hydrocarbon for biofuel generations are listed in the Table 1.1. As per requirement, feedstock can be classified on the basis of the source of origin. Feedstock can suitably be converted into biofuel and the abundance of this material in various forms are the source for different biofuel generations.

Table 1.1 Classification of biomass feedstock (Speight 2011; Sikarwar et al. 2017)

Biomass type	Examples
Forest product	Wood, logging, residue, trees, shrubs, and wood residues, sawdust, bark
Bio-renewable waste	Agricultural waste, crop residues, mill wood waste, urban wood wastes, urban organic waste
Energy crops	Short-rotation woody crops, sugar crops, forage crops, oilseed crops, switchgrass, *Miscanthus*
Aquatic plants	Algae, water weed, water hyacinth, reed, and rushes
Food crops	Grains, oil crops
Sugar crops	Sugar cane, sugar beets, molasses, *Sorghum*
Landfill	Hazardous waste, nonhazardous waste, inert waste, liquid waste
Organic wastes	Municipal solid waste (MSW), industrial organic
Algae	Prokaryotic algae, eukaryotic algae, kelps
Mosses	Bryophyta, Polytrichales
Lichen	Crustose lichens, foliose lichens, fruticose lichens

1.3 First-Generation Biorefinery

Concept of the biorefinery comes in the limelight for the researcher's interest due to the limited resources of conventional energy sources and the rising price of the oil in the global market. 1G is produced from the food crops such as corn, sugar cane, rapeseed, cassava, etc. The first-generation biofuels are biodiesel, ethanol, and biogas, which are primarily producing so far. Biodiesel is the alternative biofuel to the diesel which is produced by the transesterification of the various vegetable oils and fats. However, small modification in the properties of biodiesel can make it a better substitute for the diesel. Whereas, bioethanol is the better substitute for the gasoline and is also known as the flexi fuel. Moreover, bioethanol is produced from various sources via fermentation of the sugar or starch material. Bioethanol is used as a feedstock for ethyl butyl ether (ETBE), which can easily blend with the gasoline. Another biofuel like biogas and biomethane is produced from the anaerobic digestion of the household and municipal waste. Biodiesel, bioethanol, and biogas can easily be produced from the food crop feedstock, but the increasing demand of the edible oils makes it difficult to use food crop for the biorefinery (Lee et al. 2014b). The big challenge of the production of the first-generation biofuel is the food to feedstock, high uses of the groundwater, fertilizer, and pesticide use on the crop.

1.3.1 Transesterification

The oil obtained from the various feedstock suitably transesterified into the fatty acid methyl esters (FAME) with the help of alcohol. The alcoholic group readily reacts with the triglyceride units in the presence of a suitable catalyst (homogenous or heterogeneous) with oil group and releases biodiesel and glycerol as high-value coproduct (Kulkarni et al. 2006; Meher et al. 2006). This mechanism can suitably be applied to all kinds of oil obtained from using algae, halophilic isolate, animal fat, and vegetable oil. The transesterification reaction is an equilibrium reaction in which continuous mixing is required for getting methyl esters (biodiesel). However, biodiesel obtained from transesterification reactions (given below) has similar characteristics as compared to diesel.

1.3.2 Ethanol Production

Ethanol is produced from sugar, starch, and cellulose-containing crops. Alcohol produced from the food crops and lignocellulosic biomass is known as the bioethanol, which can be easily produced by the biochemical process (Minteer 2016). Structure of starch consists of the polymeric unit of Dglucopyranose, which is first converted into glucose molecule with the action of enzymes "amylase." Amylase produces the maltodextrin by liquefaction process. The process of saccharification of dextrin and oligosaccharides is hydrolyzed by the enzymes pullulanase and glucoamylase. The action of these enzyme converts the substrate into the smaller unit of glucose, maltose, and isomaltose. Then further, fermentation is carried by the yeast at 30 °C for ethanol production (Lee et al. 2014b).

1.3.3 Fermentation

The fermentation process converts organic compounds into simpler products with the action of the microbes and enzymes. Ethanol production from glucose utilizes fermentative consumption of pyruvate as an intermediate product (Ingram et al. 1987). The process can be classified as the aerobic and anaerobic among microbes depending up on oxygen requirement. Many microorganisms such as yeast (*Saccharomyces cerevisiae*), bacteria (*Zymomonas mobilis*), and halophilic bacteria (*Halanaerobium saccharolyticum*) convert butyrate and ethanol from glycerol and hydrogen from *senegalensis* (Kivistö et al. 2010). Most eukaryotes and prokaryotes follow Embden-Meyerhof-Parnas (EMP) also known as glycolysis pathway, and *Zymomonas mobilis* follows Entner-Doudorof (ED) pathway. *Z. mobilis* produces up to 97% of the theoretical yield of ethanol and also produces a 2.5-fold higher yield compared to *S. cerevisiae* (Weber et al. 2010). Moreover, *Z. mobilis* is suitably engineered for utilizing various types of sugar of lignocellulosic biomass such as glucose, mannose, and xylose for fermentation (Yanase et al. 2012). Another microorganism such as *Clostridia* can secrete hydrolytic complex to utilize both types of sugars (Lütke-Eversloh and Bahl 2011). However, these microorganisms are capable enough to ferment into C6 and C5 sugar or a mixture of both.

1.3.4 Anaerobic Fermentation

This process produces biogas under the absence of oxygen. However, anaerobic condition facilitates the growth of microorganism on organic biomass and converts it into methane (60–70%) and carbon dioxide (Naik et al. 2010). When we consider from the economic point, anaerobic digestion is an economically feasible process and also produces the biofertilizers for agricultural uses. The conventional process generally produces the gaseous fuel for the household purposes and electricity generation, but newer technology now focuses on the liquid fuel to overcome the separation process cost (Lee et al. 2014b). Utilizing simultaneous saccharification and fermentation (SSF) and anaerobic digestion in the case of birch wood by *Saccharomyces cerevisiae* efficiently converts sugar into ethanol, and stillage was subsequently utilized for the biomethane production (Kalyani et al. 2017).

1.3.5 Whole-Crop Utilization

In this concept, the whole crop utilization is applied for conversion of the whole plant into a useful product. Like *Jatropha*, oilseed contains 30–40% of oil. In the whole-crop utilization, different parts of the crop are treated separately. For example, oil produces the biodiesel, which can further be used for the conversion into oleochemicals. Remaining de-oiled cake is used for the gasification purposes, which can further convert into the syngas. Syngas finally is utilized for the production of various fuels.

1.4 Second-Generation Biofuels

Second-generation biofuels are produced from the nonfood crops such as agricultural residues (wheat straw, rice straw, and sugar cane bagasse, etc.) and organic waste from municipal solid waste. 2G biofuels have a key advantage over the first generation in terms of no competition with the food resources, their product formation with less emission of GHG, and less land requirement for the raw material. However, the mixed feedstock concept involved heterogeneous feed. When compared with 1G, the technological risk of operation is higher and also requires higher input cost for the process development. Process understanding in terms of the route selection requires higher technical knowledge and process modification (Margesin and Schinner 2001). Plant biomass consists of the plant cell, which comprised of the 75% of polysaccharide as a potential source of carbon (Pauly and Keegstra 2008). Lignocellulosic biomass is converted into the advanced biofuels through hydrolysis and fermentation (bioethanol and biobutanol) or gasification (FT biodiesel, dio-DME, and SNG). Raw materials for the advanced biofuels are rotational crops (poplar and *Eucalyptus*), *Miscanthus*, switchgrasses, and agricultural waste residues (Naik et al. 2010). Advanced bioethanol can be a suitable alternative of gasoline, and the sugars obtained from the hydrolysis are converted into Fischer-Tropsch diesel or biomass to liquids for the full substitution of the conventional diesel. In this, the lignocellulosic biomass with the help of suitable gasification is converted into syngas. Further, these syngas is utilized as the source of liquid hydrocarbon (kerosene, diesel, or bio-SNG), or this syngas can have transformed into the methane or dimethyl ether (DME) (Naik et al. 2010). 2G biofuels involve two main route strategies. First one is the thermochemical conversion route in which biomass conversion utilizes the thermal decay and chemical reformation in the presence of different concentrations of the oxygen. In the second one, biochemical conversion utilizes the conversion with the help of microbes using catalysts/biocatalysts. The major advantage of the thermochemical conversion over the biochemical is the conversion of the whole organic component into biofuel, but in the case of the biochemical conversion, it is only the polysaccharide that is converted into biofuel (Gomez et al. 2008).

1.4.1 Physical Process

1.4.1.1 Mechanical Extraction

In the mechanical extraction processes, crude oil from the oilseeds is recovered from the action of an external mechanical pressure generated by screw press. However, a small amount of the de-oiled cake (around 15–20% of oil) from the oilseed also comes out with the recovered oil. In some mechanical extraction processes, pre-pressing is followed by the solvent extraction for the high oil recovery around 30–40% (Stevens and Verhé 2004). On the other hand, in the full pressing system, high pressure of 9.5 MPa is applied to take out as possible oil content.

1.4.1.2 Briquetting

Various biomass resources obtained from agricultural, forest residues and other waste biomass come under second-generation fuel and are difficult to use as a direct biofuel due to uneven bulk. Therefore, this problem is overcome by the method of densification of the material to give near uniform compact shape. These densification methods are employed as pressing and maceration or combinations of both methods. In press, the density of the material increases by applying mechanical pressure in the early stage of compression. Further, an increase in the pressure has a negative impact on the density of material until it reaches the density of water, whereas in maceration, no such correlation of pressure exists, but chopping, grinding, and pulverization of the large material (like branches of the tree) are more effective methods. These methods are more effective compared to ultrafine grind material (Stevens and Verhé 2004; Osamu and Carl 1989).

1.4.1.3 Distillation

In this process, raw materials are crushed followed by steam distillation. Further, collective vapors are condensed separately which then turn into liquid form. Advanced distillation process known as molecular distillation is applied where conventional methods cannot be applied for the productions of the fragrances (Stevens and Verhé 2004).

1.4.2 Thermochemical Conversion

Thermochemical conversion involves heating of biomass followed by chemical conversion into biofuel. This process includes combustion, gasification, liquefaction, and pyrolysis under oxygen-deficient or oxygen-controlled conditions and results in the generation of syngas majorly comprising of H_2 and CO. Further, this syngas is processed into gaseous or liquid products (Lee et al. 2014b).

1.4.2.1 Combustion

In the combustion process, the chemical reactions usually take place in the air where fuel and air react with each other. Therefore, the products formed during this process are carbon dioxide and water, and it liberates energy in the form of heat (Lee et al. 2014b). Combustion is the simplest way to use biomass into energy with the low yield of products.

1.4.2.2 Gasification

Gasification is a type of pyrolysis carried out at a very high temperature and produces a mixture of gases known as producer gas (CO, H_2, CH_4, CO_2, and N_2). In the case of biomass, gasification producers are methanol, ammonia, and ethylene. Gasification utilizes conversion of the lignocellulosic biomass into the biofuel since the past 30 years. Gasification has many processes which simultaneously go on, such as drying, pyrolysis, and partial oxidation (Bridgwater and Boocock 2013). The syngas produced from the gasification process can be through catalytic and non-catalytic processes. In the non-catalytic process, higher temperature ranges 1000–1300 °C for the syngas production; in the catalytic process, it is operated at around 900 °C due to more advances in catalysis (Lee et al. 2014b). The key in gasification is controlling H_2/CO ratio for getting the different product in higher concentration (Melin et al. 2015).

1.4.2.2.1 Biomethanol

The reaction of the biomass requires oxygen to produce producer gas, which is the mixture of various gases such as CO, CO_2, H_2, CH_4, and N_2 (known as syngas). These producer gases have the potential to produce stationary power generation. Gases are first cooled down to 300 °C. Further, gases are fed to shift reactor with the controlled condition of H_2/CO ratio provided to 2.1 for the methanol synthesis. Remaining acid gases are then removed by scrubber or dissolved in solvent (Vamvuka et al. 2004). Further, high pressure of 6.1 MPa is used to compress purified syngas to increase methanol synthesis:

$$CO + 2H_2 \leftrightarrow CH_3OH$$
$$CO_2 + 3H_2 \leftrightarrow CH_3OH + H_2O$$

1.4.2.2.2 Methane

All conditions for the preparation of the syngas follow the same procedure. Here, the use of catalyst which catalyzes tar with methane decomposition requires higher H_2/CO ratio of 3.1, which forms methane at 31 bar of pressure and temperature between 300–600°C, whereas the chemical reactions are:

$$CO + 3H_2 \leftrightarrow CH_4 + H_2O$$
$$CO_2 + 4H_2 \leftrightarrow CH_4 + 2H_2O$$

1.4.2.2.3 Bioethanol Production

Biomass is majorly used for the power steam generation. Wherein, SO_2 feed with the steam explosion around 205 °C for a short time of 5 min. Then, steam is recovered by forming the product. This product flash and vapor formed during flash is condensed to recover furfural. After this steam recovery, the liquid is recovering with the help of additional water use for the simultaneous fermentation. Saccharification to form the simplest sugar is done at low temperature up to 40 °C. Therefore, saccharification forms the simplest sugar of hexoses and pentoses, which further are converted into bioethanol via the fermentation process:

$$C_6H_{10}O_5 + H_2O \rightarrow C_6H_2O_6$$

$$C_5H_8O_4 + H_2O \rightarrow C_5H_{10}O$$

$$C_6H_{12}O_6 \rightarrow 2CO_2 + 2C_2H_5OH$$

$$2C_5H_{10}O_5 \rightarrow 3CO_2 + 3C_2H_5OH$$

Combine production of the methanol and ethanol is achieved by using the following strategy. First, hydrolyzed biomass of cellulose and hemicellulose is used for ethanol production. The remaining part such as lignin with the remaining residue of the hydrolyzed biomass is used for the methanol production (Melin et al. 2015). In other strategies, biomass is hydrolyzed and coupled with hydrogenation with the use of a catalyst such as lignin with the remaining residue of the hydrolyzed biomass used for the methanol production (Melin et al. 2015). In other strategies, biomass is hydrolyzed and coupled with hydrogenation with the use of a catalyst such as Ru at 160 °C and 50 bar pressure. The major advantage of this method is the stability of sugar which produces alcohols:

$$C_6H_{12}O_6 + H_2 \rightarrow C_6H_{14}O_6$$

1.4.3 Liquefaction

Biomass is subjected to the presence of alkali, glycerine, propanol, butanol, or direct liquefaction (Demirbas 2004). This technique product is water insoluble, with high viscosity, which requires solvents, carbon monoxide (reducing gases) or hydrogen, or catalyst to be present in biomass.

In thermochemical conversion, lignocellulosic biomass is directly converted into the liquid as heavy liquid oils.

1.4.4 Pyrolysis of Biomass

The basic thermochemical process converts biomass into hydrocarbon-rich gas mixture in the absence/controlled oxygen conditions (Demirbaş 1998). In this process, biomass is converted into the vital product known as bio-oil, and other liquid products like acetic acid, acetone, and methanol. Solid products like charcoal, and gaseous products like non-condensable gases are also formed on heating 477 °C in the presence of air. However, around 70% of bio-oil are produced on a mass basis, whereas this process economically viable does not require pre-treatment of the biomass (Alonso et al. 2010). Therefore, the corrosive nature of the bio-oil towards equipment, and poor thermal stability has to be overcome by upgrading oxygen content, whereas, hydrogenation leads alkali removal and oil require catalytic cracking (Demirbaş 1998). Pyrolysis of lignocellulosic biomass such as wood is a zonal process, in which thermal degradations of hemicellulose, cellulose, and lignin occur (Chum 1991). Moreover, pyrolysis comprises of five stages (Demirbaş 2000):

First stage: Removal of moisture with some volatile loss
Second stage: Degradation of hemicellulose and emissions of CO and CO_2
Third stage: Exothermic reaction and emission of CH_4, H_2, C_2H_6
Fourth stage: External energy supply
Fifth stage: Completion of the process

1.4.4.1 Fast Pyrolysis

This process is performed in the higher temperature range of 577–977 °C with fast heating rate achieved in short residence time of 0.5–10 s. In fast pyrolysis, biomass is converted into vapors, aerosols, and char. With further cooling and condensation of vapor, the aerosol's dark color liquid is formed, whereas in terms of yield, 60–75% bio-oil, 15–25% char, and 10–15% non-condensed gases are formed (Shafizadeh 1982).

1.4.4.2 Flash Pyrolysis

Flash pyrolysis occurs in higher temperature than fast pyrolysis of 777–1027 °C with the fast heating rate of more than 723 °C/s (<0.5 s short retention time) and very fine particle size of <0.2 mm. Bio-oil is formed from flash pyrolysis, is mixed with char, and makes a mixture of bioslurry. Further, this bioslurry is fed to the gasifier at 26 bar pressure with 654–954 °C to convert into syngas with the conversion efficiency of 70%. This bio-oil formed from flash pyrolysis is used for the engines and turbines for the power generation (Demirbas 2004).

Biomass-derived syngas is used for different biofuel productions contaminated with various particulates, tars, nitrogen, sulfur contaminants, and halides with trace elements. These contaminants are removed before entering into the catalytic reactor system (Sikarwar et al. 2017).

Production of the 2G biofuel not an issue for the feed to food compared with 1G biofuel, but fiber crops used as raw material, compete with the food crops. Moreover, the pesticides and fertilizer use also the primary concern for the production of the second-generation biofuel.

1.5 Third-Generation Biofuels

In the early stage of development, algae are used for the generation of the 3G biofuel. Algae include the microalgae, macroalgae from seaweed, and cyanobacteria, also known as blue-green algae (BGA). More than 800,000 algal species are exist on the earth, and algae can rapidly grow and harvested in a month, and minimal nutrient is required for their growth; even economic production in wastewater is an added advantage (Bowyer et al. 2018). The potential of oil production from algae is 100 times greater than in area basis than the soya bean and canola oils (Hu et al. 2008). For the better growth of the algae, the optimal temperature range is between 20 and 30 °C and slow growth observed is lower than 16 °C, whereas above 35 °C is deleterious for the algae growth (Menetrez 2012). Microalgae produce lipid

Table 1.2 Various lipid extraction methods with their advantages and disadvantages Halim et al. (2011), Harun et al. (2010), Wang et al. (2015), Kanda et al. (2013) and Mubarak et al. (2015)

Method	Adv Wang antages	Disadvantages
Oil expeller	Easy to operate No solvent required	Large biomass size required Slow operation Higher power utilization and scale up problem
Ultrasonication	Reduced extraction time Low solvent required Release of cell content higher in solvent	Filtration and centrifugation add extra cost for removal of solid residue Efficiency poor, target compound, volatile in nature
Microwave	Economical Environmental friendly Reduced solvent use, time Improved extraction yield	Extraction time long Large volume solvent required, which is toxic and highly flammable Solvent recovery is highly energy intensive High operational and infrastructure cost
Solvent extraction	Simple and economical viable High efficiency, better for small scale	
Supercritical CO_2	Reduced time Low toxicity of solvents Flexibility to change pressure and temperature and improves selectivity Favorable for mass transfer Solvent-free extract	
Wet extraction	Energy saving for drying biomass Reduced solvent uses	Low-quality lipid extracted

through photosynthesis, and lipid content reaches around 20–80% of dry weight depending on the species. However, changes in the nutritional requirement can change the product yield of the lipid, protein, and carbohydrate (Shen et al. 2009). After that, lipid extracted by different extraction processes is used for the production of the biodiesel, and the leftovers, carbohydrate and starches (residual), are further processed and can be used for the production of ethanol (Shen et al. 2009, Coppola et al. 2009). Lipid extraction process includes mechanical and chemical extractions (Halim et al. 2012). Table 1.2 summarizes the various method of extraction.

1.5.1 Open Pond

A natural water reservoirs like lake and lagoon and artificial reservoirs system like ponds, shallow ponds, and tanks are the open pond system. Advantage of the open system is that it is easy to construct and operate, but poor utilization of the light, evaporation loss, loss of the CO_2 to the environment, and chances of contamination are the significant key challenges of the open pond system. This system can be suitably installed with the wastewater treatment plant (Beal et al. 2012). But the major drawbacks of this system lacks control of various parameters.

1.5.2 Photobioreactor (PBRs)

The system is closed, therefore it provides a better internally maintained control system for the growth of pure algal culture as they maintain various parameters of inadequate amount such as CO_2, water, temperature, light intensity, mixing density, and pH of the system (Lee 1999; Mussgnug et al. 2007). In PBRs, algae are cultured in glass or plastic tube, where they are grown in light of the particular wavelength which supports the optimal growth of the algae. Approximately, 7.6–10 cm light penetrates inside the PBRs due to the turbidity of the algal growth into PBRs (Lee 1999; Mussgnug et al. 2007; Müller et al. 1993; Rapala and Sivonen 1998). For the easy operations of the PBRs, various conditions optimized like volume size to the area for the optimum light penetration, spatial distribution of light, CO_2 transfer rate, and monitoring of the growth inside the reactor (Menetrez 2012). Their installation and operational costs are higher, but product yield in terms of oil is obtained around 40–55% (Hu et al. 2008; Chisti 2007). Due to the less expensive operation of the open system, it can be suitably hybrid with the closed type of PBRs for optimizing cost and product yield (Menetrez 2012).

Algae are an excellent source to produce various biofuels such as biodiesel, ethanol, and petroleum. From various reports, productivity from the algae is 602 times higher than various crops like corn and switchgrasses (Menetrez 2012). The significant advantage of algae is higher productivity, and small land for the cultivation is required. Transesterification of oil recovered from algae produces biodiesel chemically and biologically. Moreover, biological production utilizes a lipase enzyme which converts lipids into biodiesel at room temperature. This operation at room temperature cuts the energy requirement for biodiesel production (Fig. 1.1).

1.6 Fourth-Generation Biofuel

The use of the new synthetic biological tools has emerged as the fourth-generation biofuel. In this generation, the breakthrough has made promises to photobiological biofuels and electro biofuels (Aro 2016). Synthetic biology of the algae and cyanobacteria involves various designs of the new promising strains by using recombinant

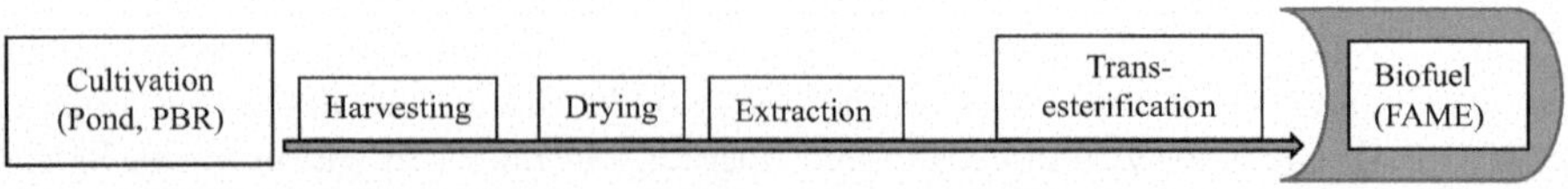

Fig. 1.1 Biofuel from algae

Table 1.3 Comparisons of different biofuel generations and its challenges

Generation	Source	Current conditions	Problems
First	Food crop sources Starch and sugar Oil crops	Commercially available. Technically developed	Competition with food supply and land uses Benefit of greenhouse gases is dependent on the feedstock
Second	Lignocellulosic biomass Agro-waste and energy crop	Advanced status of development and demonstration step	Higher cost, selection cheap, and sustainable raw material Additional cost of pre-treatment Enzyme cost is higher
Third	Algae	Fast-growing, cultivation land not required, Various potential algae strains are available, "under research and development"	High cost of downstream processing
Fourth (extension of third generation)	Algae and cyanobacteria	Primary stage of genetic engineering and new recombinant strains development, more promising than 3 G	Under research.

deoxyribonucleic acid (DNA) technologies. This technology uses the synthesis of the new strain by using new devices and the integration of the new biological parts or redesigning of the biological system for the biofuel generation.

The fourth-generation biofuels are inexhaustible, cheap, and readily available resources compared in (Table 1.3) with other generation of biofuel. During photosynthesis process, water molecule splits into hydrogen and oxygen by the action of solar energy with the interaction of photosystems I and II which is the key for the biofuel production at a large scale (Inganäs and Sundström 2016). Moreover, artificial photolysis is emerging as the new door for the biofuel generation (Inganäs and Sundström 2016). Even CO_2 water can also use as an alternative strategy for biofuel production, which utilizes carbon-based solar fuel.

1.6.1 Direct Process for Solar Fuel

This technology utilizes the combination of sensitizers and catalyst for performing individual reactions and generating approximately 30% of the solar fuel from water, but no complete device to date is made for direct fuel productions (Inganäs and Sundström 2016). In this process, oxidation of the water and reduction of the substrate are principle mechanisms. Further, proton from substrate utilized for the hydrogen fuel generation. The catalyst used is generally made up of abundant metals like cobalt, iron, and manganese, etc. In another way, an artificial photosynthesis follows the molecular mechanism; it involves the enzymes which are catalytically active used as biocatalyst like hydrogenase enzyme, which is used for hydrogen production by dark fermentation (Tamagnini et al. 2007). In direct process, with the help of the nanotechnologies, hydrogen production follows the non-molecular mechanism light-driven catalysis on the surface. This surface area can be metal, semiconductors, or fabrication of nanostructured carbon-based material. Limitations of the non-molecular mechanism are expensive, and rare metals are used as the catalyst. Involvement of molecular and non-molecular mechanism involved makes the process complex thus less understood. Limitations of the non-molecular mechanism are the use of rare metals as the catalyst which makes it expensive.

1.7 Microbial Conversion

Production of the biofuel is still a challenge because the production cost is more than the raw material. In bioprocessing, microbial conversion step converts lignocellulosic biomass into biofuel by microbial action (no action of enzyme). This method has gained more interest from researchers over the first-generation biofuel production in terms of the cost (Lynd et al. 2005). The new trend of bioprocess states that the potential of ethanol production through microbial route or its combinations mainly utilizes the cellulose machinery with other components (hemicellulose and lignin) of the pre-treated biomass with the high yield of the ethanol. Various aerobic fungi can degrade pre-treated feedstocks directly into the CO_2, and only a minor amount of ethanol is produced. However, the electrons yielded are transferred to the respiratory chain for oxidative phosphorylation, and the metabolic pathway does not utilize the substrate level phosphorylation for ethanol production. The rule of thumb of alcohol production from sugar is that 2 moles of alcohol is produced from each mole of hexose sugar. Yeast like *Saccharomyces cerevisiae* uses the glycolytic pathway to form alcohol. All the industrial processes are operated at higher pH and temperature range, which suitably use consortia of potent fungal strain like *Sarocladium strictum* with thermophiles like *Halomonas* for the improved degradation of wheat straw (Cortes-Tolalpa et al. 2018) (Table 1.4).

1.8 Enzymatic Conversion to Biofuel

Biofuel production from lignocellulosic biomass has given the new option to fulfill global energy demand but has a greater challenge of global warming. Production of biological-derived energy known as "bioenergy" is a major concern with the biocatalysis (utilizes the various enzymes to catalyze biomass conversion) (Rubin 2008). The complex structure of the lignocellulosic biomass first gets hydrolyzed

Table 1.4 Microorganism for producing biofuels (Kumar and Kumar 2017)

Microorganism	Biofuel	Biofuel yield (g/L)	References
Clostridium acetobutylicum	Butanol	3	Lütke-Eversloh and Bahl (2011)
Clostridium thermocellum	Isobutanol	5.4	Lin et al. (2015)
Escherichia coli	Butanol	30	Shen et al. (2011)
Escherichia coli	Ethanol	25	Romero-García et al. (2016)
Saccharomyces cerevisiae	Fatty acids	0.38	Yu et al. (2016)
Saccharomyces cerevisiae	Isoprenoid based biofuel	40	Westfall et al. (2012)
Pseudomonas putida	Butanol	0.05	Nielsen et al. (2009)
Cryptococcus	Lipids	7.8	Deeba et al. (2016)
Zymomonas mobilis	2,3 butandiol	10	Yang et al. (2016)
Zymomonas mobilis	Ethanol	–	Kremer et al. (2015)
Caldicellulosiruptor bescii	Ethanol	0.7	Chung et al. (2014)
Trichoderma reesei	Ethanol	10	Huang et al. (2014)
Yarrowia lipolytica	Fatty acids	55	Beopoulos et al. (2009)
Synechococcus sp.	Limonene	0.04	Davies et al. (2014)
Synechococcus elongatus	1,3 propanediol	0.28	Hirokawa et al. (2016)

into its simplest fractions with various hydrolyzing enzyme cocktails. However, the use of enzyme cocktail proved the maximum conversion into industrial product formation from lignocellulosic biomass conversion. Enzyme consortium is another way used to deconstruct recalcitrant substrate into a useful product (Lopes et al. 2018). This enzyme consortium conversion requires detailed studies on various enzymes and their action to work synergistically. To date, Sprizyme® from Novozyme, a commercial enzymatic cocktail is available in three versions (in the market) for the complete conversion of sugar. Moreover, Cellic®CTec3 (2017) from Novozyme is using cellulose conversion into ethanol. Enzyme cocktails enable the enzyme to hydrolyze cell wall structure due to their specific action. Moreover, the product of one enzyme acts as the substrate of another enzyme and also gives an option to the individual enzyme to act on an available substrate. During enzyme-substrate reaction, enzyme inhibitors are also produced, which can inhibit the hydrolysis process. Therefore, a complete enzyme mechanism can be studied to increase lignocellulosic biomass conversion. Wherein, each substrate in cellulosic biomass has its characteristic and feature which require the distinct condition to convert it into its simplest form.

1.8.1 Cellulases

There are a group of three enzymes which act synergistically on cellulose and completely convert it into glucose. These enzymes are endoglucanase, exoglucanase, and β-glucosidase.

Endoglucanase: This enzyme randomly hydrolyzes β 1-4 glucosidase linkages at the amorphous region and forms 2 Carbon units of cellobiose from cellulose chain.

Exoglucanase: Other names are cellobiohydrolase, avicelase, and exocellulases, which cleaves at exterior ends of the crystalline part of cellulose and produces cellobiose as a product.

β-glucosidase: This enzyme cleaves 2C chain molecule (cellobiose) into glucose.

These all three enzymes are collectively known as cellulytic enzyme, majorly produced from *Trichoderma reesei*. Various thermophilic potent strains such as *Thermophilic clostridia* and *Thermoanerobacterium* are also reported in producing cellulases (Lamed and Zeikus 1980).

1.8.2 Xylanases

Xylan sugar is an important component of hemicellulose which is made up of 5C chain unit (pentosans). Xylanases are the enzymes, which are used to degrade xylan unit linked with the lignin and cellulose component. Removing the xylan unit from lignocellulosic biomass gives easy accessibility to other hydrolyzing enzymes or chemicals for the increasing saccharification of cellulose. Various microbial species like *Trichoderma reesei* and *Humicola insolens* are the potent strain which degrades xylan at higher temperatures of 40–60 °C (Binod et al. 2018).

1.8.3 Lignolytic Enzymes

These are lignin-degrading enzymes which utilize lignin phenolic and non-phenolic group as substrates comprising of lignin peroxidase (LiP), manganese peroxidase (MnP), laccases, versatile peroxidase (VP) and dye-oxidizing enzymes (DyP A and DyP B). The action mechanism of LiP acts on the surface of the phenolic and non-phenolic aromatic component, and MnP acts on phenolic substrates by the action of Mn^{3+}, and VP has characterstics of both LiP and MnP (Schoemaker and Piontek 1996). Lignin-degrading enzymes remove lignin in the pre-treatment step of lignocellulosic biomass and expose site for easy degradation of cellulose.

1.8.4 Cellobiose Dehydrogenase (CBDH)

This enzyme is the oxidative enzyme used to oxidize various sugars such as lactose, glucose, and cellobiose. Wherein, the substrate is oxidized by the action of CBDH and reduction of flavin into flavin adenine dinucleotide (Knöös et al. 2014). In another way, CBDH activates cellulases by product inhibition of cellobiose. Moreover, another enzyme of this class is lytic polysaccharide monooxygenase copper (LPMO) enzyme, and it needs an external electron donor for its activity to cleave C1, C4, or both glycosidic bonds of cellulose. LPMO acts on first C-H bond of cellulose followed by the oxygen-dependent chain for the degradation into the product (Eibinger et al. 2014).

1.9 Effect of Surfactant on Enzymatic Hydrolysis

Surfactant positively enhances enzymatic hydrolysis and reduces hydrolysis time and dosage of enzymes (Helle et al. 1993). Surfactant hydrophobically interacts with the lignin part of lignocellulosic biomass and doubles the yield by increasing surface modifications or disruption; even surfactant acts as enzyme stabilizer (Kim et al. 1982). However, nonionic surfactant used in enzymatic hydrolysis such as Tween (80, 20) and polyethylene glycol (PEG) has more effect than the ionic surfactant as it increases higher surface for hydrolysis (Binod et al. 2018) (Table 1.5).

1.10 Biofuel from Nanotechnology

Nanotechnological advancement for improving bioprocess requires nontoxic NPs for microbial growth, which is economically produced and environment friendly. However, nanomaterial provides high surface area, high catalytic activity, stability, storage, durability, and higher potentials of 3R principle (recovery, reusability, and

Table 1.5 Advantages and disadvantages from biofuels in terms of GHG emissions (Nylund et al. 2008)

Biofuel	Advantage	Disadvantage
Ethanol	Reduction CO, HC, particulate matter (PM), NOx, 1,3 butadiene, aromatics	Increase acetaldehyde emissions
FAME	Reduction in PM	Increases in NOx and aldehyde emissions
Synthetic fuel	PM, NOx and aldehyde, 1,3 butadiene, benzene, polycyclic aromatic hydrocarbons (PAHs), and mutagenicity	
Natural gases	Low PM, NOx Methane performance improved by adding hydrogen	Formaldehyde, PAHs
DME	Clean burning	Challenging as a vehicle fuel
Hydrogen	Zero emissions	
CNG, LPG, biogas, and hybrids	GHG reductions with hybrids like CNG and LPG	

recycling). Previous reports showed the successful implementation of nanotechnology with all types of biofuel-producing processes such as transesterification, anaerobic digestion, gasification, FAME, and renewable hydrocarbons shows efficient and economic feasible only at lab scale (Zhang et al. 2010). Various nanocatalysts such as titanium dioxide, magnesium oxide calcium oxide, and strontium oxide are developed with high catalytic performance. Nanotechnology is also applied with the enzyme (biocatalyst) immobilization in nanoencapsulation, entrapment with silaffin, and adsorption for providing higher surface area, higher stability, higher enzyme loading and also gives the chance of reusability. Nanoparticles are used for the extraction of lipid from algae without harming cells. Nanotechnology can suitably bring breakthrough in advanced fermentation, pyrolysis, jet fuels, catalytic conversion gasification, biofuel cell, carbon capture storage, and nano-based precision forming technology (Nizami and Rehan 2018). Nanoparticles recently gained lots of interest due to their feasibility to enhance effect metabolic engineering by improving performance or product yield. Biohydrogen produced by the microbes shows enhanced production in anaerobic conditions using NPs. Whereas, these NPs improve reaction kinetics with increased transfer of electrons. Though several NPs have been reported to enhance dark fermentation for hydrogen production with incorporation of gold NPs, which shows stimulatory effect on substrate utilization by 56% and increases yield by 46%, this enhanced surface area to volume ratio provides better accessibility of binding site with bacteria as well as enzyme (Zhang and Shen 2007; Sekoai et al. 2019). Various mesoporous NPs have been developed to overcome the barrier of higher dosing of NPs for improving the production of biofuel. When NPs supplemented (zero-valent metals, metal oxide, and carbon-based NPs) with the anaerobic fermenting bacteria which further increase the hydrolysis process of organic material by increasing substrate utilization, lipase catalyzing biodiesel production by transesterification reaction produces biodiesel; this enzyme suitably nano-immobilized, produces biodiesel, and increases the reusability of the enzyme. Therefore, the use of nanotechnology also improves bioethanol production by immobilizing cellulases using manganese oxide dependent (NPs), it increases hydrolysis and catalytic efficiency of cellulase. Immobilization of enzyme by NPs acts as enzyme protector from the intermediate inhibitor. Nevertheless, NPs modified enzyme can sustain in extreme environmental conditions and supports higher product formation with reusability. Since nanotechnological implementation is still under laboratory scale, it requires higher knowledge and techno-economic assessments for process modification before its commercialization.

1.11 Lignin Strategy to Biofuel

1.11.1 Lignin Structure

Lignin matrix consists of heterogeneous structure of aromatic alcohol such as p-coumaryl, coniferyl, and sinapyl alcohols. The aromatic monolignols from lignin form guaiacol (G) from coniferyl alcohol, p-hydroxy phenyl (H) from coumaryl alcohol, and syringyl from sinapyl alcohol (S) gives lignin with distinctive characteristics from each other. Variation in monolignol compositions forms distinctive properties of wood. Where G/S-Units ratio are highly present in softwood lignin, and a variable composition ratio of G/S/H is present in hardwood lignin (Laskar et al. 2013; Lee et al. 2014a). Aromatic structure of lignin transferred to hydrocarbon (HC) fuels via chemical treatment produces products having similar properties like gasoline and diesel (Ragauskas et al. 2006).

1.11.2 Lignin Valorization

In the cellulosic biorefinery, the cellulosic part, which is consist of 45–60%, is utilized for the ethanol production; the rest of the ~20% part of lignin leftover is the key motivation for the lignin valorization into biofuel and generation of value-added by-product. This strategy involves depolymerization of lignin by using hydrogen as the reducing agent to form aromatic product with low oxygen content with higher stability. Due to the presence of different lignin monomers, degree of interaction varies to form useful product (Ragauskas et al. 2014). This barrier of interaction overcome by different pre-treatment strategies of lignocellulosic biomass. In case of hydrogenation operated at mild conditions with the help of a suitable catalyst, which ensures the desired product formation in complex reaction. In an in situ catalytic characterization, hydrogenation process yields guaiacol conversion with cyclohexanol selectivity of 99% and ~94% in 7 h, 220 °C, and initial pressure of 2 MPa with the suitable ration of water/methanol/lignin as feedstock (20:5:0.5) (Yu et al. 2013). Presently, in highly selective hydrogenation at mild conditions, step-by-step precipitation is done with the help of catalyst (Ni/SiO_2). This highly selective hydrogenation process is operated in 7h, 120 °C, and 2 MPa for complete conversion into guaiacol (Shu et al. 2016). Moreover, in wet impregnation technique of hydrogenation palladium is suitably doped with alumina as a catalyst, where 3 weight % of catalyst converts 4-ethylphenyl(cyclohexanol) in aqueous phase reaction operated for 6h, 60 °C (Yi et al. 2016). In lignin valorization other than the chemical catalyst, biological catalyst like lignolytic enzyme is also utilized for future trend of biofuel generation. The value added HC fuels produced from lignin are able to converted into C7-C18 hydrocarbon, which have suitable applications in jet fuel (Wang et al. 2015). This lignin to biofuel strategy involves degradation of lignin in such a way that O_2/C and H/C ratios match to conventional fuel. In case of hydrodeoxygenation (HDO) technique, lignin monomers increase the ratio of H/C while decreasing their O/C ratio. This process includes cleavage of CO bond in monomeric lignin unit and oxygen removal achieved by high-pressure hydrogen, which forms water vapor with the decrease in O/C ratio. Further more, double bond of aromatic carbon is saturated with the help of hydrogen using catalyst (palladium and active carbon) (Ge et al. 2017).

1.12 Sustainability Criteria

1.12.1 Food and Feedstock

Crop food used for the biofuel generation in 1G affects the food requirement in developing countries; even increasing the price of agricultural product, cereal, and sugar is the key public agenda. Therefore, biofuel expansion requires both increased agricultural productivity for food and feedstock for animal and biomass material for biofuel. This would lead conversion of forest land, and further, this would affect the increase in GHG and a potential risk to biodiversity. Moreover, 3G biofuel gives hope for the high product yield, and land requirement, but the downstream cost is higher to be reduced.

1.12.2 Water Requirement

Even 2G biofuel requires nonfood crop, but it competes with land for cultivation of lignocellulosic biomass such as *Jatropha*, which also requires other resources like water uses (400–700 times higher than other energy sources) and 90% more than the production of feedstocks (Mandil and Shihab-Eldin 2010).

1.12.3 Emissions

As expected biofuel reduces GHG emissions by 60–94% compared to the fossil fuels, but when we compare the scenario with climatic change and lifecycle assessment, biofuel is worse than gasoline (Highina et al. 2014; Holma et al. 2013).

1.12.4 Biodiversity

Biofuel production and conversion of land affect the environment, sustainability, and biodiversity. Conversion of land and forest may affect a wide range of flora and fauna of the ecosystem. No methods are available, which can tell about the direct impact on biodiversity and its damage (Araújo et al. 2017). Therefore, biodiversity is considered foremost, integrated with the biofuel generation planning and agricultural farming (Morgera et al. 2009).

1.12.5 Policies

Policies and its implementation for biofuel generation play a pivotal role in the development of sustainable development, reducing barriers, greater economics, better implementation of technologies, and funding agencies. However, many

Table 1.6 Showing the biofuel blend strategy of countries (Araújo et al. 2017)

Country	Ethanol blend aim	Biodiesel blend aim	Year of implementation
Brazil	27%	10%	2019
China		15% (10% mandate)	2020
EU	10%	20%	2020
India		20% petroleum	2017
USA		0.1	Implemented

countries like Brazil, the USA, China, and India have their own regulation policies for the development of biofuel and blending with the flexi fuel. These policies suitably help for reducing subsidies to conventional fuel and mandate blending of biofuel to curb out CO_2 emissions (Table 1.6).

1.13 Conclusions

In this book chapter, biofuel production disscussed on different process methodologies, uses of different feedstocks, its key challenges. The production of biofuel plays a pivotal role in the current scenario. In case, 1G technology impacts on food prices, and competition with the food crop is the key challenge. Wherein, the concept of 2G comes into the biofuel production, which is economically better than 1G, but again its competition with the land uses and other resources makes it difficult. Further, concept of 3G uses of algae and BGA strains, which require low area and no competition with food feedstock and high product yield opens a new door, but its downstream processing cost is still a challenge. This 3G technology is still in the developmental phase for increasing higher product yield. The genetic manipulation and engineered strain reduces its downstream cost and yield known as 4G of biofuel. Apart from 1G to 4G new advancement in nanotechnology, has potential with technological advancement increases the effectiveness of processes. Generation of biofuel from the chemical process is commercially viable but still have challenge to environment. Therefore, biologically and biochemically preccesses are promising and environmentally friendly. Furthermore, several attempts are made for making the enzymatic process economically viable, and still industrial important enzymes are required. Biofuel production requires several issues for careful analysis such as; feedstock selection, process design, reduction of cost, cultivation practices, land uses and its practices, GHG, cocktail preparation, effect on soil, and mineral, etc. Shortly, the processes compacted and designed by removing pre-treatment strategy. Lignin valorization strategy is also gaining interest to convert lignin as biofuel in recent trends. The suitable blending of biofuel with existing fuel is the alternate best option. However, large-scale production is the ultimate requirement shortly to fulfill our energy need. Nevertheless, there is still a need for technological understanding with advancement and integration of the concept of chemistry, bioprocess, and chemical engineering in biofuel generation.

1.14 Summary

This book chapter gives the idea about how biofuels are formed from various feedstocks using multiple technology strategies. Biofuel production requires various chemical, biological, biochemical processes, and nanotechnologies. Moreover, biofuel formed from these technology showed similar characteristic properties as fuel from fossil fuel would be blended suitably to open the doors of sustainable process development. In the future complete development of biofuel technology will reduce dependence on fossil fuels.

References

Abdullah B, Muhammad SAFAS, Mahmood NAN (2017) Production of biofuel via hydrogenation of lignin from biomass. In: New advances in hydrogenation processes-fundamentals and applications. InTech, London

Alonso DM, Bond JQ, Dumesic JA (2010) Catalytic conversion of biomass to biofuels. Green Chem 12:1493–1513

Araújo K, Mahajan D, Kerr R, Silva MD (2017) Global biofuels at the crossroads: an overview of technical, policy, and investment complexities in the sustainability of biofuel development. Agriculture 7:32

Aro E-M (2016) From first generation biofuels to advanced solar biofuels. Ambio 45(Suppl 1):S24–S31

Beal CM, Hebner RE, Webber ME, Ruoff RS, Seibert AF, King CW (2012) Comprehensive evaluation of algal biofuel production: experimental and target results. Energies 5:1943–1981

Beopoulos A, Cescut J, Haddouche R, Uribelarrea J-L, Molina-Jouve C, Nicaud J-M (2009) Yarrowia lipolytica as a model for bio-oil production. Prog Lipid Res 48:375–387

Binod P, Gnansounou E, Sindhu R, Pandey A (2018) Enzymes for second generation biofuels: recent developments and future perspectives. Bioresour Technol Rep 5:317–325

Bowyer J, Howe J, Levins RA, Groot H, Fernholz K, Pepke E, Henderson C (2018) Third generation biofuels implications for wood-derived fuels. Available online: http://www.dovetailinc.org/report_pdfs/2018/dovetail3gbiofuel0218.pdf. Accessed on 28 Feb 2019

Bridgwater A, Boocock D (2013) Developments in thermochemical biomass conversion: volume 1. Springer, Dordrecht

Buragohain B, Mahanta P, Moholkar VS (2010) Biomass gasification for decentralized power generation: the Indian perspective. Renew Sust Energ Rev 14:73–92

Chen H (2014) Chemical composition and structure of natural lignocellulose. In: Biotechnology of lignocellulose. Springer, Dordrecht/Heidelberg

Chisti Y (2007) Biodiesel from microalgae. Biotechnol Adv 25:294–306

Chum HL (1991) Polymers from biobased materials. Noyes Data Corp, Park Ridge

Chung D, Cha M, Guss AM, Westpheling J (2014) Direct conversion of plant biomass to ethanol by engineered Caldicellulosiruptor bescii. Proc Natl Acad Sci 111(24):8931–8936

Coppola F, Simonciniand E, Pulselli R (2009) Bioethanol potentials from marine residual biomass: an energy evaluation. Energy Environ 122:379–387

Cortes-Tolalpa L, Norder J, van Elsas JD, Falcao Salles J (2018) Halotolerant microbial consortia able to degrade highly recalcitrant plant biomass substrate. Appl Microbiol Biotechnol 102:2913–2927

Davies FK, Work VH, Beliaev AS, Posewitz MC (2014) Engineering limonene and bisabolene production in wild type and a glycogen-deficient mutant of Synechococcus sp. PCC 7002. Front Bioeng Biotechnol 2:21

Davis R, Tao L, Tan E, Biddy M, Beckham G, Scarlata C, Jacobson J, Cafferty K, Ross J, Lukas J (2013) Process design and economics for the conversion of lignocellulosic biomass to hydrocarbons: dilute-acid and enzymatic deconstruction of biomass to sugars and biological conversion of sugars to hydrocarbons. National Renewable Energy Lab.(NREL), Golden

Deeba F, Pruthi V, Negi YS (2016) Converting paper mill sludge into neutral lipids by oleaginous yeast Cryptococcus vishniaccii for biodiesel production. Bioresour Technol 213:96–102

Demartini JD, Pattathil S, Miller JS, Li H, Hahn MG, Wyman CE (2013) Investigating plant cell wall components that affect biomass recalcitrance in poplar and switchgrass. Energy Environ Sci 6:898–909

Demirbaş A (1998) Yields of oil products from thermochemical biomass conversion processes. Energy Convers Manag 39:685–690

Demirbaş A (2000) Mechanisms of liquefaction and pyrolysis reactions of biomass. Energy Convers Manag 41:633–646

Demirbas A (2004) Combustion characteristics of different biomass fuels. Prog Energy Combust Sci 30:219–230

Demirbas A (2011) Competitive liquid biofuels from biomass. Appl Energy 88:17–28

Eibinger M, Ganner T, Bubner P, Rosker S, Kracher D, Haltrich D, Ludwig R, Plank H, Nidetzky B (2014) Cellulose surface degradation by a lytic polysaccharide monooxygenase and its effect on cellulase hydrolytic efficiency. J Biol Chem 289(52):35929–35938. jbc. M114. 602227

Ge Y, Dababneh F, Li L (2017) Economic evaluation of lignocellulosic biofuel manufacturing considering integrated lignin waste conversion to hydrocarbon fuels. Procedia Manuf 10:112–122

Gomez LD, Steele-King CG, Mcqueen-Mason SJ (2008) Sustainable liquid biofuels from biomass: the writing's on the walls. New Phytol 178:473–485

Hahn-Hägerdal B, Galbe M, Gorwa-Grauslund MF, Lidén G, Zacchi G (2006) Bio-ethanol – the fuel of tomorrow from the residues of today. Trends Biotechnol 24:549–556

Halim R, Gladman B, Danquah MK, Webley PA (2011) Oil extraction from microalgae for biodiesel production. Bioresour Technol 102:178–185

Halim R, Danquah MK, Webley PA (2012) Extraction of oil from microalgae for biodiesel production: a review. Biotechnol Adv 30:709–732

Harun R, Singh M, Forde GM, Danquah MK (2010) Bioprocess engineering of microalgae to produce a variety of consumer products. Renew Sust Energ Rev 14:1037–1047

Helle SS, Duff SJ, Cooper DG (1993) Effect of surfactants on cellulose hydrolysis. Biotechnol Bioeng 42:611–617

Highina B, Bugaje I, Umar B (2014) A review on second generation biofuel: a comparison of its carbon footprints. Eur J Eng Technol 2(2):1–7

Himmel ME, Ding S-Y, Johnson DK, Adney WS, Nimlos MR, Brady JW, Foust TD (2007) Biomass recalcitrance: engineering plants and enzymes for biofuels production. Science 315:804–807

Hirokawa Y, Maki Y, Tatsuke T, Hanai T (2016) Cyanobacterial production of 1, 3-propanediol directly from carbon dioxide using a synthetic metabolic pathway. Metab Eng 34:97–103

Holma A, Koponen K, Antikainen R, Lardon L, Leskinen P, Roux P (2013) Current limits of life cycle assessment framework in evaluating environmental sustainability – case of two evolving biofuel technologies. J Clean Prod 54:215–228

Hu Q, Sommerfeld M, Jarvis E, Ghirardi M, Posewitz M, Seibert M, Darzins A (2008) Microalgal triacylglycerols as feedstocks for biofuel production: perspectives and advances. Plant J 54:621–639

Huang J, Chen D, Wei Y, Wang Q, Li Z, Chen Y, Huang R (2014) Direct ethanol production from lignocellulosic sugars and sugarcane bagasse by a recombinant Trichoderma reesei strain HJ48. Sci World J 2014:1–8

Inganäs O, Sundström V (2016) Solar energy for electricity and fuels. Ambio 45:15–23

Ingram L, Conway T, Clark D, Sewell G, Preston J (1987) Genetic engineering of ethanol production in Escherichia coli. Appl Environ Microbiol 53:2420–2425

Kalyani DC, Zamanzadeh M, Müller G, Horn SJ (2017) Biofuel production from birch wood by combining high solid loading simultaneous saccharification and fermentation and anaerobic digestion. Appl Energy 193:210–219

Kanda H, Li P, Yoshimura T, Okada S (2013) Wet extraction of hydrocarbons from Botryococcus braunii by dimethyl ether as compared with dry extraction by hexane. Fuel 105:535–539
Kim M, Lee S, Ryu DD, Reese E (1982) Surface deactivation of cellulase and its prevention. Enzym Microb Technol 4:99–103
Kivistö A, Santala V, Karp M (2010) Hydrogen production from glycerol using halophilic fermentative bacteria. Bioresour Technol 101:8671–8677
Knöös P, Schulz C, Piculell L, Ludwig R, Gorton L, Wahlgren M (2014) Quantifying the release of lactose from polymer matrix tablets with an amperometric biosensor utilizing cellobiose dehydrogenase. Int J Pharm 468:121–132
Kremer TA, Lasarre B, Posto AL, Mckinlay JB (2015) N2 gas is an effective fertilizer for bioethanol production by Zymomonas mobilis. Proc Natl Acad Sci 112:2222–2226
Kulkarni MG, Gopinath R, Meher LC, Dalai AK (2006) Solid acid catalyzed biodiesel production by simultaneous esterification and transesterification. Green Chem 8:1056–1062
Kumar R, Kumar P (2017) Future microbial applications for bioenergy production: a perspective. Front Microbiol 8:450
Lamed R, Zeikus J (1980) Glucose fermentation pathway of Thermoanaerobium brockii. J Bacteriol 141:1251–1257
Laskar DD, Yang B, Wang H, Lee J (2013) Pathways for biomass-derived lignin to hydrocarbon fuels. Biofuels Bioprod Biorefin 7:602–626
Lee C-G (1999) Calculation of light penetration depth in photobioreactors. Biotechnol Bioprocess Eng 4:78–81
Lee H, Hamid S, Zain S (2014a) Conversion of lignocellulosic biomass to nanocellulose: structure and chemical process. Sci World J 2014:1–20
Lee S, Speight JG, Loyalka SK (2014b) Handbook of alternative fuel technologies. CRC Press, Boca Raton
Limayem A, Ricke SC (2012) Lignocellulosic biomass for bioethanol production: current perspectives, potential issues and future prospects. Prog Energy Combust Sci 38:449–467
Lin PP, Mi L, Morioka AH, Yoshino KM, Konishi S, Xu SC, Papanek BA, Riley LA, Guss AM, Liao JC (2015) Consolidated bioprocessing of cellulose to isobutanol using Clostridium thermocellum. Metab Eng 31:44–52
Lopes A, Ferreira Filho EX, Moreira L (2018) An update on enzymatic cocktails for lignocellulose breakdown. J Appl Microbiol 125:632–645
Lü J, Sheahan C, Fu P (2011) Metabolic engineering of algae for fourth generation biofuels production. Energy Environ Sci 4:2451–2466
Lütke-Eversloh T, Bahl H (2011) Metabolic engineering of Clostridium acetobutylicum: recent advances to improve butanol production. Curr Opin Biotechnol 22:634–647
Lynd LR, van Zyl WH, Mcbride JE, Laser M (2005) Consolidated bioprocessing of cellulosic biomass: an update. Curr Opin Biotechnol 16:577–583
Mandil C, Shihab-Eldin A (2010) Assessment of biofuels potential and limitations. Geopolit Energy 32:6–11
Margesin R, Schinner F (2001) Potential of halotolerant and halophilic microorganisms for biotechnology. Extremophiles 5:73–83
Meher L, Sagar DV, Naik S (2006) Technical aspects of biodiesel production by transesterification – a review. Renew Sust Energ Rev 10:248–268
Melin K, Kohl T, Koskinen J, Hurme M (2015) Performance of biofuel processes utilising separate lignin and carbohydrate processing. Bioresour Technol 192:397–409
Menetrez MY (2012) An overview of algae biofuel production and potential environmental impact. Environ Sci Technol 46:7073–7085
Minteer S (2016) Alcoholic fuels. CRC Press, Boca Raton
Morgera E, Kulovesi K, Gobena A (2009) Case studies on bioenergy policy and law: options for sustainability. In: FAO legislative study. FAO, Rome
Mubarak M, Shaija A, Suchithra T (2015) A review on the extraction of lipid from microalgae for biodiesel production. Algal Res 7:117–123

Müller C, Reuter W, Wehrmeyer W, Dau H, Senger H (1993) Adaptation of the photosynthetic apparatus of Anacystis nidulans to irradiance and CO2-concentration. Bot Acta 106:480–487
Mussgnug JH, Thomas-Hall S, Rupprecht J, Foo A, Klassen V, Mcdowall A, Schenk PM, Kruse O, Hankamer B (2007) Engineering photosynthetic light capture: impacts on improved solar energy to biomass conversion. Plant Biotechnol J 5:802–814
Naik SN, Goud VV, Rout PK, Dalai AK (2010) Production of first and second generation biofuels: a comprehensive review. Renew Sust Energ Rev 14:578–597
Nielsen DR, Leonard E, Yoon S-H, Tseng H-C, Yuan C, Prather KLJ (2009) Engineering alternative butanol production platforms in heterologous bacteria. Metab Eng 11:262–273
Nizami A-S, Rehan M (2018) Towards nanotechnology-based biofuel industry. Biofuel Res J 5:798–799
Nylund N-O, Aakko-Saksa P, Sipilä K (2008) Status and outlook for biofuels, other alternative fuels and new vehicles. VTT, Espoo
Osamu K, Carl H (1989) Biomass handbook. Gordon Breach Science Publisher, New York
Pauly M, Keegstra K (2008) Cell-wall carbohydrates and their modification as a resource for biofuels. Plant J 54:559–568
Ragauskas AJ, Williams CK, Davison BH, Britovsek G, Cairney J, Eckert CA, Frederick WJ, Hallett JP, Leak DJ, Liotta CL (2006) The path forward for biofuels and biomaterials. Science 311:484–489
Ragauskas AJ, Beckham GT, Biddy MJ, Chandra R, Chen F, Davis MF, Davison BH, Dixon RA, Gilna P, Keller M (2014) Lignin valorization: improving lignin processing in the biorefinery. Science 344:1246843
Ragnar M, Henriksson G, Lindström ME, Wimby M, Blechschmidt J, Heinemann S (2000) Pulp. In: Ullmann's encyclopedia of industrial chemistry. WileyVCH, Weinheim, pp 1–92
Rapala J, Sivonen K (1998) Assessment of environmental conditions that favor hepatotoxic and neurotoxic Anabaena spp. strains cultured under light limitation at different temperatures. Microb Ecol 36:181–192
Richards E (2013) Careers in biofuels. US Bureau of Labor Statistics
Romero-García J, Martínez-Patiño C, Ruiz E, Romero I, Castro E (2016) Ethanol production from olive stone hydrolysates by xylose fermenting microorganisms. Bioethanol 2:51–65
Rubin EM (2008) Genomics of cellulosic biofuels. Nature 454:841
Schoemaker HE, Piontek K (1996) On the interaction of lignin peroxidase with lignin. Pure Appl Chem 68:2089–2096
Schutyser W, Renders T, van den Bosch S, Koelewijn SF, Beckham GT, Sels BF (2018) Chemicals from lignin: an interplay of lignocellulose fractionation, depolymerisation, and upgrading. Chem Soc Rev 47:852–908
Sekoai PT, Ouma CNM, du Preez SP, Modisha P, Engelbrecht N, Bessarabov DG, Ghimire A (2019) Application of nanoparticles in biofuels: an overview. Fuel 237:380–397
Shafizadeh F (1982) Introduction to pyrolysis of biomass. J Anal Appl Pyrolysis 3:283–305
Shen Y, Pei Z, Yuan W, Mao E (2009) Effect of nitrogen and extraction method on algae lipid yield. Int J Agric Biol Eng 2:51–57
Shen CR, Lan EI, Dekishima Y, Baez A, Cho KM, Liao JC (2011) High titer anaerobic 1-butanol synthesis in Escherichia coli enabled by driving forces. Appl Environ Microbiol 77(9):2905–2915
Shu R, Zhang Q, Xu Y, Long J, Ma L, Wang T, Chen P, Wu Q (2016) Hydrogenation of lignin-derived phenolic compounds over step by step precipitated Ni/SiO 2. RSC Adv 6:5214–5222
Sikarwar VS, Zhao M, Fennell PS, Shah N, Anthony EJ (2017) Progress in biofuel production from gasification. Prog Energy Combust Sci 61:189–248
Speight JG (2011) The biofuels handbook. Royal Society of Chemistry, Cambridge
Stevens C, Verhé R (2004) Renewable bioresources: scope and modification for non-food applications. Wiley, Chichester
Tamagnini P, Leitão E, Oliveira P, Ferreira D, Pinto F, Harris DJ, Heidorn T, Lindblad P (2007) Cyanobacterial hydrogenases: diversity, regulation and applications. FEMS Microbiol Rev 31:692–720

Valdivia M, Galan JL, Laffarga J, Ramos JL (2016) Biofuels 2020: biorefineries based on lignocellulosic materials. Microb Biotechnol 9:585–594

Vamvuka D, Arvanitidis C, Zachariadis D (2004) Flue gas desulfurization at high temperatures: a review. Environ Eng Sci 21:525–548

Wang H, Ruan H, Pei H, Wang H, Chen X, Tucker MP, Cort JR, Yang B (2015) Biomass-derived lignin to jet fuel range hydrocarbons via aqueous phase hydrodeoxygenation. Green Chem 17:5131–5135

Weber C, Farwick A, Benisch F, Brat D, Dietz H, Subtil T, Boles E (2010) Trends and challenges in the microbial production of lignocellulosic bioalcohol fuels. Appl Microbiol Biotechnol 87:1303–1315

Westfall PJ, Pitera DJ, Lenihan JR, Eng D, Woolard FX, Regentin R, Horning T, Tsuruta H, Melis DJ, Owens A (2012) Production of amorphadiene in yeast, and its conversion to dihydroartemisinic acid, precursor to the antimalarial agent artemisinin. Proc Natl Acad Sci 109:E111–E118

Yanase H, Miyawaki H, Sakurai M, Kawakami A, Matsumoto M, Haga K, Kojima M, Okamoto K (2012) Ethanol production from wood hydrolysate using genetically engineered Zymomonas mobilis. Appl Microbiol Biotechnol 94:1667–1678

Yang S, Mohagheghi A, Franden MA, Chou Y-C, Chen X, Dowe N, Himmel ME, Zhang M (2016) Metabolic engineering of Zymomonas mobilis for 2, 3-butanediol production from lignocellulosic biomass sugars. Biotechnol Biofuels 9:189

Yi J, Luo Y, He T, Jiang Z, Li J, Hu C (2016) High efficient hydrogenation of lignin-derived monophenols to cyclohexanols over Pd/γ-Al2O3 under mild conditions. Catalysts 6:12

Yu Y-X, Ying X, Wang T-J, Zhang Q, Zhang X-H, Zhang X (2013) In-situ hydrogenation of lignin depolymerization model compounds to cyclohexanol. J Fuel Chem Technol 41:443–447

Yu A-Q, Juwono NKP, Foo JL, Leong SSJ, Chang MW (2016) Metabolic engineering of Saccharomyces cerevisiae for the overproduction of short branched-chain fatty acids. Metab Eng 34:36–43

Zhang Y, Shen J (2007) Enhancement effect of gold nanoparticles on biohydrogen production from artificial wastewater. Int J Hydrog Energy 32:17–23

Zhang XL, Yan S, Tyagi RD, Surampalli RY, Zhang TC (2010) Application of nanotechnology and nanomaterials for bioenergy and biofuel production. In: In Bioenergy and biofuel from biowastes and biomass. American Society of Civil Engineers (ASCE), pp 478–496. https://doi.org/10.1061/9780784410899.ch21

Applications of Plant-Based Natural Products to Synthesize Nanomaterial

2

Muhammad Irfan, Mamoona Saeed, Bushra Iqbal, and Misbah Ghazanfar

Abstract

Plants are the important means of various kinds of phytochemicals with several applications in nanotechnology. Nanotechnology refers to the building and use of materials whose components exist at the nanoscale up to 100 nm in size. These biotechnological tools have been synthesized by the use of different kinds of plants. Plants have numerous natural products like tannins, saponins, flavonoids, steroids, alkaloids, and other nutritional products that can be obtained from several plant parts like seed, barks, leaves, roots, shoots, flowers, and stems. It has been reported that the extracts from plants act as a powerful pioneer for a nanomaterial production in safe procedures. As the plant extracts have numerous secondary metabolites, it plays part as stabilizing and reducing factors for the bioreduction reaction to form new metallic nanoparticles. These plant-based nanoparticles have various applications in different fields especially in biofuel production.

Keywords

Nanoparticles · Plants · Natural · Nanomaterials · Applications · Substances

2.1 Introduction

In the present days, nanoparticles and its various items have become principal components of our items in routine, so metallic nanoparticles like silver in antiperspirants and particles with improved discharge properties into meds are known as nanoparticles and the whole distance to "nano impregnations" of spray lodges,

M. Irfan (✉) · M. Saeed · B. Iqbal · M. Ghazanfar
Department of Biotechnology, University of Sargodha, Sargodha, Pakistan

M. Srivastava et al. (eds.), *Nanomaterials in Biofuels Research*,
Clean Energy Production Technologies, https://doi.org/10.1007/978-981-13-9333-4_2

baths, and cleaning bowls (Moss and Siccardi 2014; Vincent and Loeve 2014; Kettler et al. 2016). As these nanomaterials, which include particles having diameters in the range of hundred nanometers, have now significant role in our routine life, there are also concerns about a possible toxic effect of these nanomaterials on humans and environment as well (Roy et al. 2017). However, such evaluations usually disregard the truth that nature is a proficient nanotechnologist itself, with various precedents of general nanomaterials emerging from sources that are natural, like mineral springs, volcanoes but specifically from living beings also. Actually, life rotates around the cell which itself is microscopic in size and metabolize molecules which are picoscopic (Kajander et al. 2001; Ciftcioglu et al. 2006; Urbano and Urbano 2007).

Presently, we sometimes fortuitously judge the emerging field of nanomaterials. Here, we will shortly discuss the rising area of natural nanoparticles. Actually, there is a demand for a difference among natural nanomaterials, such as already found or synthesized in the environment without man involvement, and substances which are "bio" and "nano" as well. In another case, the predicate "natural" corresponds to "biological," as in "natural products," which mentions mainly biological substances (Mukherjee et al. 2001; Patel et al. 2015a, b; Pasula and Lim 2017). Through the earliest starting point, all of us ought to complement to grounds of nanoparticles (NPs) as is broad and incidental. Figure 2.1 illustrates the synthesis of nanomaterials from plants and their applications in various fields.

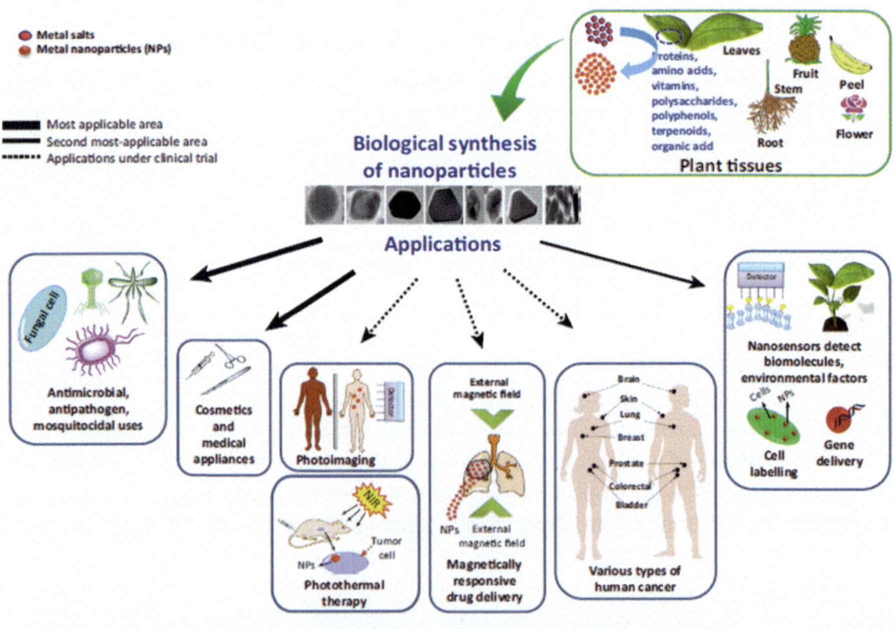

Fig. 2.1 Biological synthesis and applications of metallic nanoparticles in environmental and biomedical areas. (Modified from Singh et al. 2016a, b)

2.2 Inorganic Nanoparticles Derived from Natural Sources

In the earth, so many inorganic nanoparticles are present which are derived from natural sources. So such particles are common, so far more often than nonnatural. Volcanic powder mists contain a comprehensive assorted diversity in polydispersity miniaturized scale plus nanomaterials. Such particles extend in size somehow from 100 to 200 nm and are synthetically basically made out of composites of iron and silicates. These particles are deliberately floating in air and if only once heaved can prompt attentive pulmonary disorders. "Carbon nanotube" sediment gathered from the cremation of Texas pine, for example, in point, involves many layers (15–70 nanometers of size) of carbon nanotubes. Such carbon-based particles deliberately turned out to be airborne and present serious well-being perils to creatures and the human populace (Murrand Guerrero 2006).

Miraculously, drinking water, when seen under microscope, which could be loaded with variously sized nanoparticles just like minute strong materials of every shape. Generally, these particles are not of good quality and cannot easily be molded and are very much characterized. There are inexhaustible instances of characteristic nanomaterials structured thusly, for example, the $CaSO_4$ and silicate particles in spring water (Wu et al. 2016). In reality, these inorganic substances meanwhile urged counterparts to produce a scope of similar particles dependent on normally happening particles, for example, refined nanoparticles of Fe_3O_4 and MnO_2 (Song et al. 2010; Cho et al. 2017).

Likewise, in such "straightforward" physical occasions, there are comparable however progressively controlled molecule-inducing forms, frequently made on regular oxidation. One constantly noticed precedent of hydrogen sulfide gas oxidation (HS) condensed in volcanic eruptions or/and in wells which are common in numerous locality of our planet (Ezoe et al. 2002; Berlo et al. 2014). All things considered must be well characterized—at any rate with respect to synthetic course of action—and could get significant consideration.

2.3 Biological Synthesis of Nanomaterials

To be sure, the living cell is habitually managing "nanotechnology," in other words with objects of a nanoscopic rate. Just to wind up an inclination for sizes, basic strand of DNA in width is almost equal to 2.5 nm (Shors 2011). Henceforth some cells that lie in microbes are specific and are included in the field of nanotechnology. Some procedures are somewhat very much perceived and contemplated with respect to some metals like selenium and sulfur just as metallic NPs (Kuppusamy et al. 2016). The particles created by such biogenic production lines are generally of decent quality, for example, little circular states of a practically uniform size. Not astoundingly, such natural procedures might be inappropriate ultimately to create materials of good quality and then to deliver.

To be sure, the subtraction of ecological contaminants (e.g., overwhelming metals, natural and inorganic toxins), from defiled destinations utilizing nanomaterials or nanoparticles shaped in or by plants, parasites, and microbes with the aid of nanotechnology, frequently referenced to as nanobioremediation (NBR), is a growing, earth-responsive, and practical substitute to customary compound strategies (Singh and Walker 2006; Yadav et al. 2017). At this point, the three fundamental methodologies of present-day bioremediation grasp the utilization of organisms, plants, and sequestered chemicals, for event, nitrate reductase, or laccase (Sanghi et al. 2011). All things being equal, there is likewise an extra plus point of that strategy. Presently, the nanoparticles made by those life forms are not all encompassing spotted as contaminants however basically as esteemed nanoparticles of a pretty much common birthplace. Inside this specific situation, a portion of these bioreductively framed "normal" nanomaterials have been found currently with focus on likely restorative and horticultural entries (Salata 2004; Li et al. 2011; Duhan et al. 2017; Iavicoli et al. 2017). It is conceivable, for example, to tenderfoot innocuous microorganisms, for example, *Staphylococcus carnosus* and *Saccharomyces cerevisiae*, to create genuinely homogeneous selenium nanoparticles from selenite, an ion with normal measurements of 60 and 80 nanometers, correspondingly (Zhang et al. 2012; Estevam et al. 2017). Such particles are assembled from the microbes and yeasts after rupture of the cell. The investigators have wandered about conceivable implementations as nourishment increments and most likely as antimicrobial operators as a portion of such particles uncover an unambiguous antimicrobial movement (Zhang et al. 2012; Estevam et al. 2017; Skalickova et al. 2017).

In the area of farming, likely implementations have considerably much surfaces and a conceivable "cap trap" of immediately enhancing dirt with selenium for braced sustenance items, of giving plants components for their characteristic obstruction frameworks, and of dispensing with plant pathogens appears to be doable (Estevam et al. 2017). Inside this unique situation, one needs to push such normally delivered particles are not proportionate to modernly created materials. They are not "artificially unadulterated" and every now and again likewise contain a "characteristic" covering of proteins whose design is an impression of the yeasts or microscopic organisms they have been made in. Later the organic action of such characteristic particles may come from the mass material of the molecule itself, for example, selenium, from different mixes kept or restricted to the molecule and furthermore from the covering, which is as often as possible wealthy in proteins (Prakash et al. 2009; Wang et al. 2010; Yazdi et al. 2012; Estevam et al. 2017). In cases like these, a broad "intracellular diagnostics" is needed to explain the correct target(s) and point-by-point method(s) of activity (Mániková et al. 2014). Eventually, one may foresee a complex procedure by which microorganisms are created or tainted and by remediating that dirt delivers unmistakable nanoparticles which might be reaped and push off in prescription, horticulture, or several more appropriate applications. The significant merits of these techniques might be impressive and are not unrealistic additionally, as germane contaminants, for example, extensive metals, regularly likewise describe the establishment of mostly intriguing particles. In any case, there might be some further advantages, particularly with regard to pathogenic organisms and microorganisms. Various examinations have built up that the arrangement of nanomaterials by or inside pathogenic microscopic organisms is a compelling instrument to cancel those living beings. It has been seen, for example, that pathogenic strains of *Staphylococcus aureus*, for example, HEMSA and HEMSA 5 M, reduce SeO_3^{2-} to basic selenium when tested with incredibly extraordinary convergences of this anion (around 2 mM) in a shallow endeavor to manage this experience (Estevam et al. 2015). At last, this defensive plan comes up short, and the stores of selenium framed inside the microscopic organisms kill these cells. This sort of "self-destructive normal nanotechnology" is found between numerous microorganisms and organisms, just as pathogenic ones. It somewhat clarifies to some degree the antimicrobial activity frequently related with reactive selenium species (RSeS), such as SeO_3^{2-}, SeO_4^{2-}, TeO_3^{2-}, and TeO_4^{2-}. These activities might be explicit for specific life forms, giving these operators and chaperon forms with specific "sensor/effector" properties. Later on, this sort of common nanotechnology thusly can give an intriguing chance to trade off, debilitate, harm, or potentially even execute such pathogenic living beings (Estevam et al. 2017).

2.4 Processing Natural Materials

Various methodologies had great impact on processing of natural materials. The subsequent particles of such regular items are of an elite nature, as they are common items, yet being twisted to an uncommon, unnatural shape and size. It is along these lines barely amazing that numerous normal items have been nanosized in the most recent few years. Cancer prevention agents, for example, rutin, have been transformed into alleged "nanocrystals" utilizing an expressive procedure which incorporates wet globule processing (WBM) and high-weight homogenization (HPH) (Müller and Keck 2008; Mauludin et al. 2009). Here, nanotechnology can be utilized to create nanoparticles with a drastically enhanced solvency, phenomenal discharge energy, and later a decent bioavailability and organic action. This technique is especially alluring in the area of—frequently sparingly dissolvable—cell reinforcements and plant items wealthy in such cancer prevention agents, i.e., materials which initially have deficient discharge energy on the lipid/fluid skin surface; however on account of the new innovation, it can nowadays be utilized effectively, for example, in beauty care products. To be sure, the guideline of nanosizing coarse materials to advance their organic movement is extremely straightforward and is for the most part dependent on the Noyes-Whitney condition, one of the real conditions in biopharmacy.

For this situation, the immersion dissolvability increments because of a higher disintegration weight, which is elucidated by a greater shape of the particles (Kelvin condition) and the diffusional separate likewise diminished (Prantl condition). Nanosizing prompts an imperative increment in the general speed of disintegration, which is particularly intriguing if dynamic constituents break up gradually or not properly soluble in water. Besides, nanosizing enhances the bioactivity of inadequately solvent dynamic fixings. Because of the expansion in dissolvability, the fixation slope, when contrasted with bigger estimated materials, is expanded (Keck and Müller 2006). Because of these more prominent structures and the simplicity of generation, nanosizing, i.e., the creation of nanocrystals, has turned into a noteworthy readiness guideline in pharmaceutics to propel the bioactivity of dynamic fixings (Müller and Keck 2012; Scholz and Keck 2015). A portion of these processed plant materials have been assessed as likely nourishment supplementation and even as characteristic medications and antimicrobial specialists (Griffin et al. 2016). NPs of normal items must be utilized in the grounds of sustenance, medication, and beautifying agents or on account of substantial scale getting together, in "green" horticulture. The exercises trial for those nanosized materials are regularly encouraging, yet there is a well parity of contentions which should be estimated. Therapeutic plant is equivalently clear and altogether simpler than withdrawal, refinement, and detailing of the dynamic component(s) kept in that. It additionally delivers no or minimal waste. Also, NPs that are basically "normal," in any event to the extent their concoction organization is unstable, that contain every one of the elements in plant will not face any broad alterations and would not be treated with natural solvents. In a perfect world, they even portray a characteristic moderate discharge assembly of biologically available and naturally dynamic fixings. On account of high-pressure homogenization (HPH), such materials are additionally sanitized as progressively current investments.

2.5 Plant-Based Synthesis of Metallic NPs and Their Applications

NPs have exceptionally fascinating applications covering many areas (Chandran et al. 2006). Biological ways of making natural elements capable of reusing are shown in Fig. 2.2.

2.5.1 Traditional Strategies of Metals

Anciently metals were considered symbol of strength, and even nowadays some metals are considered as influential like gold and silver. In the eighteenth era, gold was used for mental and spiritual purification. Egyptians utilize gold

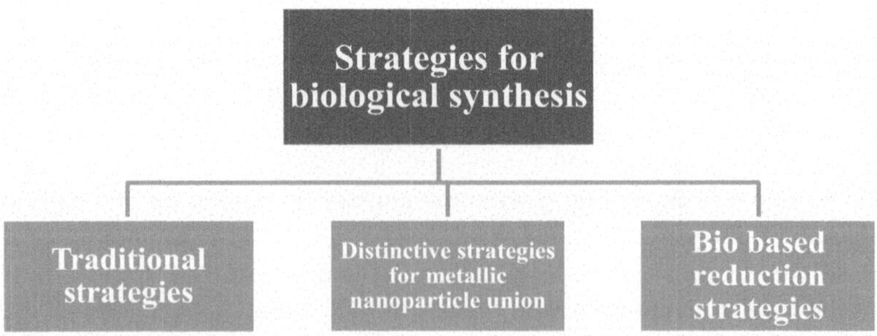

Fig. 2.2 Biological ways for synthesis of metallic nanoparticles

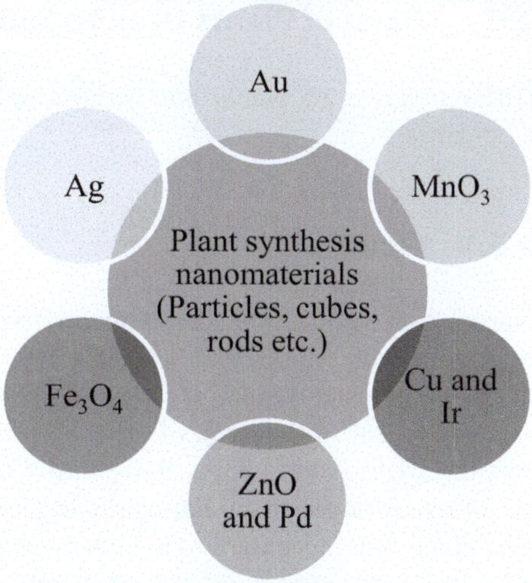

Fig. 2.3 Different plant-derived metallic nanoparticles

metal-solubilized water. The significance of gold is as yet respected in countryside areas, as the employees there prepare their rice dishes with a gold pellet to supersede the mineral lack in the body by nourishment admission. Silver is used for the healing of wounds and ulcer removal (Singh et al. 2013). Truth be told, the silver nanoparticles (AgNPs) which are colloidal in nature were utilized against antimicrobes, as wound dressing material, as tooth bond, and as a water purifier (Narayanan and Sakthivel 2011).

2.5.2 Distinctive Strategies for Union Metallic Nanoparticle

Some techniques are employed for the union of nanoparticles (nanoparticles, such as natural, compound, enzymatic, and physical). Physical strategies include ball processing, warm dissipate, beat laser desorption, atomic shaft, and flame amalgamation dispersion of NPs (Joerger et al. 2000). Synthetic techniques are utilizing radiation of high frequency and settling operators which are hurtful to human well-being. Figure 2.3 exhibits various kinds of metallic nanoparticles formed from plant means.

2.5.3 Bio-based Reduction Strategies

Silver is the biochemical feedback of silver nitrate that basically prompts the arrangement of AgNPs with the help of plant stock (Tripathy et al. 2010). The plants possess many vital biomolecules, like proteins, amino acids, and chemicals; moreover, they have some metals too. Then all these biomolecules must engage with bioreduction mechanism. So the conclusion in this metal like gold was the conversion and reduction of Au^+ into metallic Au_0 nanoparticles in the redox catalysts (Thakkar et al. 2010).

2.6 Parts of Plants Used to Synthesize Nanomaterials

Different parts of plants are being employed in the formation of eco-friendly nanoparticles.

2.6.1 Flowers

Noruzi et al. (2011) considered an environment-friendly disposed strategy amalgamation of gold nanoparticles (GNPs) by utilizing flower petals. The concentrated media contains sufficient proteins and sugars. A portion of these utilitarian mixes are the critical wellsprings of tetrachloroaurate salt, which declines into the main part of GNPs. Additionally, flowers of *Clitoria ternatea* and *Catharanthus roseus* are utilized for the metal nanoparticle combination with required shape and sizes. The plant material-incorporated nanoparticles are viably handling disease-causing microbes, and correspondingly the therapeutic utilizable *Nyctanthes arbor-tristis* flowers for GNPs extracts are obtained through green science technique (Das et al. 2011). Fluid concentrate of flowers of *Mirabilis jalapa* goes about as a reducing agent and delivered GNPs with eco-friendly technique (Vankar and Bajpai 2010). Table 2.1 delineates the plant metabolites which speak to in the bioreduction response to union of metal NPs and their applications in pharmacology.

Table 2.1 Different types of plant-based metallic nanoparticles (Kuppusamy et al. 2016)

Plant used	Part of plants	Nanoparticles	Shape	Size (nm)	Plants involved in bioremediations
Acalypha indica	Leaf	Ag, Au	Spherical	20–30	Quercetin, plant pigment
Alternanthera sessilis	Whole	Ag	Spherical	40	Amine, carboxyl group
Aloe vera	Leaf	In_2O_3	Spherical	50	Biomolecules
A. paniculata	Leaves	Ag	Spherical	67–88	Alkaloid, flavonoids
A. mexicana	Leaves	Ag	Spherical	20–50	Proteins
Boswellia serrata	Gum	Ag	Spherical	70–90	Secondary metabolites
Caria papaya	Fruit	Ag	Spherical	15	Catechins
Cassia fistula	Stem	Au	Spherical	55–98	Hydroxyl group
Cinnamon zeylanicum	Leaves	Ag	Spherical	45	Water-soluble organics
Citrullus colocynthis	Calli	Ag	Triangle	5–70	Polyphenols
Citrus sinensis	Peel	Ag	Spherical	35	Water-soluble compound
Dillenia indica	Fruit	Ag	Spherical	11–24	Biomolecules
Dioscorea bulbifera	Tuber	Ag	Rod-triangular	8–24	Ascorbic acid
Euphorbia prostrata	Leaves	Ag	Rod, spherical	52	Proteins, phenols
Mirabilis jalapa	Leaves	Au	Spherical	90–150	Polysaccharides

2.6.2 Stem

Shameli et al. in 2012 investigated the stem of *Callicarpa maingayi* with removed methanolic group utilized for amalgamation of AgNPs and shaped [Ag (Callicarpamaingayi)] + complex. The plant extracts have aldehyde collections, and it's for the most part associated with the decrease of silver particles into metallic Ag nanoparticles. The distinctive useful gathering demonstrates amide and polypeptides that are dependable mixes with topping of ionic substances into metal NPs. The atomic examinations on biosynthesis of silver precious stones are intricate and not yet completely comprehended. Be that as it may, some past investigations are proposed demonstrating components of nanoparticle cooperation with pathogenic living beings. The biologically produced silver nanoparticles have protein external cell mass of microscopic organisms, growths, or entities of viruses that degrade the lipoproteins of microbe cell divider. At last the division of cell was ceased and cell prompts passing. At room temperature silver nanoparticle photosynthesis utilized extracts of *Cissus quadrangularis* (Vanaja et al. 2013). The removed stem of some portion of plants demonstrates the distinctive utilitarian gatherings, especially the amine, phenolic, carboxyl, and aggravates that are associated with the decrease of silver particles. Thus, incorporated silver nanoparticles indicate greater action against *Bacillus subtilis* and *Klebsiella planticola* pathogenic microscopic organisms. Along these lines, the biosynthesized metal nanoparticles went about as great antibacterial specialists. Various plant parts involved in formation of nanoparticles are shown in Fig. 2.4.

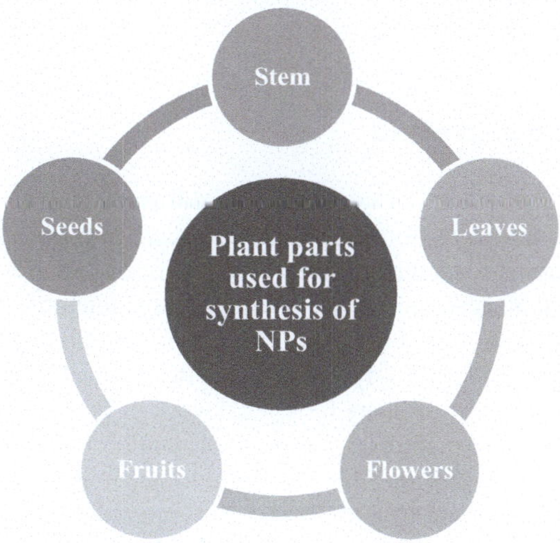

Fig. 2.4 Different types of parts of plants involved in synthesis of nanomaterials

2.6.3 Seeds

A few seeds that extricate like that of fenugreek contain high substance of secondary metabolites like flavonoids and other bioactive items, for example, lignin and nutrients. *Chloroauric corrosive* could be utilized for breakdown of solid parts of the fenugreek seeds. The COO gathering (carboxylic) and C=N and C=C useful gatherings are separately lying in the seed. The useful gathering of metabolites goes about as a surfactant of GNPs, and the flavonoids can settle the electrostatic adjustment of GNPs (Mittal et al. 2013). The fluid drawouts of *Macrotyloma uniflorum* impact the decrease of silver particles. Caffeic corrosive might be expected in the concentrate. Subsequently, decreased caffeic corrosive response could happen in one moment.

2.6.4 Fruits

Gopinath et al. (2012) utilized plant organic product of *Tribulus terrestris* extricate with expansion of various molar convergences of silver nitrate arrangement so as to incorporate eco-accommodating silver NPs with certain morphological highlights. The concentrate contains dynamic phytochemical that are subject for the single step decrease response. The round states of AgNPs were delivered by the *T. terrestris* extricates and proved commendable antimicrobial action against multidrug safe human pathogens. There is comparable cover utilizing polyphenol from grapes to combine palladium NPs and act viably against bacterial maladies (Amarnath et al. 2012). Not withstanding this, *Rumex hymenosepalus* serves as a balancing out operator for AgNPs. The utilization of ideal physic-concoction strategies to create nanomaterials is exceptionally valuable in pharmacological proposition to treat numerous endemic illnesses.

2.6.5 Leaves

Leaves that extricate are utilized as mediators to shape nanoparticles. Leaves of *Murraya koenigii*, *Centella asiatica*, and *Alternanthera sessilis*, including various leaves separate, are studied. Most recently leaves of *Piper nigrum* contained a critical bioactive material which is taking an interest in the nanoparticle arrangement by eco-philic strategy. The naturally orchestrated silver nanoparticles of 100 lg/ml concentration were dynamic medication, which focus on HEp-2 and HeLa cell line to control the ordinary biological work in disease cells. The AgNPs are powerful medication in malignant growth treatment to fix oncology in addition to terrifying disorders. *P. nigrum* removes longumine, it goes about as a topping operator for the arrangement of AgNPs and may upgrade the cytotoxic impacts of the tumor cells (Jacob et al. 2012). A green blend of AgNPs utilizing plant *Artemisia nilagirica* leave extracts has been depicted by Vijayakumar et al. (2013). It is by all accounts a crucial device for antimicrobial operators for now and close fates like that of silver nanoparticle amalgamation from the plants that control diverse pathogens in human.

2.7 Plant-Derived Formation of Silver Nanoparticles

Both physical and chemical methods had been used for the synthesis of silver nanoparticles for an extended period, but now for this purpose, developments have found the principal role of biological systems (Fig. 2.5).

Chemical and physical procedures are energy-exhaustive operations which indicate high production cost. Synthesis of AgNPs by chemical methods such as ethylene glycol, hydrazine hydrate, sodium borohydride, and dimethyleformamide can cause absorption of dangerous chemicals on the nanoparticle surfaces leading to toxicity problems (Iravani 2011). Furthermore, aqua chemical paths form nanocrystalline silver colloids which show accumulation with time, which means accommodation with the size factor upon storage. Plant derivatives have been explored as much better nominees over other biological systems, e.g., fungi and microorganisms, among the bio-based formation of silver nanoparticles, as they do not demand

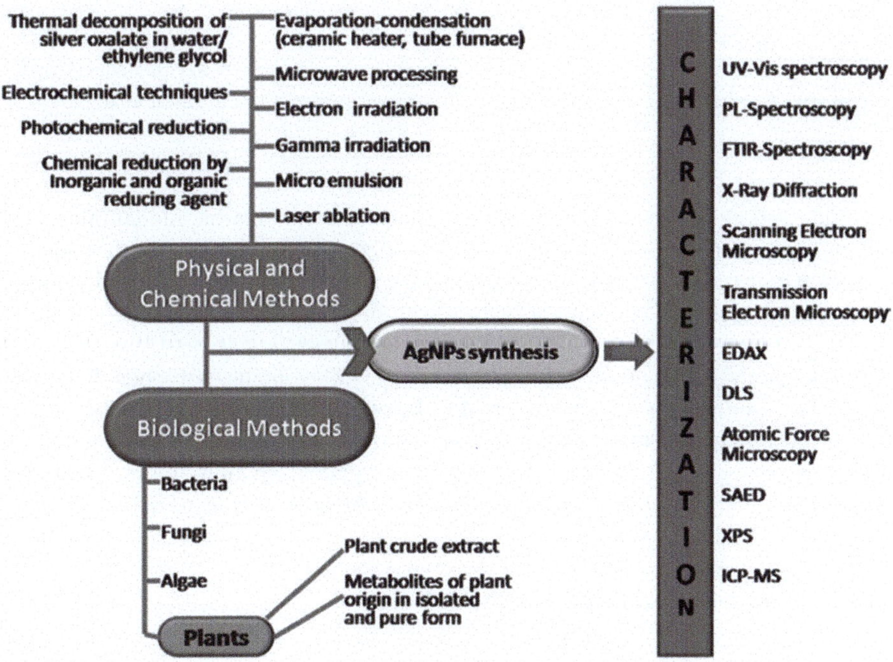

Fig. 2.5 Chemical, physical, and biological methods for the synthesis of AgNPs and the techniques employed for the characterization of produced nanoparticles (Patel et al. 2015a, b)

Table 2.2 Investigations by several researchers on plant-based AgNP synthesis and characteristics of formed nanoparticles

Plants	AgNP specifications	Plant parts	References
Citrus sinensis	10–35 nm	Peel	Kaviya et al. (2011)
Allium sativum	Spherical, 7.3 nm	Garlic cloves	Rastogi and Arunachalam (2011)
Allium cepa	Spherical, 33.6 nm	Leaves	Saxena et al. (2010)
Capsicum annuum	Spherical and crystalline, 10–70 nm	Fruit	Li et al. (2007)
Daucus carota	31–52 nm	Taproot spherical	Mukunthan and Balaji (2012)
Eucalyptus globulus	5–50 nm	Bark	Astalakshmi et al. (2013)
Mangifera indica	Triangular, hexagonal, and nearly spherical, 20 nm	Leaves	Philip (2010)
Ocimum sanctum	5–10 nm	Root and stem	Ahmad et al. (2010)
Piper betle	17–120 nm	Leaves	Rani and Rajasekharreddy (2011)
Zingiber officinale	6–20 nm	Rhizome	Kumar et al. (2012)
Trachyspermum ammi and *Papaver somniferum*	60–80 nm	Seeds	Vijayaraghavan et al. (2012)
Rosa rugose	12 nm	Leaves spherical	Dubey et al. (2010)
Svensoniahyderobadensis	24 nm	Roots	Rao and Savithramma (2013)
Parthenium hysterophorus	20–70 nm	Leaves	Anwar et al. (2015)

harmful capping and reducing factors, high temperature, radiation, fungal/microbial strains and expensive media for growth of microbes/fungi as well as for the production of nanoparticles. During formation and implementation zones, they also minimize the possibility of contamination or infection (Borase et al. 2014).

Many other researchers have also investigated the participation of several plant-based metabolites like amines (Prasad et al. 2011), flavonoids, polyphenols, terpenoids (Marimuthu et al. 2011), aldehydes, ketones (Chandran et al. 2006), starch (Vigneshwaran et al. 2006), arabinose and galactose (Kora et al. 2010), and saponins (Elavazhagan and Arunachalam 2011) in AgNP formation. Table 2.2 shows utilization of various plant parts in the formation of silver nanomaterials.

2.8 Plant-Based Gold Nanoparticle

Gold-derived nanoparticles have several applications in various fields. Au(3+) ion is reduced to the Au atom by binding of the atom to the cell surface, whereas further reduced Au also combines and clumps to produce GNPs. Au ions have good chances

Table 2.3 Plant-based synthesis of gold nanoparticles

Plants	AuNP size	Plant parts	References
Citrus maxima	25.7710 nm	Fruit	Yu et al. (2016)
Ixora coccinea	5–10 nm	Flower	Nagaraj et al. (2011)
Coleus forskohlii	5–18 nm	Root	Naraginti et al. (2016)
Cassia fistula	55.2–98.4 nm	Stem bark	Daisy and Saipriya (2012)
Punica granatum	70 nm	Fruit peel	Ganeshkumar et al. (2013)
Argemone mexicana	22–26 nm	Leaf	Varun et al. (2015)
Hygrophila spinosa	50–80 nm	Leaf	Koperuncholan (2015)
Abelmoschus esculentus	45–75 nm	Seed	Jayaseelan et al. (2013)
Morinda citrifolia	12.17–38.26 nm	Root	Suman et al. (2014)
Hibiscus sabdariffa	10–60 nm	Leaf and stem	Mishra et al. (2014)

at the starting phases to attach with the atom and produce clumps (Cai et al. 2011). GNPs have applications in the field of medicine (Honary et al. 2012). The bio-based synthesis of nanoparticles has specific morphology and determined shapes such as hierarchical tubes, triangle, hexagonrods, decahedrons, icosahedrons, and nanotriangles as well as nodous ribbons. Different nanoparticles have different efficient applications in various fields; thus researchers are now focusing on their shapes (Du et al. 2007). GNPs are being produced by utilizing several plant parts, e.g., stem, leaf, root, and fruit as showed in Table 2.3.

2.9 Plant-Based Zinc Oxide Nanoparticles

Zinc oxide (ZnO) nanoparticles are currently of greater interest due to their certain characteristics, e.g., their production is cheap, safe to use, and can be synthesized easily (Elumalai et al. 2015). The reason of the attention of researchers toward zinc oxide NPs is their vast applications in the sector of optics, electronics, and biomedical systems (Azizi et al. 2014). ZnO nanoparticles show enormous applications as anti-inflammatory (Ramesh et al. 2015) and wound-healing characteristics (Nagajyothi et al. 2013). They have ultraviolet filtering characteristics thus being used in cosmetics, e.g., sunscreen creams and lotions. They have also vast applications in medicines such as anticancer, drug delivery, antifungal, antibacterial, antidiabetic, and agricultural characteristics (Ramesh et al. 2015). They have also been used in manufacturing rubber and paint, in applications in dentistry, and in eliminating arsenic and sulfur from water (Ali et al. 2016). Anbuvannan et al. (2015) reported ZnO nanoparticles have several morphologies like nanorods, nanowires, nanoflowers, nanoflakes, and nanobelts. Table 2.4 shows plant-based synthesis of ZnO NPs derived from different plant parts like flowers, stem, leaves, and fruit peels.

Table 2.4 Plant-based synthesis of ZnO nanoparticles

Plants	Shape	Plant parts	References
Rosa canina	Spherical	Fruit extract	Jafarirad et al. (2016)
Solanum nigrum	Hexagonal and quasispherical	Leaf extract	Ramesh et al. (2015)
Cocos nucifera	Spherical and hexagonal	Coconut water	Krupa and Vimala (2016)
Gossypium	Spherical and nanorod	Cellulosic fiber	Aladpoosh and Montazer (2015)
Calotropis gigantea	Spherical	Fresh leaves	Vidya et al. (2013)
Nephelium lappaceum	Spherical and hexagonal	Fruit peels	Yuvakkumar et al. (2015)
Coptidis rhizoma	Spherical and rod shaped	Dried rhizome	Nagajyothi et al. (2014)
Vitex negundo	Hexagonal	Flowers	Ambika and Sundrarajan (2015)

2.10 Biofuel Applications of Nanoparticles

Commercialization process of biofuels is limited due to hinders in their production processes. Productions of high-energy yield and cost-effective biofuels are the main challenges. Nanoparticles in this regard may play a significant role in forming this process economically viable.

2.10.1 Role in Pretreatment

Most frequently found biopolymer is the lignocellulosic biomass rich in carbohydrate content. Two-thirds of the lignocellulosic biomass consists of cellulose and hemicellulose and are among the cheapest sources for production of biofuel (Srivastava et al. 2015a). Lignin barrier is removed prior to saccharification by pretreatments to expose cellulose and hemicellulose (Alvira et al. 2010). Biological and chemical pretreatment methods are common among various pretreatment procedures for breakdown of lignocellulosic matter (Srivastava et al. 2015a). Although pretreatment is considered as the expensive phase in the transformation of cellulose, it has the ability to enhance the efficiency remarkably and lower the net cost of biofuel production (Alvira et al. 2010; Srivastava et al. 2015a). Degrading enzymes, e.g., cellulases and hemicellulases, are employed after the pretreatment to liberate the fermentable sugars (Rawat et al. 2014). Thus many investigators are working on pretreatment process to improve the bioconversion efficiency of lignocellulosic biomass. Wei et al. (2015) observed that production of sugar from corn stover was improved in the existence of iron oxide nanoparticle by employing the acid pretreatment. Furthermore, nanoparticle-acid pretreated substrate exhibited nearly 13–19% more xylose and glucose relative to control (acid-treated substrate without nanoparticles). This study found that iron oxide nanoparticle can aid the pretreatment process, as results exhibited improved sugar formation over the metallic iron nanoparticles. The experiment was conducted at 100 °C which showed its economic viability, and the reported data exhibited a positive link among the concentrations of released sugar and iron during the biomass pretreatment. In another investigation done by Yang et al. (2015), breakdown of cellulosic biomass was improved in the existence of decreased graphene oxide functionalized with iron oxide nanoparticles. The iron oxide reduced graphene oxide-SO_3H (Fe_3O_4-RGO-SO_3H); nanocomposite was successfully produced employing the reduced graphene oxide (RGO), containing iron oxide nanoparticles and benzene sulfonic acid which were directly harbored on the surface of RGO through C-C covalent bonds. The distinctive structure of Fe_3O_4-RGO-SO_3H nanocomplex along with increased dispersion in water sustains the reachability of cellulose to the active sites and enhances the production of sugar which can further be employed for the biofuel production. Although these investigations have explored a new zone for the production of biofuels employing the nanoparticles, currently this avenue is at the very initial phase. So, many more attempts to improve the economic sensibility are needed.

2.10.2 Role in Cellulase Production and Stability

After the pretreatment process, enzymatic hydrolysis is done by cellulases to degrade lignocellulosic biomasses. For enhanced enzymatic hydrolysis, greatly efficient cellulases are needed, which are capable of working in unrelenting conditions. Furthermore uses of several cofactors, e.g., metal ions, are reported in the many investigations other than various substitute methods to make enhanced cellulase efficiency and its production in industries (Srivastava et al. 2014). To increase the stability of enzyme application of nanoparticles is a new avenue in the area of bioenergy production (Srivastava et al. 2014; Singh et al. 2016a, b). Few but encouraging and potential investigations have been outlined in this field currently. The study by Dutta et al. (2014) noticed an enhanced production of cellulase in the existence of hydroxyapatite nanoparticle employing the bacterial strain. Highly thermostable enzyme was used in this research which kept its half-life at 80 °C. Besides enhanced thermal stability, these authors also observed an enhancement in reducing sugars when substrates, e.g., rice husk and rice straw, were used. Similarly Srivastava et al. (2015a) reported an improved cellulase production, its thermal strength, and sugar production in the existence of iron oxide/alginate nanocomplex. A higher sugar yield was observed in this study by employing thermotolerant fungal sp. *Aspergillus fumigatus* AA001 in the solid-state fermentation using the same nanocomplex. Furthermore, thermal stability of cellulase enzyme was also improved besides its production in the existence of uncovered iron oxide nanoparticle as relative to control. The production of cellulase was enhanced in the existence of pure iron oxide nanoparticle and iron oxide/alginate nanocomposite by 35% and 40%, respectively, when compared to control. In addition, iron oxide/alginate nanocomplex treated cellulase exhibited its thermal stability at 70 °C for 8 h by maintaining its 56% of relative activity, whereas crude cellulase could keep only 19%. These investigations distinctly evince that the nanoparticles may play a significant part to change the whole process of bioconversion. Srivastava et al. (2014) suggested that the major cause of enhancement in the thermal stability of cellulase was cellulase immobilization on to the nanoparticles. Besides, thermal stability and an improved cellulase production have been noticed in the existence of nickel cobaltite nanoparticle by thermotolerant *Aspergillus fumigatus* NS (class: Eurotiomycetes) through solid-state fermentation. Furthermore, untreated cellulase showed thermal stability at 80 °C for 7 h in the presence of $NiCo_2O_4$ nanoparticles, whereas at the same temperature, untreated cellulase was stable up to 4 h. Moreover it was also observed that production time was lessening in the existences of nanoparticles in above studies. Therefore, in the near future, nanoparticles may prove their capability in biofuel production process. Verma et al. (2013) reported improvement in the above stated investigations; there are also many researches that have concluded improvements in the hydrolysis efficiency of cellulases besides its better production and thermostability in the presence of various kinds of nanomaterials, e.g., iron oxide, zinc oxide nanoparticles, etc. Ansari and Husain (2012) immobilize cellulase on iron oxide magnetic nanoparticles. They reported that the nanoparticles may act as a carrier and provide resistance against the unfavorable pH and inhibitor besides thermal stability. Verma et al. (2013) observed that the thermostability of the enzyme β-glucosidase was enhanced in the occurrence of iron oxide magnetic nanoparticles and showed the half-life of that enzyme at 70 °C. As nanoparticles have explored a new path to enhance the production of cellulase and its thermal stability, their exact mechanism is not well known. So, further studies are focusing on this specific area for its industrial and economic viability.

2.10.3 Role in Saccharification

Saccharification or hydrolysis of lignocellulosic biomass is the next phase after pretreatment to liberate sugars using cellulase enzyme. Temperature 45–50 °C needed for enzymatic hydrolysis makes the whole process slow, more vulnerable to contamination by microbes, and usually incomplete, ending in decreased production of fermentable sugars and demand greater enzyme quantity. Hence, it is suggested that these bottlenecks can be overcome by employing the thermostable cellulase enzymes and thermophilic microbes (Yeoman et al. 2010). Cellulase that can tolerate increased temperatures can also work at higher hydrolysis temperatures. The presence of nanoparticles enhances the thermal stability of cellulase, and these thermostable cellulases may perform significantly at higher degradation temperatures. Some current investigations concluded the enhanced thermostability and degrading efficacy at increased temperatures (Dutta et al. 2014; Srivastava et al. 2015a, b).

Dutta et al. (2014) employed rice husk/rice straw as substrates, and improved thermal stability of cellulase enzyme was observed resulting in enhanced production of sugar in saccharification process. In this study calcium hydroxyapatite nanoparticles were used, and substrates were treated at 80 °C to get reducing sugars. Srivastava et al. (2015b) used iron oxide/alginate nanocomposite and found betterment in degradation efficacy of cellulase enzyme and sugar yield at 70 °C. Under the solid-state fermentation, greater sugar yield was reported at higher temperature employing *Aspergillus fumigatus* AA001 in the occurrence of iron oxide/alginate nanocomplex. Though nanoparticles have greater capability to enhance the degrading efficacy of cellulase enzyme, their mechanism is not very clear; so attention must be paid in this aspect for industrial scale production.

2.11 Optional Metabolite Impact on Bio-decrease Response

A considerable portion of the optional metabolites and a few proteins have reasonably advanced the combination of NPs which are metallic in nature from the other ionic mixes. The decreased response principally included plant biomolecules (optional metabolites, e.g., sugars (polysaccharides), proteins, natural mixes, shades, and plant tars. Plant normal items are engaged with the decrease response to incorporate green nanoparticles. Plants are especially taking an interest in guard components to create different substance mixes, for example, polyphenols, saponins, and cell reinforcement chemicals, like alkaloids. The proposed decrease response illustrated the primary components for the biological synthesis of metallic NPs. The plant removes various useful collections, for example, alkenyl, phenolic and liquor, amine, and carboxylic group. It is for the most part indicated as plant auxiliary metabolites and may be miniaturized scale and large-scale biomolecules (Jha et al. 2009). These substances are completely taking an interest for the NP generation. For example, *R. hymenosepalus* plant removes the advances of NP unions with quick response energy. Consequently, the dissolvable concentrate of *R. hymenosepalus* is rich in polyphenols, for example, stilbenes and catechins, atoms that go about as diminishing and balancing out specialists for silver NPs creation (Awwad et al. 2013).

2.12 Business Uses of Biosynthesized Nanoparticles

2.12.1 NPs in Waste Treatment NPs

Nanoproducts have various applications in everyday life. There are different ecologically accommodating products accessible in business showcase which are highly efficient, for example, bone and teeth concrete and handcrafted items (Kouvaris et al. 2012). For example, Au, Ag, and platinum NPs have wide applications in beauty products and they are utilized as fixings in different items, like, to protect against the UV rays sun blockers, toothpastes, mouthwash, and fragrances and shampoos (Kumar and Yadav 2009). NPs of silica act in fixing business items. The altered silica NPs are utilized as pesticide, and it is utilized in an assortment of non-farming implementations.

2.12.2 Beautifiers

Metallic NPs are utilized as additive specialists in restorative enterprises. New element of metallic nanoparticles is utilized for various business applications, for the most part beauty care products, medical materials, and sustenance additives (Songand and Kim 2009; Kokura et al. 2010). The metallic NPs like the platinum, gold, and Ag are connected for some business items, for example, soap, detergent, shampoo, and shoes.

2.12.3 NPs in Food Industry

Some metals like silver are quick warmth directing; therefore, nano-Ag is utilized in different mechanical frameworks. It is predominantly utilized in warmth-inclined instrumentation like in PCR cover (Weiss et al. 2006). In food businesses, sustenance items have very high microbial sullying because of different procedures, for example, assembling and handling of crude materials. Consequently, requirement is there to build up a savvy biological sensor to decide the nature of the items.

2.13 Component Blend of Metallic NPs

Difference highlighted in hydrogen particles is basically a reaction by variation in size and states of NP arrangement. Shankar et al. (2003) revealed that the extracts of *Aloe vera* provided Au-Ag center NPs in different sizes by changing the pH of the medium which is dissolvable. So, biologically synthesis of NPs by hay plant concentrate have variable size.

2.14 Conclusion

Environmental hazards caused by chemical synthesis of nanoparticles have urged synthesis of nanoparticles using plant means, as it includes eco-friendly methods for the synthesizing nanoparticles. This chapter focused on the plant-based nanoparticles and methods of their synthesis. Various plant parts, e.g., leaves, flowers, roots, stems, fruit and its peels, etc., are being used for the green synthesis of nanoparticles especially metallic nanoparticles. This chapter also overviews the applications of these plant-based nanoparticles in different fields, e.g., medical, agriculture, food industry, and cosmetic industry.

References

Ahmad N, Sharma S, Alam MK et al (2010) A rapid synthesis of silver nanoparticles using dried medicinal plant of basil. Colloids Surf B 81:81–86

Aladpoosh R, Montazer M (2015) The role of cellulosic chains of cotton in biosynthesis of ZnOnanorods producing multifunctional properties: mechanism, characterizations and features. Carbohydr Polym 126:122–129

Ali K, Dwivedi S, Azam A, Saquib Q, Al-Said MS, Alkhedhairy AA, Musarrat J (2016) Aloe vera extract functionalized zinc oxide nanoparticles as nanoantibiotics against multi-drug resistant clinical bacterial isolates. J Colloid Interface Sci 472:145–156

Alvira P, Tomás-Pejó E, Ballesteros M, Negro MJ (2010) Pretreatment technologies for an efficient bioethanol production process based on enzymatic hydrolysis: a review. Bioresour Technol 101(13):4851–4861

Amarnath K, Kumar J, Reddy T, Mahesh V, Ayyappan SR, Nellore J (2012) Synthesis and characterization of chitosan and grape polyphenols stabilized palladium nanoparticles and their antibacterial activity. Colloids Surf B: Biointerfaces 92:254–261

Ambika S, Sundrarajan M (2015) Antibacterial behaviour of Vitex negundo extract assisted ZnO nanoparticles against pathogenic bacteria. J Photochem Photobiol B Biol 146:52–57

Anbuvannan M, Ramesh M, Viruthagiri G, Shanmugam N, Kannadasan N (2015) Synthesis, characterization and photocatalytic activity of ZnO nanoparticles prepared by biological method. Spectrochim Acta A Mol Biomol Spectrosc 143:304–308

Ansari SA, Husain Q (2012) Potential applications of enzymes immobilized on/in nanomaterials: a review. Biotechnol Adv 30(3):512–523

Anwar MF, Yadav D, Kapoor S, Chander J, Samim M (2015) Comparison of antibacterial activity of Ag nanoparticles synthesized from leaf extract of Parthenium hystrophorus L in aqueous media and gentamicin sulphate: in-vitro. Drug Development and Industrial Pharmacy 41:43–50

Astalakshmi A, Nima P, Ganesan V (2013) A green approach in the synthesis of silver nanoparticles using bark of Eucalyptus globulus, Labill. Int J Pharm Sci Rev Res 23:47–52

Awwad AM, Salem NM, Abdeen AO (2013) Green synthesis of silver nanoparticles using carob leaf extract and its antibacterial activity. Int J Ind Chem 4(1):29

Azizi S, Ahmad MB, Namvar F, Mohamad R (2014) Green biosynthesis and characterization of zinc oxide nanoparticles using brown marine macroalga Sargassum muticum aqueous extract. Mater Lett 116:275–277

Berlo K, van Hinsberg VJ, Vigouroux N, Gagnon JE, Williams-Jones AE (2014) Sulfide breakdown controls metal signature in volcanic gas at Kawah Ijen volcano, Indonesia. Chem Geol 371:115–127

Borase HP, Salunke BK, Salunkhe RB, Patil CD, Hallsworth JE, Kim BS, Patil SV (2014) Plant extract: a promising biomatrix for ecofriendly, controlled synthesis of silver nanoparticles. Appl Biochem Biotechnol 173(1):1–29

Cai F, Li J, Sun J, Ji Y (2011) Biosynthesis of gold nanoparticles by biosorption using Magnetospirillum gryphiswaldense MSR-1. Chem Eng J 175:70–75

Chandran SP, Chaudhary M, Pasricha R, Ahmad A, Sastry M (2006) Synthesis of gold nanotriangles and silver nanoparticles using *Aloe vera plant extract*. Biotechnol Prog 22(2):577–583

Cho MH, Choi ES, Kim S, Goh SH, Choi Y (2017) Redox-responsive manganese dioxide nanoparticles for enhanced MR imaging and radiotherapy of lung cancer. Front Chem 5:109

Ciftcioglu N, McKay DS, Mathew G, Kajander OE (2006) Nanobacteria: fact or fiction? Characteristics, detection, and medical importance of novel self-replicating, calcifying nanoparticles. J Investig Med 54(7):385–394

Daisy P, Saipriya K (2012) Biochemical analysis of Cassia fistula aqueous extract and phytochemically synthesized gold nanoparticles as hypoglycemic treatment for diabetes mellitus. Int J Nanomedicine 7:1189

Das RK, Gogoi N, Bora U (2011) Green synthesis of gold nanoparticles using *Nyctanthes arbortristis* flower extract. Bioprocess Biosyst Eng 34(5):615–619

Du L, Jiang H, Liu X, Wang E (2007) Biosynthesis of gold nanoparticles assisted by Escherichia coli DH5α and its application on direct electrochemistry of hemoglobin. Electrochem Commun 9(5):1165–1170

Dubey SP, Lahtinen M, Sillanpää M (2010) Green synthesis and characterizations of silver and gold nanoparticles using leaf extract of Rosa rugosa. Colloids Surf A Physicochem Eng Asp 364(1–3):34–41

Duhan JS, Kumar R, Kumar N, Kaur P, Nehra K, Duhan S (2017) Nanotechnology: the new perspective in precision agriculture. Biotechnol Rep 15:11–23

Dutta N, Mukhopadhyay A, Dasgupta AK, Chakrabarti K (2014) Improved production of reducing sugars from rice husk and rice straw using bacterial cellulase and xylanase activated with hydroxyapatite nanoparticles. Bioresour Technol 153:269–277

Elavazhagan T, Arunachalam KD (2011) Memecylon edule leaf extract mediated green synthesis of silver and gold nanoparticles. Int J Nanomedicine 6:1265

Elumalai K, Velmurugan S (2015) Green synthesis, characterization and antimicrobial activities of zinc oxide nanoparticles from the leaf extract of Azadirachta indica (L.). Appl Surf Sci 345:329–336

Estevam EC, Witek K, Faulstich L, Nasim MJ, Latacz G, Domínguez-Álvarez E et al (2015) Aspects of a distinct cytotoxicity of selenium salts and organic selenides in living cells with possible implications for drug design. Molecules 20(8):13894–13912

Estevam EC, Griffin S, Nasim MJ, Denezhkin P, Schneider R, Lilischkis R et al (2017) Natural selenium particles from *Staphylococcus carnosus*: hazards or particles with particular promise? J Hazard Mater 324:22–30

Ezoe Y, Lin CH, Noto M, Watanabe Y, Yoshimura K (2002) Evolution of water chemistry in natural acidic environments in Yangmingshan, Taiwan. J Environ Monit 4(4):533–540

Ganeshkumar M, Sathishkumar M, Ponrasu T, Dinesh MG, Suguna L (2013) Spontaneous ultrafast synthesis of gold nanoparticles using Punica granatum for cancer targeted drug delivery. Colloids Surf B: Biointerfaces 106:208–216

Gopinath V, MubarakAli D, Priyadarshini S, Priyadharsshini NM, Thajuddin N, Velusamy P (2012) Biosynthesis of silver nanoparticles from Tribulus terrestris and its antimicrobial activity: a novel biological approach. Colloids Surf B: Biointerfaces 96:69–74

Griffin S, Tittikpina NK, Al-Marby A, Alkhayer R, Denezhkin P, Witek K et al (2016) Turning waste into value: Nanosized natural plant materials of *Solanum incanumL.* and *Pterocarpus erinaceus* Poir with promising antimicrobial activities. Pharmaceutics 8(2):11

Honary S, Gharaei-Fathabad E, Paji ZK, Eslamifar M (2012) A novel biological synthesis of gold nanoparticle by Enterobacteriaceae family. Trop J Pharm Res 11(6):887–891

Iavicoli I, Leso V, Beezhold DH, Shvedova AA (2017) Nanotechnology in agriculture: opportunities, toxicological implications, and occupational risks. Toxicol Appl Pharmacol 329:96–111

Iravani S (2011) Green synthesis of metal nanoparticles using plants. Green Chem 13(10):2638–2650

Jacob C (2011) Redox signalling via the cellular thiolstat. Biochem Soc Trans 39:1247–1253

Jacob SJP, Finub JS, Narayanan A (2012) Synthesis of silver nanoparticles using Piper longum leaf extracts and its cytotoxic activity against Hep-2 cell line. Colloids Surf B: Biointerfaces 91:212–214

Jafarirad S, Mehrabi M, Divband B, Kosari-Nasab M (2016) Biofabrication of zinc oxide nanoparticles using fruit extract of Rosa canina and their toxic potential against bacteria: a mechanistic approach. Mater Sci Eng C 59:296–302

Jayaseelan C, Ramkumar R, Rahuman AA, Perumal P (2013) Green synthesis of gold nanoparticles using seed aqueous extract of Abelmoschus esculentus and its antifungal activity. Ind Crop Prod 45:423–429

Jha AK, Prasad K, Kumar V, Prasad K (2009) Biosynthesis of silver nanoparticles utilizing *Eclipta* leaf. Biotechnol Prog 25:1475–1477

Joerger R, Klaus T, Granqvist CG (2000) Biologically produced silver–carbon composite materials for optically functional thin-film coatings. Adv Mater 12(6):407–409

Kajander EO, Ciftcioglu N, Miller-Hjelle MA, Hjelle JT (2001) Nanobacteria: controversial pathogens in nephrolithiasis and polycystic kidney disease. Curr Opin Nephrol Hypertens 10(3):445–452

Kaviya S, Santhanalakshmi J, Viswanathan B, Muthumary J, Srinivasan K (2011) Biosynthesis of silver nanoparticles using Citrus sinensis peel extract and its antibacterial activity. Spectrochim Acta A Mol Biomol Spectrosc 79(3):594–598

Keck CM, Müller RH (2006) Drug nanocrystals of poorly soluble drugs produced by high pressure homogenisation. Eur J Pharm Biopharm 62(1):3–16

Kettler K, Krystek P, Giannakou C, Hendriks AJ, de Jong WH (2016) Exploring the effect of silver nanoparticle size and medium composition on uptake into pulmonary epithelial 16HBE14o-cells. J Nanopart Res 18(7):182

Kokura S, Handa O, Takagi T, Ishikawa T, Naito Y, Yoshikawa T (2010) Silver nanoparticles as a safe preservative for use in cosmetics. Nanomed Nanotechnol Biol Med 6(4):570–574

Koperuncholan M (2015) Bioreduction of chloroauric acid (HAuCl4) for the synthesis of gold nanoparticles (GNPs): a special empathies of pharmacological activity. Int J Phytopharm 5(4):72–80

Kora AJ, Sashidhar RB, Arunachalam J (2010) Gum kondagogu (Cochlospermum gossypium): a template for the green synthesis and stabilization of silver nanoparticles with antibacterial application. Carbohydr Polym 82(3):670–679

Kouvaris P, Delimitis A, Zaspalis V, Papadopoulos D, Tsipas SA, Michailidis N (2012) Green synthesis and characterization of silver nanoparticles produced using Arbutus unedo leaf extract. Mater Lett 76:18–20

Krupa AND, Vimala R (2016) Evaluation of tetraethoxysilane (TEOS) sol–gel coatings, modified with green synthesized zinc oxide nanoparticles for combating microfouling. Mater Sci Eng C 61:728–735

Kumar V, Yadav SK (2009) Plant-mediated synthesis of silver and gold nanoparticles and their applications. J Chem Technol Biotechnol 84(2):151–157

Kumar KP, Paul W, Sharma CP (2012) Green synthesis of silver nanoparticles with Zingiber officinale extract and study of its blood compatibility. BioNanoScience 2(3):144–152

Kuppusamy P, Yusoff MM, Maniam GP, Govindan N (2016) Biosynthesis of metallic nanoparticles using plant derivatives and their new avenues in pharmacological applications–An updated report. Saudi Pharm J 24(4):473–484

Li S, Shen Y, Xie A, Yu X, Qiu L, Zhang L, Zhang Q (2007) Green synthesis of silver nanoparticles using Capsicum annuum L. extract. Green Chem 9(8):852–858

Li X, Xu H, Chen ZS, Chen G (2011) Biosynthesis of nanoparticles by microorganisms and their applications. J Nanomater 2011:16

Mániková D, Letavayová LM, Vlasáková D, Košík P, Estevam EC, Nasim MJ et al (2014) Intracellular diagnostics: hunting for the mode of action of redox-modulating selenium compounds in selected model systems. Molecules 19(8):12258–12279

Marimuthu S, Rahuman AA, Rajakumar G, Santhoshkumar T, Kirthi AV, Jayaseelan C et al (2011) Evaluation of green synthesized silver nanoparticles against parasites. Parasitol Res 108(6):1541–1549

Mauludin R, Müller RH, Keck CM (2009) Development of an oral rutin nanocrystal formulation. Int J Pharm 370(1–2):202–209

Mishra V, Mishra RK, Dikshit A, Pandey AC (2014) Interactions of nanoparticles with plants: an emerging prospective in the agriculture industry. In: Emerging technologies and management of crop stress tolerance. Academic, San Diego, pp 159–180

Mittal AK, Chisti Y, Banerjee UC (2013) Synthesis of metallic nanoparticles using plant extracts. Biotechnol Adv 31(2):346–356

Moss DM, Siccardi M (2014) Optimizing nanomedicine pharmacokinetics using physiologically based pharmacokinetics modelling. Br J Pharmacol 171(17):3963–3979

Mukherjee P, Ahmad A, Mandal D, Senapati S, Sainkar SR, Khan MI et al (2001) Bioreduction of AuCl4− ions by the fungus, *Verticillium sp.* and surface trapping of the gold nanoparticles formed. Angew Chem Int Ed 40(19):3585–3588

Mukunthan KS, Balaji S (2012) Silver nanoparticles shoot up from the root of Daucus carota (L.). Int J Green Nanotechnol 4(1):54–61

Müller RH, Keck CM (2008) Second generation of drug nanocrystals for delivery of poorly soluble drugs: smart crystal technology: 134. Eur J Pharm Sci 34(1):S20–S21
Müller RH, Keck CM (2012) Twenty years of drug nanocrystals: where are we, and where do we go?.*European*. J Pharm Biopharm 80(1):1–3
Murr LE, Guerrero PA (2006) Carbon nanotubes in wood soot. Atmos Sci Lett 7(4):93–95
Nagajyothi PC, An TM, Sreekanth TVM, Lee JI, Lee DJ, Lee KD (2013) Green route biosynthesis: characterization and catalytic activity of ZnO nanoparticles. Mater Lett 108:160–163
Nagajyothi PC, Sreekanth TVM, Tettey CO, Jun YI, Mook SH (2014) Characterization, antibacterial, antioxidant, and cytotoxic activities of ZnO nanoparticles using Coptidis Rhizoma. Bioorg Med Chem Lett 24(17):4298–4303
Nagaraj B, Krishnamurthy NB, Liny P, Divya TK, Dinesh R (2011) Biosynthesis of gold nanoparticles of Ixora coccinea flower extract & their antimicrobial activities. Int J Pharm Bio Sci 2(4):557–565
Naraginti S, Kumari PL, Das RK, Sivakumar A, Patil SH, Andhalkar VV (2016) Amelioration of excision wounds by topical application of green synthesized, formulated silver and gold nanoparticles in albino Wistar rats. Mater Sci Eng C 62:293–300
Narayanan KB, Sakthivel N (2011) Green synthesis of biogenic metal nanoparticles by terrestrial and aquatic phototrophic and heterotrophic eukaryotes and biocompatible agents. Adv Colloid Interf Sci 169(2):59–79
Noruzi M, Zare D, Khoshnevisan K, Davoodi D (2011) Rapid green synthesis of gold nanoparticles using Rosa hybrida petal extract at room temperature. Spectrochim Acta A Mol Biomol Spectrosc 79(5):1461–1465
Pasula RR, Lim S (2017) Engineering nanoparticle synthesis using microbial factories. Eng Biol 1(1):12–17
Patel P, Agarwal P, Kanawaria S, Kachhwaha S, Kothari SL (2015a) Plant-based synthesis of silver nanoparticles and their characterization. In: Nanotechnology and plant sciences. Springer, Cham, pp 271–288
Patel V, Berthold D, Puranik P, Gantar M (2015b) Screening of cyanobacteria and microalgae for their ability to synthesize silver nanoparticles with antibacterial activity. Biotechnol Rep 5:112–119
Pawlak J, Łodyga-Chruścińska E, Chrustowicz J (2014) Fate of platinum metals in the environment. J Trace Elem Med Biol 28(3):247–254
Philip D (2010) Green synthesis of gold and silver nanoparticles using Hibiscus rosa-sinensis. Phys E 42(5):1417–1424
Prakash NT, Sharma N, Prakash R, Raina KK, Fellowes J, Pearce CI, Pattrick RA (2009) Aerobic microbial manufacture of nanoscale selenium: exploiting nature's bio-nanomineralization potential. Biotechnol Lett 31(12):1857
Prasad KS, Pathak D, Patel A, Dalwadi P, Prasad R, Patel P, Selvaraj K (2011) Biogenic synthesis of silver nanoparticles using *Nicotianatobaccum* leaf extract and study of their antibacterial effect. Afr J Biotechnol 10(41):8122–8130
Ramesh V, Armash A (2015) Green synthesis of gold nanoparticles against pathogens and cancer cells. Int J Pharm Res 5(10):250–256
Rani PU, Rajasekharreddy P (2011) Green synthesis of silver-protein (core–shell) nanoparticles using Piper betle L. leaf extract and its ecotoxicological studies on Daphnia magna. Colloids Surf A Physicochem Eng Asp 389(1–3):188–194
Rao ML, Savithramma N (2013) Biological synthesis and validation of silver nano particles from roots of Svensonia hyderobadensis (Walp.) Mold- A rare medicinal plant taxon. Int J Adv Sci Tech Res 5:524–541
Rastogi L, Arunachalam J (2011) Sunlight based irradiation strategy for rapid green synthesis of highly stable silver nanoparticles using aqueous garlic (Allium sativum) extract and their antibacterial potential. Mater Chem Phys 129(1–2):558–563
Rawat R, Srivastava N, Chadha BS, Oberoi HS (2014) Generating fermentable sugars from rice straw using functionally active cellulolytic enzymes from Aspergillus niger HO. Energy Fuel 28(8):5067–5075

Roy D, Goswami R, Pal A (2017) Nanomaterial and toxicity: what can proteomics tell us about the nanotoxicology? Xenobiotica 47(7):632–643
Salata OV (2004) Applications of nanoparticles in biology and medicine. J Nanobiotechnol 2(1):3
Sanghi R, Verma P, Puri S (2011) Enzymatic formation of gold nanoparticles using *Phanerochaete chrysosporium*. Adv Chem Eng Sci 1(03):154
Saxena A, Tripathi RM, Singh RP (2010) Biological synthesis of silver nanoparticles by using onion (Allium cepa) extract and their antibacterial activity. Dig J Nanomater Biostruct 5(2):427–432
Scholz P, Keck M, C. (2015) Nanocrystals: from raw material to the final formulated oral dosage form-a review. Curr Pharm Des 21(29):4217–4228
Shameli K, Bin Ahmad M, Jaffar Al-Mulla EA, Ibrahim NA, Shabanzadeh P, Rustaiyan A et al (2012) Green biosynthesis of silver nanoparticles using Callicarpa maingayi stem bark extraction. Molecules 17(7):8506–8517
Shankar SS, Ahmad A, Sastry M (2003) Geranium leaf assisted biosynthesis of silver nanoparticles. Biotechnol Prog 19(6):1627–1631
Shors T (2011) Understanding viruses. Jones & Bartlett Publishers, Burlington
Singh BK, Walker A (2006) Microbial degradation of organophosphorus compounds. FEMS Microbiol Rev 30(3):428–471
Singh S, Saikia JP, Buragohain AK (2013) A novel 'green' synthesis of colloidal silver nanoparticles (SNP) using *Dillenia indica* fruit extract. Colloids Surf B: Biointerfaces 102:83–85
Singh P, Kim YJ, Zhang D, Yang DC (2016a) Biological synthesis of nanoparticles from plants and microorganisms. Trends Biotechnol 34(7):588–599
Singh P, Singh R, Borthakur A, Srivastava P, Srivastava N, Tiwary D, Mishra PK (2016b) Effect of nanoscale TiO 2-activated carbon composite on Solanum lycopersicum (L.) and Vignaradiata (L.) seeds germination. Energy Ecol Environ 1(3):131–140
Skalickova S, Milosavljevic V, Cihalova K, Horky P, Richtera L, Adam V (2017) Selenium nanoparticles as a nutritional supplement. Nutrition 33:83–90
Song S, Rao R, Yang H, Liu H, Zhang A (2010) Facile synthesis of Fe3O4/MWCNTs by spontaneous redox and their catalytic performance. Nanotechnology 21(18):185602
Songand JY, Kim BS (2009) Biological synthesis of metal nanoparticles. In: Biocatalysis and agricultural biotechnology. CRC Press, Boca Raton, pp 399–407
Srivastava N, Rawat R, Sharma R, Oberoi HS, Srivastava M, Singh J (2014) Effect of nickel–cobaltite nanoparticles on production and thermostability of cellulases from newly isolated thermotolerant Aspergillus fumigatus NS (Class: Eurotiomycetes). Appl Biochem Biotechnol 174(3):1092–1103
Srivastava N, Srivastava M, Mishra PK, Singh P, Ramteke PW (2015a) Application of cellulases in biofuels industries: an overview. J Biofuels Bioenergy 1(1):55–63
Srivastava N, Singh J, Ramteke PW, Mishra PK, Srivastava M (2015b) Improved production of reducing sugars from rice straw using crude cellulase activated with Fe3O4/Alginate nanocomposite. Bioresour Technol 183:262–266
Suman TY, Rajasree SR, Ramkumar R, Rajthilak C, Perumal P (2014) The green synthesis of gold nanoparticles using an aqueous root extract of Morinda citrifolia L. Spectrochim Acta A Mol Biomol Spectrosc 118:11–16
Thakkar KN, Mhatre SS, Parikh RY (2010) Biological synthesis of metallic nanoparticles. Nanomed Nanotechnol Biol Med 6(2):257–262
Tripathy A, Raichur AM, Chandrasekaran N, Prathna TC, Mukherjee A (2010) Process variables in biomimetic synthesis of silver nanoparticles by aqueous extract of *Azadirachtaindica* (Neem) leaves. J Nanopart Res 12(1):237–246
Urbano P, Urbano F (2007) Nanobacteria: facts or fancies? PLoS Pathog 3(5):e55
Vanaja M, Rajeshkumar S, Paulkumar K, Gnanajobitha G, Malarkodi C, Annadurai G (2013) Phytosynthesis and characterization of silver nanoparticles using stem extract of *Coleus aromaticus*. Int J Mater Biomater Appl 3(1):1–4
Vankar PS, Bajpai D (2010) Preparation of gold nanoparticles from *Mirabilis jalapa* flowers. Indian J Biochem Biophys 47(3):157–160

Varun S, Sellappa S, RafiqKhan M, Vijayakumar S (2015) Green synthesis of gold nanoparticles using Argemonemexicana L. Leaf extract and its characterization. Int J Pharm Sci Rev Res 32:42–44

Verma ML, Chaudhary R, Tsuzuki T, Barrow CJ, Puri M (2013) Immobilization of β-glucosidase on a magnetic nanoparticle improves thermostability: application in cellobiose hydrolysis. Bioresour Technol 135:2–6

Vidya C, Hiremath S, Chandraprabha MN, Antonyraj ML, Gopal IV, Jain A, Bansal K (2013) Green synthesis of ZnO nanoparticles by *Calotropis gigantea.* Int J Curr Eng Technol 1:118–120

Vigneshwaran N, Nachane RP, Balasubramanya RH, Varadarajan PV (2006) A novel one-pot 'green' synthesis of stable silver nanoparticles using soluble starch. Carbohydr Res 341(12):2012–2018

Vijayakumar M, Priya K, Nancy FT, Noorlidah A, Ahmed ABA (2013) Biosynthesis, characterisation and anti-bacterial effect of plant-mediated silver nanoparticles using *Artemisia nilagirica.* Ind Crop Prod 41:235–240

Vijayaraghavan K, Nalini SK, Prakash NU, Madhankumar D (2012) One step green synthesis of silver nano/microparticles using extracts of Trachyspermum ammi and Papaver somniferum. Colloids Surf B: Biointerfaces 94:114–117

Vincent BB, Loeve S (2014) Metaphors in nanomedicine: the case of targeted drug delivery. NanoEthics 8(1):1–17

Wang T, Yang L, Zhang B, Liu J (2010) Extracellular biosynthesis and transformation of selenium nanoparticles and application in H_2O_2biosensor. Colloids Surf B: Biointerfaces 80(1):94–102

Wei Y, Li X, Yu L, Zou D, Yuan H (2015) Mesophilic anaerobic co-digestion of cattle manure and corn Stover with biological and chemical pretreatment. Bioresour Technol 198:431–436

Weiss J, Takhistov P, McClements DJ (2006) Functional materials in food nanotechnology. J Food Sci 71(9):R107–R116

Wu CY, Martel J, Wong TY, Young D, Liu CC, Lin CW, Young JD (2016) Formation and characteristics of biomimetic mineralo-organic particles in natural surface water. Sci Rep 6:28817

Yadav KK, Singh JK, Gupta N, Kumar V (2017) A review of nanobioremediation technologies for environmental cleanup: a novel biological approach. J Mater Environ Sci 8:740–757

Yang Z, Huang R, Qi W, Tong L, Su R, He Z (2015) Hydrolysis of cellulose by sulfonated magnetic reduced graphene oxide. Chem Eng J 280:90–98

Yazdi MH, Mahdavi M, Varastehmoradi B, Faramarzi MA, Shahverdi AR (2012) The immunostimulatory effect of biogenic selenium nanoparticles on the 4T1 breast cancer model: an in vivo study. Biol Trace Elem Res 149(1):22–28

Yeoman CJ, Han Y, Dodd D, Schroeder CM, Mackie RI, Cann IK (2010) Thermostable enzymes as biocatalysts in the biofuel industry. In: Advances in applied microbiology, vol 70. Academic, Cambridge, MA, pp 1–55

Yu J, Xu D, Guan HN, Wang C, Huang LK (2016) Facile one-step green synthesis of gold nanoparticles using Citrus maxima aqueous extracts and its catalytic activity. Mater Lett 166:110–112

Yuvakkumar R, Suresh J, Saravanakumar B, Nathanael AJ, Hong SI, Rajendran V (2015) Rambutan peels promoted biomimetic synthesis of bioinspired zinc oxide nanochains for biomedical applications. Spectrochim Acta A Mol Biomol Spectrosc 137:250–258

Zhang L, Li D, Gao P (2012) Expulsion of selenium/protein nanoparticles through vesicle-like structures by Saccharomyces cerevisiae under microaerophilic environment. World J Microbiol Biotechnol 28(12):3381–3386

Zimmermann S, Sures B (2004) Significance of platinum group metals emitted from automobile exhaust gas converters for the biosphere. Environ Sci Pollut Res 11(3):194

Application of Plant-Based Natural Product to Synthesize Nanomaterial

3

Mohammad Khajeh Mehrizi and Zahra Shahi

Abstract

Nanotechnology is rapidly advancing and is used for a wide range of applications. The chemical and physical methods used to synthesize nanoparticles are expensive and often raise questions of environmental risk because of the use of toxic and hazardous chemicals. Biological methods for nanoparticle synthesis using microorganisms, enzymes, and plant extracts have been suggested as alternatives to chemical and physical methods. The methods have drawn the attention of researchers due to their eco-friendly nature and low-cost, energy-efficient, non-toxic, and easier steps for synthesis. Among these methods, plants (*Thuja occidentalis, Ocimum sanctum, Curcuma longa, Ficus benghalensis* leaf extract, etc.) are the best candidate and suitable for large-scale biosynthesis of nanoparticles. Many biomolecules in plants such as proteins, amino acids, polysaccharides, alkaloids, alcoholic compounds, and vitamins could be involved in bioreduction, formation, and stabilization of nanoparticles. So, using plants for biosynthesis of nanoparticles has become a major focus of researchers.

Keywords

Nanotechnology · Chemical synthesis · Biosynthesis · Plant

3.1 Definition of Nanoparticles

"Nano" is a Greek word synonymous to dwarf meaning extremely small. The prefix is used to indicate one billionth of a meter or 10^{-9} m. The nanosized world includes systems whose size is above molecular dimensions and below macroscopic

The original version of this chapter was revised. A correction to this chapter can be found at https://doi.org/10.1007/978-981-13-9333-4_12.

M. Khajeh Mehrizi (✉) · Z. Shahi
Textile Department, Yazd University, Yazd, Iran
e-mail: mkhajeh@yazd.ac.ir

M. Srivastava et al. (eds.), *Nanomaterials in Biofuels Research*,
Clean Energy Production Technologies, https://doi.org/10.1007/978-981-13-9333-4_3

(generally >1 nm and <100 nm). Nanoparticles play a basic role as building blocks of nanotechnology. In other words, nanotechnology is a branch of science (often referred to as the "tiny science") which deals with matter having at least one dimension size from 1 to 100 nm (Pal et al. 2011, 2018).

3.2 Physicochemical Properties and Application of Nanoparticles

Physicochemical properties of nanoparticles are large surface area, mechanically strong, optically active, high thermal conductivity, and chemically reactive. Their reactivity, toughness, and other properties are dependent on their unique size, shape, and structure. Due to these characteristics, they are suitable candidates for various commercial and domestic applications (Khan et al. 2017).

Nowadays, nanotechnology is widely applied in different fields mainly in water purification, electronics (capacitors), coatings, packaging, antibacterial, cosmetic, drug delivery, imaging, chemotherapy, and catalysis (Pal et al. 2018; Sujitha and Kannan 2013; Thakkar et al. 2010).

3.2.1 Silver Nanoparticles (Ag NPs)

Silver nanoparticles have many important applications. It is used as an antimicrobial agent in textiles, cosmetics, and household appliances (Khalil et al. 2014). Nanosilver-sprayed bandages can be used to avoid bacterial infections in wounds (Aondona 2018). Also, these nanoparticles can be exploited in medicine for burn treatment (Rai et al. 2009). Silver nanoparticles are suitable for biological sensing and imaging. Due to their high conductivity, silver nanoparticles are applied in conductive inks and pastes for a range of electronic devices. Also, these nanoparticles are used as catalysts for accelerating some chemical reactions (Khalil et al. 2014).

3.2.2 Zinc Oxide Nanoparticles (ZnO NPs)

Zinc oxide nanoparticles have been widely used in many industries such as solar cells, photocatalysts, and pharmaceutical. It is used as a UV blocker in sunscreens (Sangeetha et al. 2011). Also, these nanoparticles, with a wide band gap (3.37 eV), are semiconductors for nano-electronic applications (Qu et al. 2011).

3.2.3 Titanium Nanoparticles (TiO_2 NPs)

Titanium nanoparticles possess optical, dielectric, antimicrobial, antibacterial, chemical stability, and catalytic properties which lead to industrial applications such as pigment, fillers, and photocatalyst (Sundrarajan and Gowri 2011).

The photocatalytic activity of titanium nanoparticles is ideal in antimicrobial applications including air and water purification and also antimicrobial coatings on biomedical devices. Also, the US Food and Drug Administration (FDA) has approved the use of these nanoparticles in human food, drugs, and cosmetics (Fernando et al. 2018).

3.2.4 Copper Nanoparticles (Cu NPs)

Copper nanoparticles, due to their unique physical and chemical properties and low cost of preparation, have been of great interest recently. These nanoparticles have industrial use such as catalytic processes, high-temperature superconductors, and solar cells (Honary et al. 2012). Furthermore, copper nanoparticles, due to their antimicrobial properties, have been used for coating of medical devices (Fernando et al. 2018).

3.2.5 Gold Nanoparticles (Au NPs)

Gold nanoparticles have been used at atomic force microscopy and surface plasmon resonance imaging (Iravani 2011). Also, gold nanoparticles have been studied as potential drug and gene delivery systems in cancer therapy (Fernando et al. 2018).

Based on all of the above, the synthesis of metallic nanoparticles is an active area of academic and, more importantly, "application research" in nanotechnology, and there is a great interest in synthesized nanoparticles due to their extraordinary properties.

3.3 Synthesis of Nanoparticles

Various methods can be employed for the synthesis of nanoparticles. These methods are divided into two main classes (bottom-up and top-down approach). In top-down method, larger molecules are decomposed into smaller units, which are then converted into suitable nanoparticles (destructive). Also, bottom-up method synthesizes the material from atomic or molecular species via chemical reactions; therefore this approach is also called building up approach. Some important manufacturing methods used in nanoparticle synthesis are showed in Fig. 3.1 (Khan et al. 2017).

Physical methods (arc discharge, physical vapor condensation, energy ball milling method, etc.) generally employ high temperatures, and the disadvantages of this method include the required space, time, and high energy. The absence of solvent contamination and the uniformity of the distribution of nanoparticles are the advantages of the physical methods in comparison with chemical processes. Chemical method of synthesis can be subdivided into chemical reduction, electrochemical, and pyrolysis methods. Chemical methods have the advantage of higher yield compared to physical methods (Fernando et al. 2018; Siddiqi et al. 2018).

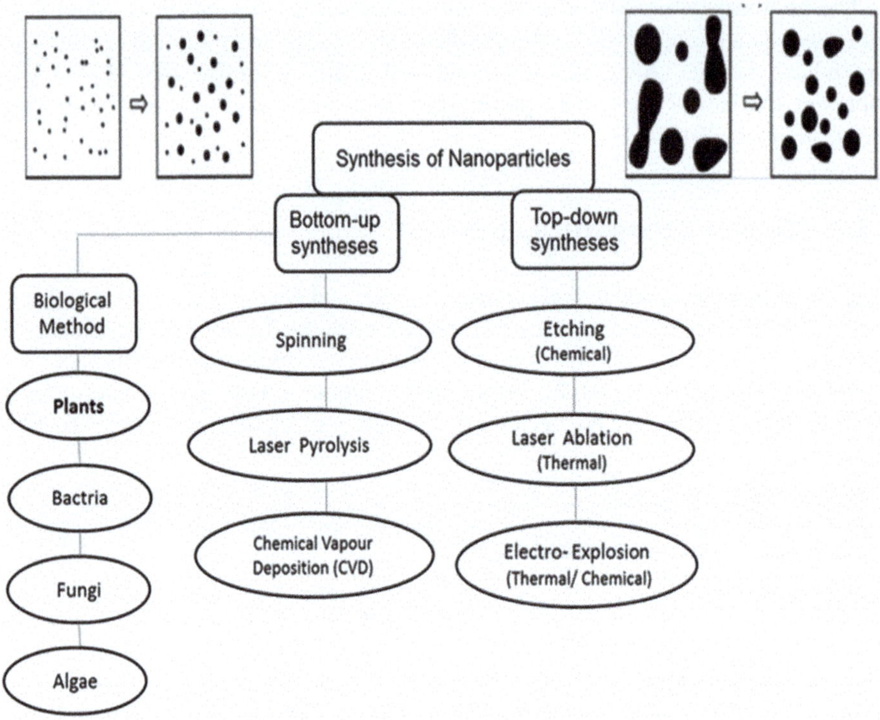

Fig. 3.1 Manufacturing methods used in nanoparticle synthesis (Khan et al. 2017)

Generally, although physical and chemical methods are more popular in the synthesis of nanoparticles, they have limitations such as increase cost of production, release of hazardous by-products, long time, high pressure, difficulty in purification, and not suitable for biological applications. Therefore, development of nontoxic and eco-friendly methods for synthesis of nanoparticles is required (Nasir et al. 2016).

3.4 Biosynthesis of Nanoparticles Using Plants

Bio-nanotechnology has emerged in developing biosynthetic and environment-friendly technology for the synthesis of nanomaterials (Rai et al. 2009). Biosynthesis of nanoparticles using biological agents such as plant extract, bacteria, fungi, yeast, and algae has gained popularity in nanotechnology (Siddiqi et al. 2018).

Developing of biosynthetic has added much importance because of its eco-friendly products, biocompatibility, and economic potential and also to avoid adverse effects during their application especially in the medical field (anticancer activity, drug delivery, etc.) compared with physiochemical nanoparticles (Sundrarajan and Gowri 2011). Additionally, plants and microorganism constituents act as stabilizing and capping agents and do not require stabilizing agents (Singh et al. 2016). Among the various biosynthetic methods, the use of plant extracts has advantages such as readily available, safety in handling, and possessing a broad variety of metabolites (Sundrarajan and Gowri 2011). Also, because of their cost-effective and environment-friendly nature, they are easily used for large-scale synthesis and in this method there is no need to use high pressure, energy, and toxic chemicals (Prasad and Elumalai 2011). Plant systems are suitable and reliable for high demand in nanoparticle synthesis with applications in the biomedical and environmental areas (Singh et al. 2016).

Various parts of plants, such as leaves, stems, roots, shoots flowers, barks, and seeds, have been successfully used for the biosynthesis of nanoparticles (Siddiqi et al. 2018). The dry form of plants compared to fresh is preferable due to different content of water in various plant parts (Mohamad et al. 2014).

Recently, nanoparticle synthesis was achieved using plant extracts of *Azadirachta indica, Camellia sinensis, Nyctanthes arbor-tristis, Coriandrum, Nelumbo nucifera, Ocimum sanctum, Alfalfa, Lemon grass, Emblica officinalis, Aloe vera,* and several others which are compatible with the green chemistry principles (Sangeetha et al. 2011; Patidar and Jain 2017).

However, there are certain limitations in using plants. The heterogeneity of the size and shape of nanoparticle synthesis of plants hinder their use in specific applications. Also, efficient extraction and purification of nanoparticles from plants is a difficult method with a low recovery (Makarov et al. 2014).

3.4.1 The Role of Plant Metabolites in the Reduction of Metal Ions

The mechanism of nanoparticle synthesis by plants includes (a) the activation phase by the reduction of metal ions, (b) the growth phase that the small nanoparticles spontaneously coalesce into larger particles, and (c) the termination phase determining the final shape (Makarov et al. 2014).

The synthesis of nanoparticles can be done by reducing metal ion using some chemical molecules of plants containing amino acids, organic acid, fats, proteins, nucleic acids, reducing sugar, and types of secondary metabolites, such as flavonoids, alkaloids, polyphenols, heterocyclic compounds, and polysaccharides, which have significant roles in metal salt reduction and, furthermore, act as capping and stabilizing agents for the synthesis of nanoparticles without producing any toxic by-product (Siddiqi et al. 2018; Makarov et al. 2014). In other words, the hydroxyl and carboxylic groups in plants may act as stabilizing agents in the synthesis of nanoparticles (Mohamad et al. 2014). Figure 3.2 shows the flowchart of the biosynthesis of nanoparticles using plants.

Examples of the main types of compounds capable of reducing metal ions are shown in Fig. 3.3.

Terpenoids are a class of organic polymers synthesized in plants. Eugenol is the main terpenoid of *Cinnamomum zeylanisum* (cinnamon) extracts that plays an essential role in the bioreduction of tetrachloroauric acid ($HAuCl_4$) and silver nitrate ($AgNO_3$) solution to nanoparticles.

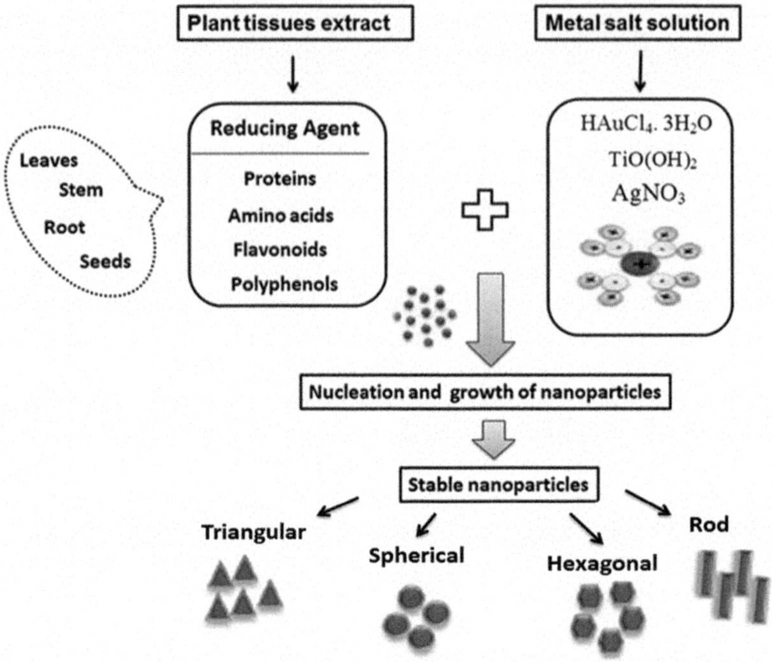

Fig. 3.2 Flowchart representing the biosynthesis of nanoparticles using plants

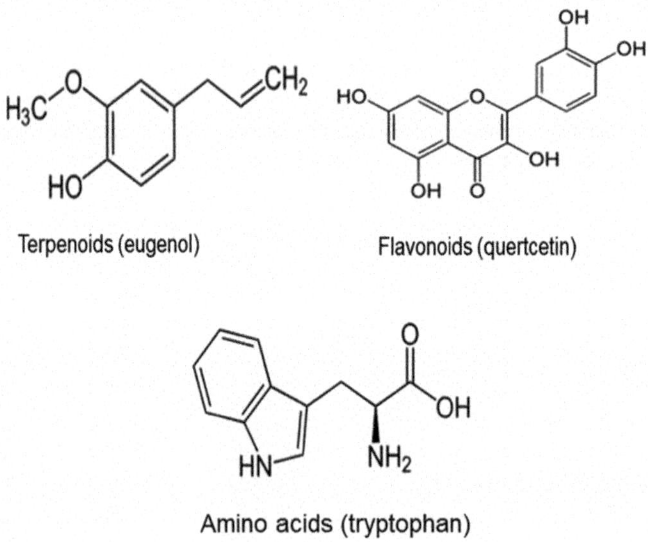

Fig. 3.3 The main types of plant metabolites involved in the synthesis of nanoparticles

Flavonoids are a large group of polyphenolic compounds that comprise several classes, such as anthocyanins isoflavonoids, flavonols, and chalcones which can reduce metal ions into nanoparticles. The tautomeric transformations of flavonoids from the enol to the keto form release a hydrogen atom that can reduce metal ions.

Amino acids can bind to metal ions through the amino and carbonyl groups of the chain, such as the carboxyl groups of glutamic acid, a nitrogen atom of the imidazole ring of histidine or thiol groups of cysteine; studies have shown amino acids such as lysine cysteine, arginine, and methionine are capable of binding silver ions and aspartate can reduce $HAuCl_4$ solution to form gold nanoparticles.

Also the sugars present in plant extracts can induce the formation of metal nanoparticles, for example, monosaccharides containing a keto group (e.g., fructose) have tautomeric transformations from a ketone to an aldehyde (Singh et al. 2016).

3.4.2 Factors Affecting the Biological Synthesis of Nanoparticles Using Plants

The reduction process of metal ions besides the nature of a plant extract containing reducing agents with different concentrations, dependent on the pH, incubation temperature, reaction time, salt concentration, mixing ratio, and electrochemical potential of a metal ion plays a key role in nanoparticle synthesis. Each of these factors is briefly discussed as follows (Singh et al. 2016; Makarov et al. 2014):

Influence of pH

The pH of a plant extract is one of the influential factors and different nanoparticles can be synthesized at different pH values. A change in pH results in a charge change in an extract, which affects their ability to bind and reduce metal ions, and also it causes change in shape, size, and yield of nanoparticles synthesized (Makarov et al. 2014).

Larger nanoparticles can be produced at a lower pH values compared to high pH values. For example, synthesized gold nanoparticles using *Avena sativa* plant were in larger size (25–85 nm) at pH 2 and smaller size (5–20 nm) at pH 3 and 4. In other words, more accessible functional groups at pH = 3–4 within the plants extract were available for nanoparticle nucleation (Shah et al. 2015).

Influence of Reactant Concentration

The studies showed that by varying the amount of leaf extract in the reaction medium could significantly influence the shape and size of the nanoparticles. In the study, by changing the *Aloe vera* leaf extract concentration, hexagonal, triangular, and spherical gold nanoparticles in different sizes (50–350 nm) have been produced (Shah et al. 2015).

Influence of Reaction Time

The reaction time is an important factor in synthesizing nanoparticles. The study revealed that increasing the reaction time (30 min and 4 h) in the biosynthesis of Ag nanoparticles using *Azadirachta indica* leaf extract tended to produce particles with increasing size (10–35 nm) (Shah et al. 2015).

3.4.2.1 Influence of Reaction Temperature

Temperature is an important physical parameter in the formation of nanoparticles in plant.

In general, temperature increases the reaction rate and efficiency of nanoparticle synthesis.

In other words, higher rate of reduction at higher temperature is due to the consumption of metal ions in the formation of nuclei. Temperature is an important factor in determining the size and shape of nanoparticles synthesized via plant. In the research, by raising reaction temperature (25–60 °C) in the synthesis of silver nanoparticles using *Citrus sinensis* peel extract, the average particle size decreased (35–10 nm) (Makarov et al. 2014; Shah et al. 2015; Punjabi et al. 2015).

3.5 Major Nanoparticles Synthesized by Plant Extracts

3.5.1 Biosynthesis of Silver Nanoparticles

Olive Leaf Extract

Khalil et al. synthesized the silver nanoparticles (20–25 nm) with the bioreduction method using aqueous *olive leaf* extract and stock solution of silver nitrate. The reduction process Ag^+ to Ag^0 nanoparticles was followed by the color change of the solution from yellow to brownish-yellow to deep brown by changing the contact time, concentration, temperature, and pH. Figure 3.4 shows UV-visible spectra of

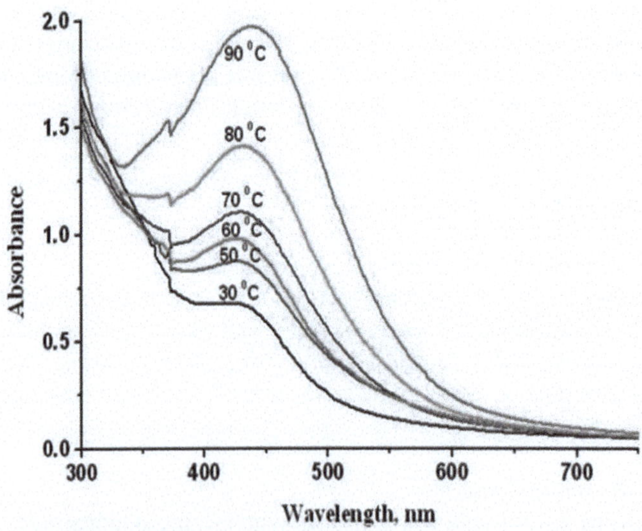

Fig. 3.4 UV-vis spectra of Ag NPs as a function of temperature

the Ag NPs prepared at different temperatures. It can be seen that the absorbance increases with raising temperature. In other words, increasing the reaction temperature led to a rapid reduction of the Ag^{+} ions.
Also, results indicated that the reduction of silver ions was increased at elevated pH due to increased activity of olive leaf extract constituent. As a result, the number of nucleus and size of the silver nanoparticles decreased with increased pH of the reactions (Khalil et al. 2014).

Telfairia occidentalis **Leaf Extract**
Aondona et al. synthesized the silver nanoparticles using dry fluted pumpkin (*Telfairia occidentalis*) leaf extract as reducing agent and 0.01 M silver nitrate solution. The produced nanoparticles had an average size of 30 nm and angular and cubicle shape (Aondona 2018). The green synthesis of nanoparticle includes three steps: (i) selection of solvent, (ii) the reducing agent, and (iii) the nontoxic capping agent (Nasir et al. 2016). The work has shown that dried fluted pumpkin leave extract can be used as a reducing agent without any capping to produce nanoparticle. Due to the eco-friendly method of nanoparticle biosynthesis, these can be used on wound bandage (Aondona 2018).

Thuja occidentalis **Leaf Extract**
Barua et al. synthesized polyethylene glycol-stabilized colloidal silver nanoparticles (7–14 nm) using the reductive potency of *Thuja occidentalis* leaf extract under ambient conditions. The silver ions in the polyethylene glycol matrix are stabilized by electrostatic interaction. The leaves contain considerable amounts of tannic acid, fenchone, and isothujone as the main components. Also, it includes carvotanacetone, myrcen, and camphen that may be directly related to the excellent reducing potential of the leaf extract. The silver ion is reduced by –OH group of the active components of the leaves, which is oxidized to carbonyl group.

The peak at around 420 nm (pH = 8) in the UV-vis spectrum was observed which indicates the formation of silver nanoparticles (Fig. 3.5) (Barua et al. 2013).

Kurian et al. synthesized the silver nanoparticles (20–50 nm) using *Curcuma longa* (turmeric) and *Zingiber officinale* rhizome extracts and silver sulfate solution. The reaction was completed at 3 h using turmeric rhizome as reducing agent and it reported one of the fastest biosynthesis route. Silver nanoparticles showed greatest antibacterial activity against *Staphylococcus aureus* pathogen (Kurian et al. 2016).

Ocimum sanctum **Leaf Extract**
Singhal et al. synthesized the stable silver nanoparticles using *Tulsi (Ocimum sanctum)* leaf extract. It was observed that *Tulsi* leaf extract can reduce silver ions into silver nanoparticles within 8 min of reaction time. Thus, this method can be used for rapid and eco-friendly biosynthesis of stable silver nanoparticles in the medical industry. The TEM images of the prepared silver nanoparticles are shown in Fig. 3.6. It was observed that silver nanoparticles were circular in shape with maximum particles in range size within 4–30 nm (Singhal et al. 2011).

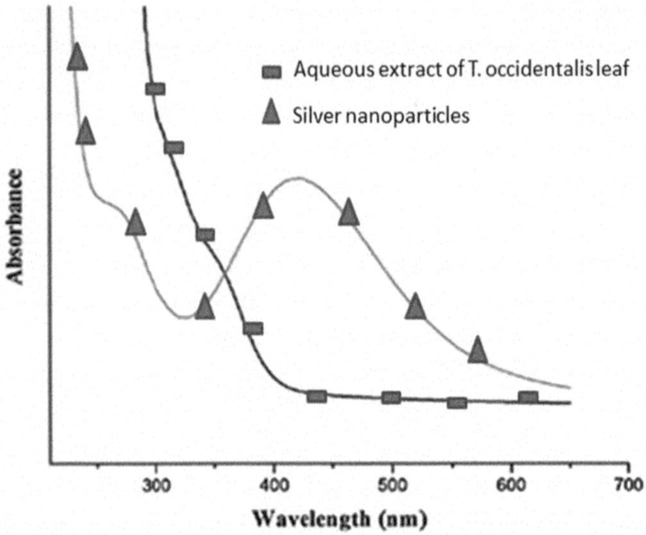

Fig. 3.5 UV-vis spectra showing the formation of the PEG/Ag nanoparticle *Curcuma longa*/*Zingiber officinale* extracts

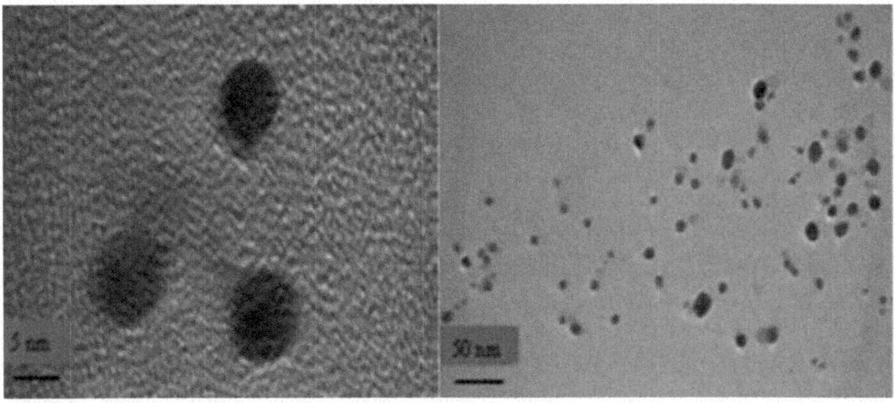

Fig. 3.6 TEM images of biosynthesized silver nanoparticles

Ahmad et al. synthesized silver nanoparticles (~10 nm) from the dried stem and root of basil plant. The size of the nanoparticles derived from the stem (~62%) is more uniform than those derived from the root (~30%). Rosmarinic acid and flavonoids like luteolin, apigenin, and isoeugenol were the most abundant components identified in basil. The formation of enol form of the luteolin liberates reactive hydrogen which is responsible for the conversion of Ag^+ to Ag^0 (Fig. 3.7a and b). Also, there is liberation of reactive hydrogen in the keto form of the rosmarinic acid

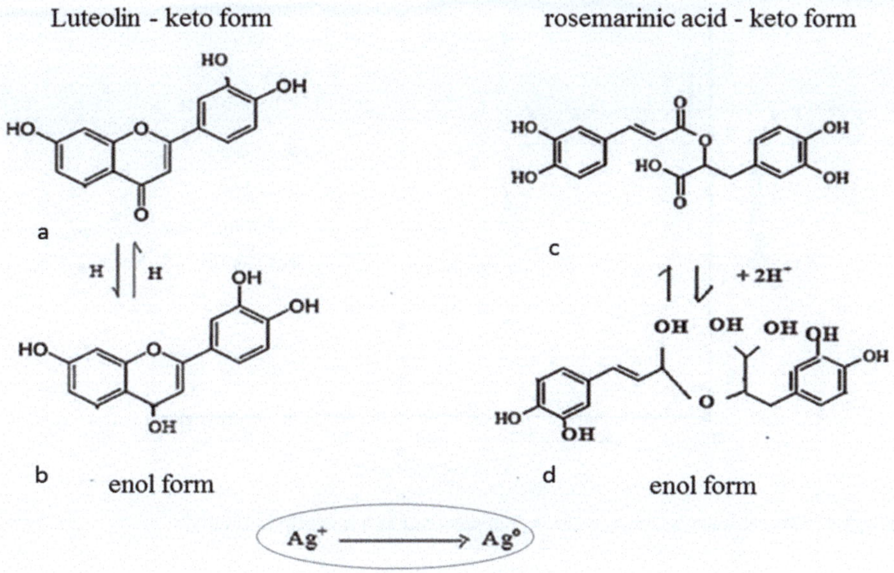

Fig. 3.7 Mechanism of biosynthesis of Ag nanoparticles

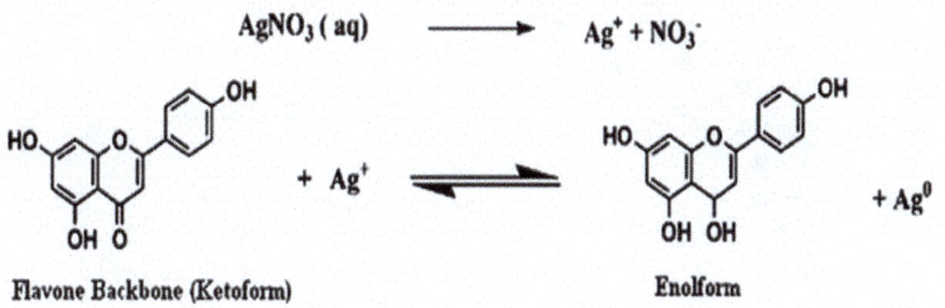

Fig. 3.8 Mechanism of biosynthesis of silver nanoparticles

to unstable enol form (Fig. 3.7c and d). Thus, in both of them the reactive hydrogen is liberated which participates in the synthesis of silver nanoparticles (Ahmad et al. 2010).

Ficus benghalensis **Leaf Extract**

Saxena et al. developed a green synthetic method for silver nanoparticles using *Ficus benghalensis* leaf extract which acts as a reducing and capping agent. *F. benghalensis* plant leaf contains high levels of flavonoids. Phenolics and other chemicals within the plant leaf extract reduce silver salts (Fig. 3.8). Also proteins and enzymes present in the extract facilitate the formation of pure nanoparticles by reduction of the metal ions.

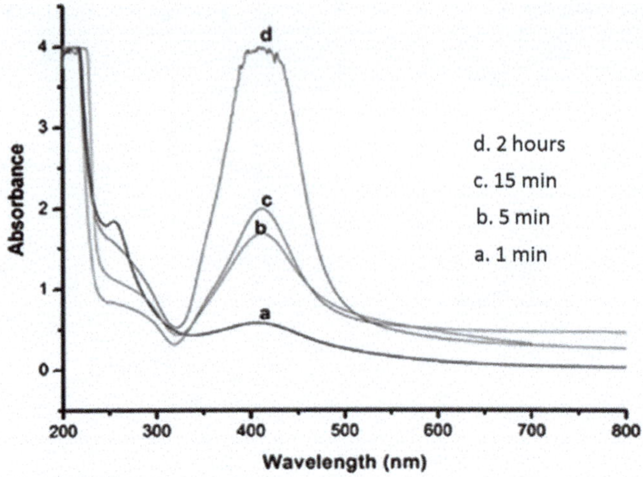

Fig. 3.9 UV-vis spectra showing absorption recorded as a function of time

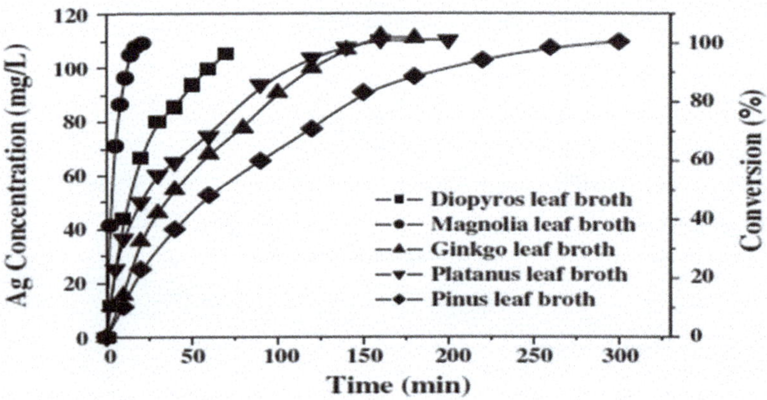

Fig. 3.10 Time courses of silver nanoparticle formation obtained from various plant leaf extracts

UV-vis spectra were recorded as function of reaction time (Fig. 3.9). The results showed that after 5 min the surface plasmon resonance of silver occurs at 410 nm. So, the use of *Ficus benghalensis* extract is a fast method for the synthesis of silver nanoparticles and reduces silver ions without using any harsh conditions (Saxena et al. 2012).

Pine, Persimmon, Ginkgo, Magnolia, and Platanus Leaf Extract
Song et al. synthesized silver nanoparticles (average particle size ranged from 15 to 500 nm) from the five plant leaf extracts (pine, persimmon, *Ginkgo, Magnolia*, and *Platanus*). *Magnolia* leaf extracts are the best reducing agent in the synthesis of silver nanoparticles (11 min, 95 °C) (Fig. 3.10).

This method of silver nanoparticle production can potentially be used in various human contacting areas such as cosmetics, foods, and medical applications (Song and Kim 2009).

3.5.2 Biosynthesis of Gold Nanoparticles

Olive Leaf Extract

Khalil et al. synthesized gold nanoparticles using olive leaf extracts as reducing agent and $HAuCl_4$. $3H_2O$ solution. The high phenolic content of the hot water extract of olive leaves helped in the reduction of Au^{3+} to Au nanoparticles. The TEM images showed that a mixture of shapes (triangular, hexagonal, and spherical) was formed at lower leaf concentration and high pH, while smaller spherical shapes were obtained at higher concentration and low pH (Fig. 3.11) (Khalil et al. 2012).

Pelargonium graveolens **Leaf Extract**

Shankar et al. reported the use of *geranium* leaves (*Pelargonium graveolens*) in the synthesis of gold nanoparticles. The reducing and capping agents in gold nanoparticles synthesized using geranium leaves appear to be terpenoids. The nanoparticles grown showed a variety of shapes that included rods, flat sheets, and triangles (Fig. 3.12).

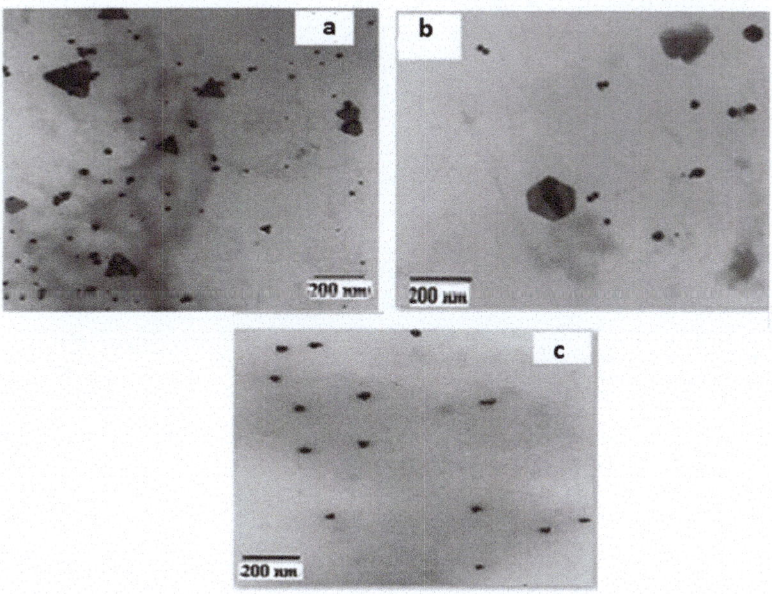

Fig. 3.11 Effect of the leaf extract quantity on the size and shape of the gold nanoparticles. (**a**) TEM image measured at 0.5 ml (triangle shapes), (**b**) 0.5 ml (hexagonal shapes), (**c**) 5 ml of extract

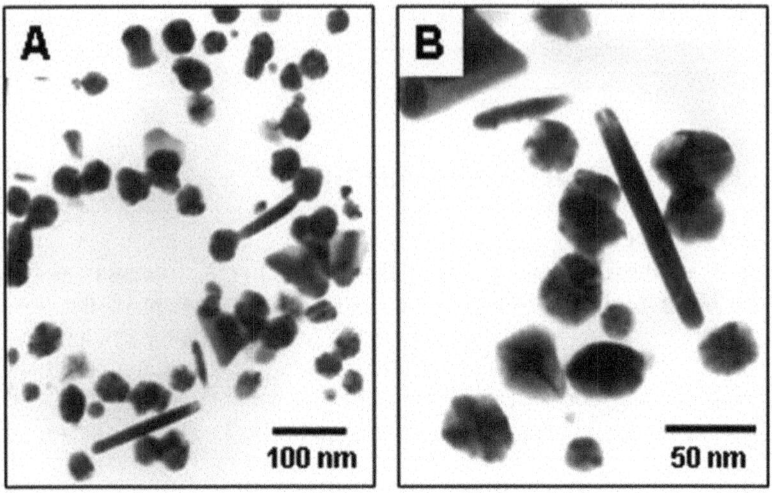

Fig. 3.12 TEM images of gold nanoparticles

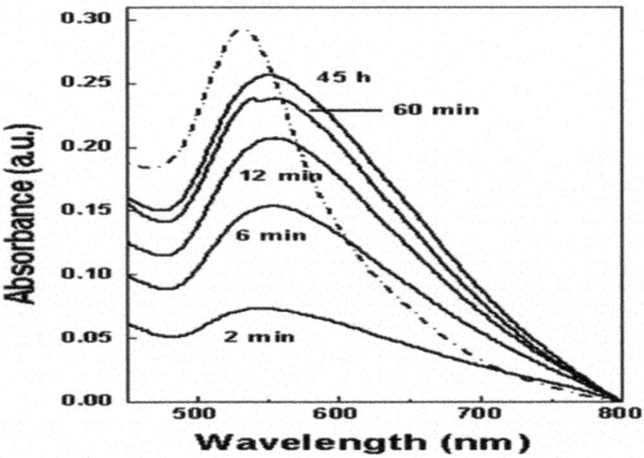

Fig. 3.13 UV-vis spectra recorded as a function of time of reaction

Also, the results showed that the gold surface plasmon resonance occurs at 547 nm after 2 min (Fig. 3.13) (Shankar et al. 2003).

Honey
Philip synthesized gold nanoparticles (~15 nm) using honey as reducing and capping agent. The reducing agent is fructose, and the capping material responsible for stabilization is proteins present in honey (Philip 2009).

Citrus Fruit Extract

Sujitha et al. reported a novel method for the biosynthesis of gold nanoparticles through the reduction of aqueous $AuCl_4$ ions using citrus fruit extract (*C. limon, C. reticulata,* and *C sinensis*). The rich source of citric acid and ascorbic acid in the citrus fruit extract may be responsible for the reduction of metal ions and stabilization of synthesized nanoparticles.

TEM studies showed the particles to be of various shapes and sizes (15–80 nm) that could be changed by altering the concentration of citrus extract. At lower concentrations of the extract, different shapes of nanoparticles, while at higher concentrations spherical particles are observed (Sujitha and Kannan 2013).

3.5.3 Biosynthesis of Palladium Nanoparticles

Cinnamomum camphora **Extract**

Yang et al. synthesized palladium nanoparticles (3.2–6.0 nm controlled by initial concentration of the palladium ions) using *Cinnamomum camphora* leaf without capping agent. The polyol components and the heterocyclic components were responsible for the reduction and stabilization of palladium nanoparticles, respectively. Reaction between the hydroxyl group of the polyols in the *C. camphora* leaf broth and the palladium ions is shown below (Yang et al. 2010).

$$\mathrm{nPd}(\mathrm{II}) + 2\mathrm{R} - (\mathrm{OH})_n \rightarrow \mathrm{nPd}(0) + 2\mathrm{nR} = \mathrm{O} + 2\mathrm{nH}^+ \tag{3.1}$$

Coffee and Tea Extract

Mallikarjuna et al. synthesized the palladium nanoparticles (in the size range of 20–60 nm) using coffee and tea extract at room temperature. Caffeine and polyphenols in the extracts can form complexes with metal ions and reduce them to the corresponding metals (Nadagouda and Varma 2008).

Cinnamomum zeylanicum **Bark Extract**

Sathishkumar et al. synthesized the palladium nanoparticles (15–20 nm) using an extract of *Cinnamomum zeylanicum* bark and palladium chloride ($PdCl_2$) solution. This palladium nanoparticle biosynthesis is performed under a moderate pH (=5) and room temperature (30 °C). Figure 3.14 shows a representative TEM picture of the palladium nanoparticles that were synthesized for 72 h (Sathishkumar et al. 2009).

3.5.4 Biosynthesis of Titanium Dioxide Nanoparticles

Eclipta prostrata **Leaf Extract**

Rajakumar et al. synthesized titanium dioxide nanoparticles using *Eclipta prostrata* leaf extract and $TiO(OH)_2$ solution. The synthesized nanoparticles are quite polydisperse and range in size from 36 to 68 nm (Fig. 3.15).

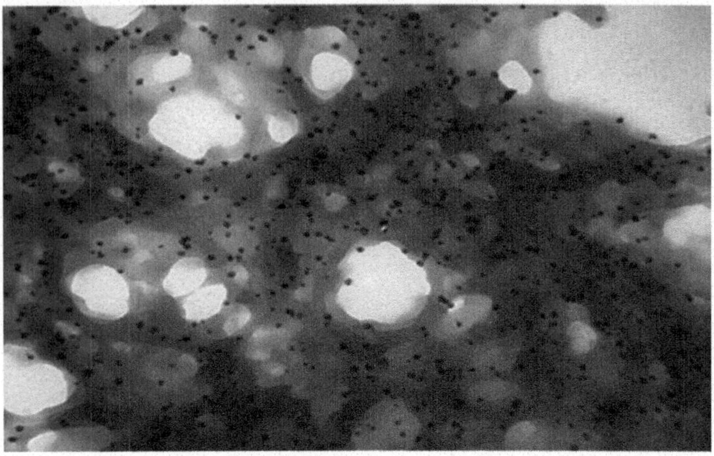

Fig. 3.14 TEM image of palladium nanoparticles

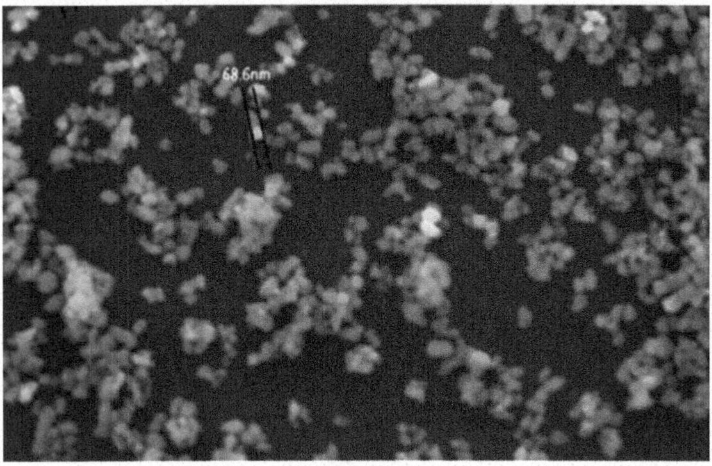

Fig. 3.15 FESEM image of the TiO_2 nanoparticles

Alcohols, phenols, alkanes, and aliphatic amines in *Eclipta* may participate in the process of nanoparticle synthesis. Also, heterocyclic compounds such as flavones are the reducing and capping ligands of the nanoparticles (bioreduction of $TiO(OH)_2$ to TiO_2 (Rajakumar et al. 2012).

Moringa oleifera **Leaf Extract**

Patidar et al. synthesized titanium dioxide nanoparticles using *Moringa oleifera* leaf extract and titanium tetraisopropoxide solution. Anatase phase TiO_2 nanoparticles with mean crystalline size of 12.22 nm and band gap of 3.9 eV at wavelength 318 nm was obtained by UV-visible spectroscopy (Sundrarajan and Gowri 2011).

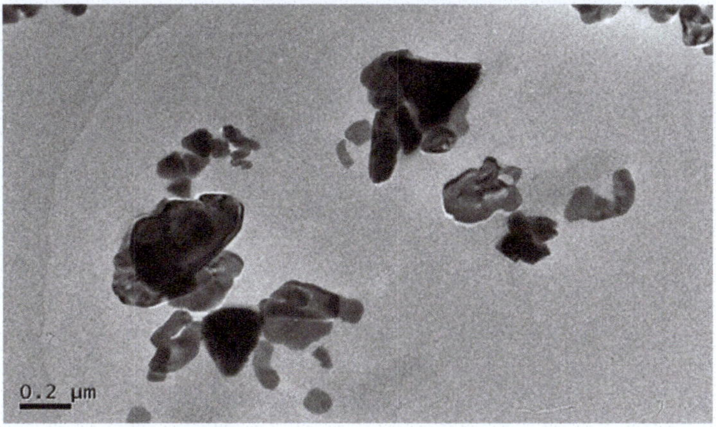

Fig. 3.16 TEM image of synthesized zinc oxide nanoparticles

Vigna radiata **Seed Extract**

Chatterjee et al. synthesized oval titanium dioxide nanoparticles using *Vigna radiata* (green gram) legume extract. *Vigna radiata* has been confirmed to be a good source for the biosynthesis of titanium nanoparticles. *Vigna radiata* seeds were chosen for the experiment because seeds have strong reducing capacity. The particles were found to be effective against pathogens and can be further used in antibiotic drugs (Singh et al. 2016).

3.5.5 Biosynthesis of Zinc Oxide Nanoparticles

Physalis alkekengi **Extract**

Qu et al. synthesized zinc oxide nanoparticles using *Physalis alkekengi* plant. The TEM images showed that nanoparticles are not uniformly distributed and the morphology of most nanoparticles is triangular and elongated with a mean size of 72.5 nm (Fig. 3.16) (Qu et al. 2011).

Aloe vera **Leaf Extract**

Sangeetha et al. synthesized zinc oxide nanoparticles (25–40 nm) using *aloe vera* leaf extract and zinc nitrate. The nanoparticles obtained have been found to be predominantly spherical and the particle size could be controlled by varying the concentrations of leaf solution. Figure 3.17 shows the effect of leaf concentration on the nanoparticle synthesis. The rate of synthesis increased with increase in concentrations. Increasing the leaf concentration to more than 25% resulted to 100% conversion to nanoparticle after 6–7 h.

The zinc oxide nanoparticles prepared from *Aloe vera* leaf are expected to have extensive applications in biomedical and cosmetic industries (Sangeetha et al. 2011).

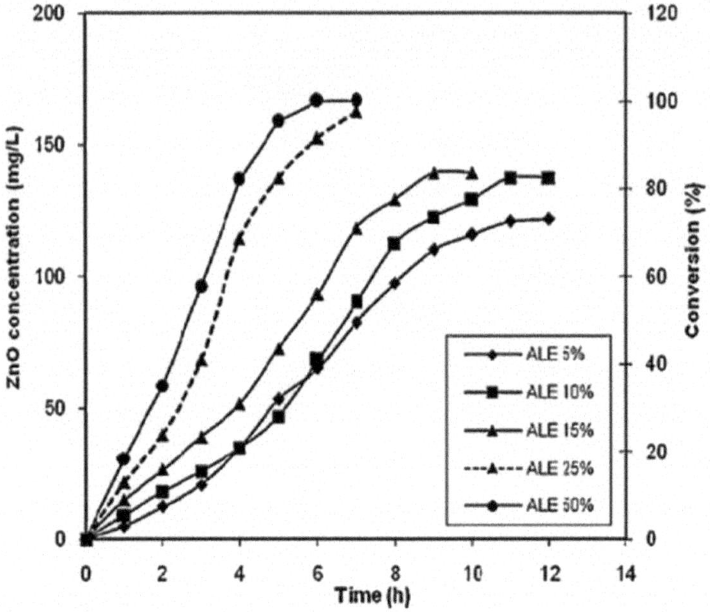

Fig. 3.17 Effect of various concentrations of *Aloe* leaf extract (ALE) on the time course of nanoparticle synthesis

3.5.6 Biosynthesis of Iron Nanoparticles

Camellia sinensis **Extract**

Hoag et al. synthesized zerovalent iron (ZVI) nanoparticles using tea (*Camellia sinensis*) polyphenols and $Fe(NO_3)_3$ solutions. The tea extracts (polyphenols) used function as both a reducing and capping agent for iron. The reaction between polyphenols and ferric nitrate occurs within a few minutes at room temperature and is indicated by color changes from yellow to dark brown in the formation of iron nanoparticles (Hoag et al. 2009).

Dodonaea viscosa **Leaf Extract**

Daniel et al. reported the biosynthesis of zerovalent iron nanoparticles (27 nm) using the leaf extract of *Dodonaea viscosa*. The *Dodonaea viscosa* leaf contains flavonoids such as santin, saponins, alizarin, and tannins as active substances. The flavonoids help in the reduction of metal salts, and the tannins and saponins may act as the capping agents which prevents the further oxidation of nanoparticles. Further oxidation of nanoparticles (Fig. 3.18) showed reduction of metal into nanoparticles (Daniel et al. 2013).

Aloe vera **Extract**

Phumying et al. synthesized magnetite (Fe_3O_4) nanoparticles using ferric acetylacetonate ($Fe(C_5H_8O_2)_3$) solution and *Aloe vera* extract. The synthesized nanoparticles are crystalline and have particle sizes of ~6–30 nm. The XRD patterns of the

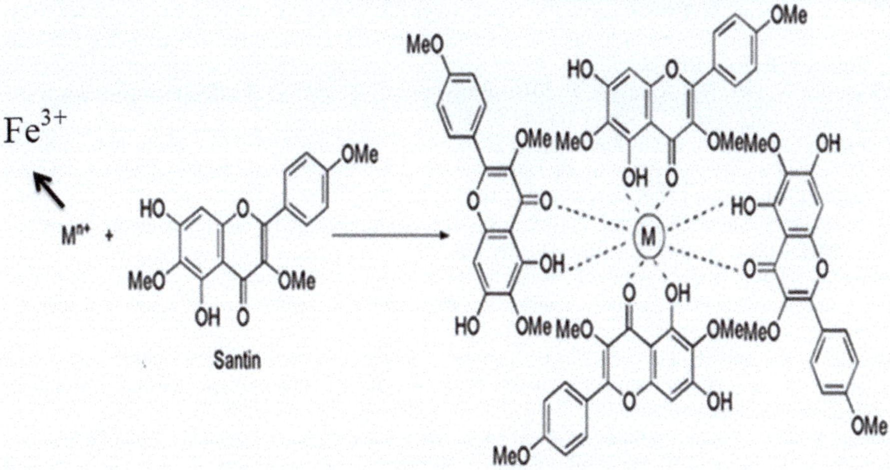

Fig. 3.18 The reduction mechanism of metal into nanoparticles

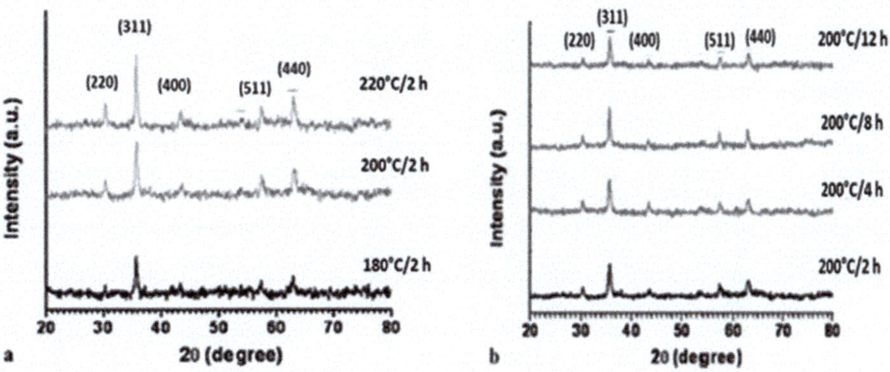

Fig. 3.19 XRD patterns of Fe_3O_4 nanoparticles: (**a**) different hydrothermal temperatures and (**b**) different hydrothermal times

nanoparticles are shown in Fig. 3.19. No diffraction peaks of impurities were observed, which indicates that a high purity of final products was successfully under the experimental conditions. It was seen that the peaks become sharper and narrower along with the increased reaction temperatures and times indicating the enhancement of crystallinity (Phumying et al. 2013).

References

Ahmad N, Sharma S, Alam MK, Alam VN, Singh SF, Shamsi BR, Mehta AF (2010) Rapid synthesis of silver nanoparticles using dried medicinal plant of basil. Colloids Surfaces Biol 81:81

Aondona IP (2018) Biosynthesis of eco-friendly silver Nano-particles using dry fluted pumpkin (*Telfairia occidentalis*) leave extract as reducing agent. Res Dev Mater Sci 4:5

Barua S, Konwarh R, Bhattacharya SS, Das P, Devi KSP, Maiti TK, Mandal M, Karak N (2013) Non-hazardous anticancerous and antibacterial colloidal 'green' silver nanoparticles. Colloids Surfaces B: Biol 105:37

Chatterjee A, DN, NS, Abraham J (2016) Biosynthesis of titanium dioxide nanoparticles using *vigna radiata*. Asian J Pharm Clin Res 9:85

Daniel SCGK, Vinothini G, Subramanian N, Nehru K, Sivakumar M (2013) Biosynthesis of cu, ZVI, and Ag nanoparticles using *Dodonaea viscosa* extract for antibacterial activity against human pathogens. J Nanopart Res 15:1319

Fernando SSN, Gunasekara TDCP, Holton J (2018) Antimicrobial nanoparticles: applications and mechanisms of action Sri Lankan J Infec Dis 8:2

Hoag GE, Collins JB, Holcomb JL, Hoag JR, Nadagouda MN, Varma RS (2009) Degradation of bromothymol blue by 'greener' nano-scale zero-valent iron synthesized using tea polyphenols. J Mater Chem 19:8671

Honary S, Barabadi H, Fathabad Gh E, Naghibi F (2012) Green synthesis of copper oxide nanoparticles using penicillium aurantiogriseum, penicillium citrinum and penicillium waksmanii. Digest J Nanomater Biol 7:3

Iravani S (2011) Green synthesis of metal nanoparticles using plants. Green Chem 13:2638

Khalil MMH, Ismail EH, Magdoub FE (2012) Biosynthesis of au nanoparticles using olive leaf extract. Arab J Chem 5:431

Khalil MMH, Ismail EH, El-Baghdady KZ, Mohamed D (2014) Green synthesis of silver nanoparticles using olive leaf extract and its antibacterial activity. Arab J Chem 7:1131

Khan I, Saeed K, Khan I (2017) Nanoparticles: properties, applications and toxicities. Arab J Chem 2017:228

Kurian M, Varghese B, Athira TS, Krishna S (2016) Novel and efficient synthesis of silver nanoparticles using curcuma longa and *Zingiber Officinale* rhizome extracts. Int J Nanosci Nanotechnol 12:3

Makarov VV, Love AJ, Sinitsyna OV, Makarova SS, Yaminsky IV, Taliansky ME, Kalinina NO (2014) "Green" nanotechnologies: synthesis of metal nanoparticles using plants. Acta Nat 6:1

Mohamad NAN, Arham NA, Jai J, Hadi A (2014) Plant extract as reducing agent in synthesis of metallic nanoparticles: a review. Adv Mater Res 832:350

Nadagouda MN, Varma RS (2008) Green synthesis of silver and palladium nanoparticles at room temperature using coffee and tea extract. Green Chem 10:859

Nasir GA, Mohammed AK, Samir HF (2016) Biosynthesis and characterization of silver nanoparticles using olive leaves extract and sorbitol. Iraqi J Biotechnol 15:1

Pal SL, Jana U, Manna PK, Mohanta GP, Manavalan R (2011) Nanoparticle: an overview of preparation and characterization. J Appl Pharm Sci 1:6

Pal S, Mondal S, Maity J, Mukherjee R (2018) Synthesis and characterization of ZnO nanoparticles using *Moringa Oleifera* leaf extract: investigation of Photocatalytic and antibacterial activity. Int J Nanosci Nanotechnol 14:165

Patidar V, Jain P (2017) Green synthesis of TiO2 nanoparticle using *Moringa Oleifera* leaf extract. Int Res J Eng Technol 4:3

Philip D (2009) Honey mediated green synthesis of gold nanoparticles. Spectrochim Acta Part A Mol Biomol Spectrosc 73:374

Phumying S, Labuayai S, Thomas C, Amornkitbamrung V, Swatsitang E, Maensiri S (2013) *Aloe vera* plant-extracted solution hydrothermal synthesis and magnetic properties of magnetite (Fe_3O_4) nanoparticles. Appl Phys A Mater Sci Process 111:1187

Prasad TNVKV, Elumalai EK (2011) Biofabrication of ag nanoparticles using *Moringa oleifera* leaf extract and their antimicrobial activity. Asian Pac J Trop Biomed 1:439

Punjabi K, Choudhary P, Samant L, Mukherjee S, Vaidya S, Chowdhary A (2015) Biosynthesis of nanoparticles : a review. Int J Pharm Sci Rev Res 30:1

Qu J, Yuan X, Wang X, Shao P (2011) Zinc accumulation and synthesis of ZnO nanoparticles using *Physalis alkekengi* L. Environ Pollut 159:1783

Rai M, Yadav A, Gade A (2009) Silver nanoparticles as a new generation of antimicrobials. Bio Adv 27:76

Rajakumar G, Rahuman AA, Priyamvada B, Khanna VG, Kumar DK, Sujin PJ (2012) *Eclipta prostrata* leaf aqueous extract mediated synthesis of titanium dioxide nanoparticles. Mater Lett 68:115

Sangeetha G, Rajeshwari S, Venckatesh R (2011) Green synthesis of zinc oxide nanoparticles by *aloe barbadensis* miller leaf extract: structure and optical properties. Mater Res Bull 46:2560

Sathishkumar M, Sneha K, Kwak IS, Mao J, Tripathy SJ, Yun YS (2009) Phyto-crystallization of palladium through reduction process using Cinnamom zeylanicum bark extract. J Hazard Mater 171:400

Saxena A, Tripathi RM, Zafar F, Singh P (2012) Green synthesis of silver nanoparticles using aqueous solution of *Ficus benghalensis* leaf extract and characterization of their antibacterial activity. Mater Lett 67:91

Shah M, Fawcett D, Sharma S, Tripathy SK, Poinern GEJ (2015) Green synthesis of metallic nanoparticles via biological entities. Materials 8:7278

Shankar SS, Ahmad A, Pasricha R, Sastry M (2003) Bioreduction of chloroaurate ions by geranium leaves and its endophytic fungus yields gold nanoparticles of different shapes. J Mater Chem 13:1822

Siddiqi KS, Husen A, Rao RAK (2018) A review on biosynthesis of silver nanoparticles and their biocidal properties. J Nanobiotechnol 16:14

Singh P, Kim YJ, Zhang D, Yang DC (2016) Biological synthesis of nanoparticles from plants and microorganisms. Trends Biotechnol 34:7

Singhal G, Bhavesh R, Kasariya K, Sharma AR, Singh RP (2011) Biosynthesis of silver nanoparticles using *Ocimum sanctum* (Tulsi) leaf extract and screening its antimicrobial activity. J Nanopart Res 13:2981

Song JY, Kim BS (2009) Rapid biological synthesis of silver nanoparticles using plant leaf extracts. Bioprocess Biosyst Eng 32:79

Sujitha MV, Kannan S (2013) Green synthesis of gold nanoparticles using Citrus fruits (*Citrus limon*, Citrus reticulata and *Citrus sinensis*) aqueous extract and its characterization. Mol Biomol Spectrosc 102:15

Sundrarajan M, Gowri S (2011) Green synthesis of titanium dioxide nanoparticles by nyctanthes arbor-tristis leaves extract. Chalcogenide Lett 8:8

Thakkar KN, Mhatre SS, Parikh RY (2010) Biological synthesis of metallic nanoparticles. Nano Biol Med 6:257

Yang X, Li Q, Wang H, Huang J, Lin L, Wang W, Sun D, Su Y, Opiyo JB, Hong L, Wang Y, He N, Jia L (2010) Green synthesis of palladium nanoparticles using broth of *Cinnamomum camphora* leaf. J Nanopart Res 12:1589

Green Synthesis Approach to Fabricate Nanomaterials

4

Ramchander Merugu

Abstract

Nanotechnology can overcome many technical hurdles due to their unique properties and small size compared to macrosize particles. Carbon and graphene are the major nanomaterial sources that are widely explored. They are widely produced by chemical, physical, and biological methods. Compared to the chemical and physical methods which are energy intensive, green approaches are economical and eco-friendly. Chemical synthesis includes the use of metal precursors and reducing agents which are generally toxic and non-eco-friendly. To prevent agglomeration of the synthesized nanoparticles, capping agents are also required. Physical methods are energy intensive and have many disadvantages. Biological methods do not require capping agents as the biomolecules present in it can act as capping agents. Different methods of synthesis using plants, microorganisms like bacteria and fungi, actinomycetes, algae, viral particles, and biomolecules have been reviewed in this chapter. Green fabrications have its own advantages and cover a lot of scope for potential applications in the future. The toxicity of the nanomaterial when it is released into the environment is the only cause of concern which needs to be researched further. In this chapter, various methods of fabrication of nanomaterials using biological methods are discussed in the light of existing literature.

Keywords

Synthesis · Characterization · Biological methods · Plant based · Microbial based · Green synthesis · Bacterial · Fungal · Biomolecules

---

R. Merugu (✉)
Department of Biochemistry, Mahatma Gandhi University, Nalgonda, Telangana, India

M. Srivastava et al. (eds.), *Nanomaterials in Biofuels Research*,
Clean Energy Production Technologies, https://doi.org/10.1007/978-981-13-9333-4_4

4.1 Introduction

Feynman (1959) introduced the term "nanotechnology," which refers to the study of materials at nanoscale. Nanotechnology is that branch of interdisciplinary areas which deals with the study of nanomaterial and analysis of matter on an atomic or molecular scale. They show unique and extraordinary properties. Many different nanomaterials form a component such as sensing devices, therapeutic systems, and consumer-friendly goods such as wrinkle-resistant clothes and tubeless tires (Singh and Singh 2012). Hassan and Abul Kalam (2013) claimed that the biofuels are ten times easily biodegraded than renewable sources. Nanomaterials exhibit unique and improved properties as they have larger surface area to volume ratio. Nanotechnology aims at miniaturization of matter to achieve enhanced performance by enhancement of their functionalities. Nanotechnology offers solutions for overcoming potential drawbacks associated with these problems. Engelmann et al. (2013) proposed that the nanomaterials exhibit new and unique properties with applications in all areas of energy. Nanotechnology can also be used for generation, storage, and distribution of energy. Verma et al. (2013a, b) have envisaged the applications of nanomaterials in bioenergy and biosensors. Nanotechnology has a hidden potential which can bring a drastic change to meet the ever-increasing energy requirements. Easily available raw materials and inputs deliver no toxicity or chemical risks to the environment. Green nanoscience involves application of green chemistry principles to the design, development, and applications of nanomaterials. Nanoparticles possess different chemical, physical, and optical characteristics when compared to other materials of different sizes. At the nanoscale, properties of the particles change unpredictably, making them behave differently with the same substance at the macro scale. Nanoscale materials can be used efficiently for the oxidation of carbon-based biomass into carbon dioxide. Nanoparticles also convert small-chain fatty acids along with minute quantities of n-heptane and n-octane which are very useful as sources of energy. Both can be used for assessing the quality of petroleum and petroleum-based products.

4.2 Synthesis and Characteristics of Nanomaterials

Generally, two methods of synthesis are widely used, namely, the top-down and bottom-up synthesis approaches.

4.2.1 Top-Down Approach

In the top-down (physical) approach, nanoparticles are obtained from their bulk materials using irradiation, laser ablation, thermal decomposition, arc discharge, etc.

4.2.2 Bottom-Up Approach

In this technique chemical and biological approaches are used where nanoparticles are obtained from their basic building blocks which generate nanoparticles of the desired shape and size.

4.2.3 Chemical Approach

Generally, chemical synthesis includes the use of metal precursors, reducing agents, and stabilizing/capping agents. These methods are generally toxic and non-eco-friendly. Capping agents are used to prevent agglomeration of the synthesized nanoparticles. Carbon at nanoscales shows increased surface area, tenacity, elasticity, strength, and electricity. Nowadays, the focus is on conversion of lignocellulosic biomass into ethanol which reduces greenhouse gas emissions by 86% over fossils fuels. Fukuda et al. (2001) reported that biofuels can be obtained commercially from feedstocks such as animal fats, vegetable oils, and starch using transesterification and fermentation. Ongoing efforts are to produce biofuels from nonfood feedstocks which have many process-related problems. To overcome these potential barriers nanotechnology and nanomaterials stand as a possible solution.

4.3 Green Synthesis Approaches

Green-based methods for fabricating nanoparticles are being investigated for increasing the production rate of biofuels. Green fabrication of nanoparticles involves the use biomolecules like proteins, amino acids, carbohydrates, enzymes, and cells of microorganisms and plants. Green synthesis has received good attention as it is an inexpensive and eco-friendly way to fabricate nanoparticles. During the green synthesis process, modulation of the particle parameters such as size, chemical composition, and surface area is possible. To reduce metal toxicity, the microorganisms reduce the metallic compounds to nanoparticles. Prokaryotic as well as eukaryotic organisms can be used as nanofactories for different kinds of nanomaterials. Factors that influence the biosynthesis approach is pH, temperature, and incubation time. Sathishkumar and Narendhirakannan (2011) opined that the method of synthesis and incubation temperature determines the nature of the nanoparticle formed. Higher temperatures speed up the reaction rates, giving a better yield when compared to lower reaction temperatures (Sathishkumar et al. 2010). Rai et al. (2006) observed that there is less agglomeration at higher temperatures when compared to lower temperatures. In biogas generation the construction of the anaerobic digester is crucial for productivity. If the digester is not properly maintained, corrosion, wear, and tear decrease the output of biogas. These limitations can be overcome using nanomaterials at the construction stage. Application of nanomaterials incorporated in the form of photovoltaics can achieve excellent absorption of sunlight in the biogas plants which will make the process quicker, efficient, and more

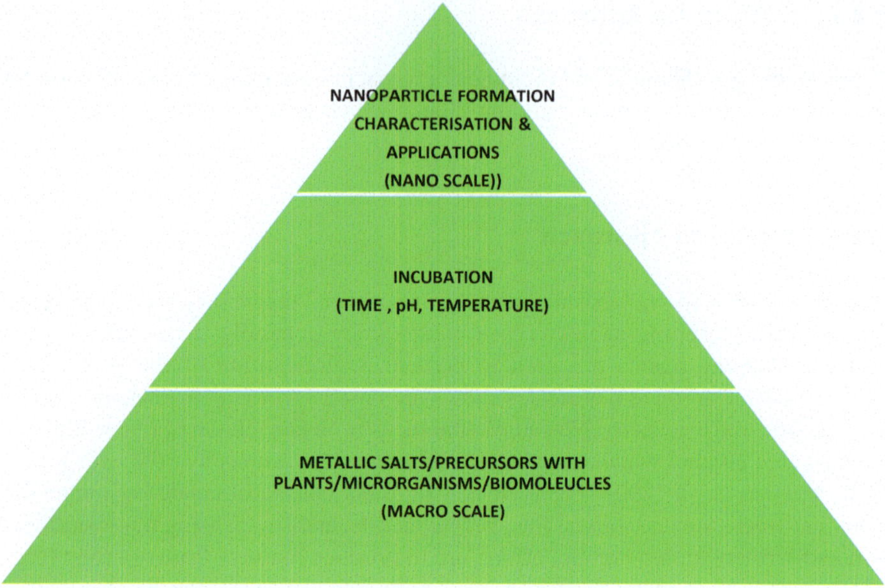

Fig. 4.1 Green fabrication of nanoparticles

reliable. Lo et al. (2012) have reported improved production of biogas from micro-nano municipal solid waste incinerator fly ash and bottom ash. Green fabrication approaches are shown in Fig. 4.1, while the methods for characterization of synthesized nanoparticles are shown in Fig. 4.2.

4.4 Plant-Based Synthesis

Phytomining is a process where plants which hyperaccumulate metals are grown on metal-rich soils. These metal ions are converted into nanoparticles due to plant metabolites present in them (Rai and Yadav 2008). But it is a time-consuming process. Imran and Rani (2015) suggested that to overcome this, whole plant biomass or plant extracts are used for the synthesis. Alkaline pH was found to favor the formation of metal oxide nanoparticles (Singh et al. 2016). Zinc oxide is one of the major biofabricated nanoparticles due to its antimicrobial activity and photocatalytic activity, etc. Iron oxide metal nanoparticles also have been synthesized (Ahmed et al. 2017). Plant-based synthesis has an advantage of lack of pathogenicity when compared to bacteria or fungi (Pantidos and Horsfall 2014).

Datta et al. (2017) investigated the *Parthenium hysterophorus* for synthesis of zinc oxide nanoparticles with a particle size ranging between 16 and 45 nm. Using the aqueous extract of tomatoes, Sutradhar and Saha (2016) prepared zinc oxide nanoparticles. Sindhura et al. (2013) reported zinc nanoparticle fabrication using *Parthenium hysterophorus*. Sangeetha et al. (2011) investigated the production of nanostructure zinc oxide particles using *Aloe vera* leaf extract. Jain and Pradeep

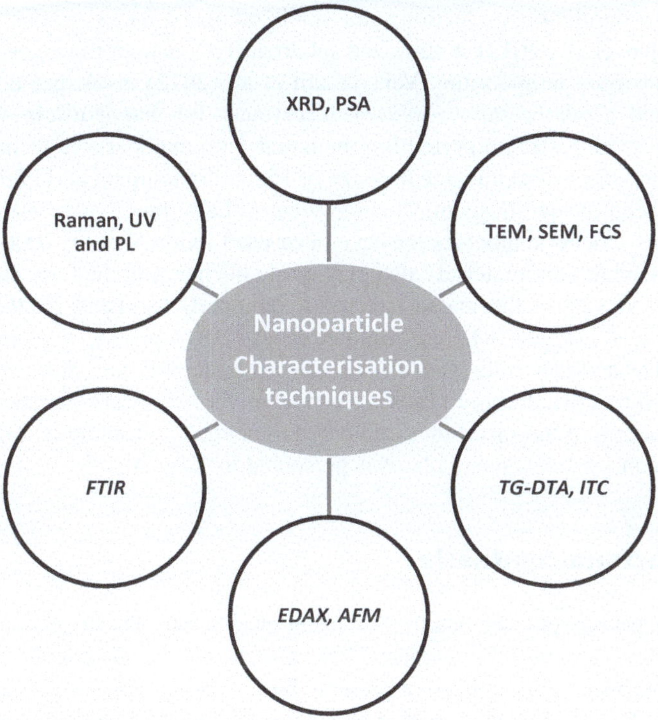

Fig. 4.2 Methods for characterization of synthesized nanoparticles

(2004) stated that green fabrication method can also be used for large-scale synthesis. Green synthesis of nanoparticles is simple where biomaterials act as reducing agent. *Brassica juncea* and *Medicago sativa* were reported to accumulate 50-nm silver nanoparticles when grown with silver nitrate (Harris and Bali 2008). In *M. sativa* 4-nm-sized gold nanoparticles were formed (Gardea-Torresdey et al. 2002). Similarly in *Iris pseudacorus* copper particles sized 2 nm were formed (Manceau et al. 2008). The main mechanism for the green synthesis of silver nanoparticles is plant-assisted reduction due to alkaloids, flavonoids, terpenoids, amino acids, saponins, and phenols carbohydrates which act as reducing, capping, and stabilizing agents. Plant-mediated synthesis of nanoparticles is advantageous over other methods compared to microbial synthesis by eliminating the elaborative and complicated processes of maintaining microorganism cultures and scale-up problems. Sadhasivam et al. (2010) suggested the use of experimental procedures for the fabrication of green nanomaterials. Nadeem et al. (2017), Asha et al. (2016), Marimuthu et al. (2013), Jalill et al. (2016), and Singh (2016) have used biosynthetic approach for nanoparticle production. Biosynthesis of TiO_2 nanoparticles were synthesized using an eco-friendly approach by Pantidos and Horsfall (2014). Bao et al. (2012) have synthesized nanoparticles with controlled size and shape. Mittal et al. (2014a, b) considered plants to be better suited for synthesis of nanoparticles. Plant-based compounds such as phenolic acids, alkaloids, proteins, and carbohydrates can be used for the regulation, reduction, and stabilization of nanoparticles (Dobrucka 2017). Roopan et al. (2012) synthesized spherical TiO_2 nanoparticles using *Annona squamosa* at room temperature. Velayutham et al. (2012) investigated *Calotropis gigantea* and *Catharanthus roseus* leaf extracts for nanoparticle fabrication. Sundrarajan et al. (2017) reported bio-mediated TiO_2 nanoparticle formation from *Morinda citrifolia* leaf extracts with a size of 15 nm. Sivaranjani and Philominathan (2016) reported the synthesis of *Moringa oleifera* Lam. leaf extract mediated TiO_2 nanoparticles. Nickel nanoparticles have been used (Kidwai et al. 2006) to reduce aldehydes. Cobalt nanomaterials also play a role in changing the fatty acid chemistry. Rajiv et al. (2013) fabricated zinc oxide nanoparticles using *Parthenium hysterophorus* leaf extracts, whereas Ghidan et al. (2016) produced nanoplates. pH changes have a direct influence on the size of nanoparticles. Soni and Prakash (2011) stated that pH alteration causes a change in the secondary metabolites which help in reduction of the metal ions compared to acidic environments. The nanofabrication of nanoparticles using plants is presented in Table 4.1.

4.5 Bacterial Synthesis

Bacteria are prokaryotic organisms with faster growth rate. They can also be used as precursors for nanoparticle synthesis. The growth conditions can be easily controlled. Bacteria are also utilized in bioremediation and bioleaching. Magnetotactic bacteria and S-layer bacteria are also reported to be used for the synthesis of these nanoparticles (Pum and Sleytr 1999; Spring and Schleifer 1995). In order to overcome metal toxicity, bacteria produce proteins which bind with the metal ions and reduce them to nanoparticles. Saifuddin et al. (2009) established a synthesis approach for silver nanoparticle synthesis using *Bacillus subtilis* and microwave irradiation in water. Extracellular silver nanoparticle synthesis (~40 nm) with supernatant of *Bacillus licheniformis* was showed by Kalishwaral et al. (2008). Wei et al. (2012) used solar irradiation of *Bacillus amyloliquefaciens* extracts to obtain nanoparticles. Satyanarayana et al. (2010) predicted that proteins with a molecular weight between 25 and 66 kDa and 66 and 116 kDa induced the formation of gold nanoparticles and silver nanoparticles. Konishi et al. (2007a, b) investigated the bacterium *S. algae* which grew anaerobically in lactate medium at pH 2.0 for the production of gold nanoparticles within the bacterial cells and outside the bacterial cells. Liangwei et al. (2015) designed a Hb-coli-nAu-glassy carbon electrode for electrochemical deposition of gold nanoparticles. The effect of pH on gold nanoparticle formation using the bacteria *E.coli* and *D. desulfuricans* was also observed by Deplanche et al. (2010) where they concluded that in acidic pH, spherical gold nanoparticles and at neutral and basic pH, triangles, hexagons, and rod-shaped nanoparticles were formed. *Plectonema boryanum* UTEX 485 was shown to produce gold nanoparticles of desired morphology at temperatures between 25 and 200 °C as reported by Lengke et al. (2006). Klaus-Joerger (2001) opined that cobalt, chromium, and nickel metals could be introduced along with magnetite crystal nanoparticles. Jahn et al. (2005) discussed the mechanism for electron transfer and

Table 4.1 Plant-mediated synthesis of nanoparticles

Name of the plant	Nanoparticle synthesized	Size	Shape	References
Aloe vera	Au and Ag	50–350	Spherical, triangular	Chandran et al. (2006)
Aloe vera	In_2O_3	5–50	Spherical	Maensiri et al. (2008)
Camellia sinensis	Ag, Au	30–40	Spherical, triangular, irregular	Nestor et al. (2008)
Curcuma longa	Pd	10–15	Spherical	Sathishkumar et al. (2009)
Eucalyptus macrocarpa	Ag	10–100	Spherical, cubes	Poinern et al. (2013a, b)
Eucalyptus macrocarpa	Au	20–100	Spherical, triangular, hexagonal	Poinern et al. (2013a, b)
Diospyros kaki	Pt	15–19	Crystalline	Song et al. (2010)
Citrullus colocynthis	Ag	31	Spherical	Satyavani et al. (2011)
Avena sativa	Gold	5–20	Rod-shaped	Armendariz et al. (2004)
Brassica juncea	Silver	2–35	Spherical	Haverkamp and Marshall (2009)
Cinnamomum camphora	Gold and silver	55–80	Spherical	Huang et al. (2007)
Carica papaya	Silver	60–80	Spherical	Mude et al. (2009)
Citrus limon	Silver	< 50	Spherical, spheroidal	Prathna et al. (2011)
Cycas sp.	Silver	2–6	Spherical	Jha and Prasad (2010)
Eucalyptus hybrid	Silver	50–150	Crystalline, spherical	Dubey et al. (2009)
Jatropha curcas	Silver	15–50	Spherical	Bar et al. (2009)
Ludwigia adscendens	Silver	100–400	Spherical	Mochochoko et al. (2013)
Mangifera indica	Silver	20	Spherical, triangular, hexagonal	Philip (2011)
Pelargonium roseum	Gold	2.5–27.5	Crystalline	Shankar et al. (2003)
Psidium guajava	Gold	25–30	Spherical	Raghunandan et al. (2009)
Pyrus sp. *(pear fruit extract)*	Gold	200–500	Triangular, hexagonal	Ghodake et al. (2010)
Psidium guajava	Gold	25–30	Spherical	Raghunandan et al. (2009)
Rhododendron dauricum	Silver	25–40	Spherical	Mittal et al. (2012)
Terminalia catappa	Gold	10–35	Spherical	Ankamwar (2010)

insoluble iron formation from iron-reducing bacteria. The marine bacterium *Desulfuromonas acetoxidans* was found to reduce Fe (III) and Mn (IV) by Roden and Lovley (1993). Kashefi and Lovley (2000) stated that *P. islandicum* reduced various metal salts at 100°C with an electron donor and electron acceptor. It was observed by these workers that this bacterium was not capable of reducing As (V) and Se (VII). Platinum nanoparticle formation of size 5 nm within 10 minutes at pH 7 and temperature 25 °C using *S. algae* was illustrated by Konishi et al. (2007a, b). The process could take place only in the presence of lactate. Yong et al. (2002) found that the platinum ions were adsorbed to the sulfate-reducing bacterium *D. desulfuricans*. A 90-kDa protein of *Tetrathiobacter kashmirensis* bioreduced selenite to elemental selenium (Hunter and Manter 2008). Both intracellular and extracellular formation of nanostructured selenium using *P. aeruginosa* SNT1 was reported by Yadav et al. (2008). *Bacillus selenitireducens* formed nanorods of elemental tellurium with 10-nm size (Baesman et al. 2007). *Bacillus* sp. BZ could produce elemental rod-shaped tellurium nanoparticles (Zare et al. 2012).

Holmes et al. (1995) have investigated the exposure of cadmium ions to *Klebsiella aerogenes* which produced intracellular CdS nanoparticle formation with size range between 20 and 200 nm. Similarly, Cunningham and Lundie (1993) observed that *Clostridium thermoaceticum* was able to form cadmium extracellularly in the presence of cysteine after 12-h exposure. Labrenz et al. (2000) found zinc sulfide particle formation of 2–5-nm diameter by sulfate-reducing bacteria. Periplasmic hydrogenases could be responsible for Pd (II) reduction resulting in the formation of palladium nanoparticles which were found to be inhibited by Cu (II) (Lioyd et al. 1998). Microbial Au (III) reduction was also investigated in *Thermotoga maritima, Geothermobacterium ferrireducens*, *Shewanella alga*, *Desulfovibrio vulgaris*, *Desulfovibrio desulfuricans*, and *Escherichia coli*. The natural product piperitone inhibited the formation of silver nanoparticles in *Klebsiella pneumoniae* which confirmed that nitroreductase enzymes play a major role in the formation of silver nanoparticles. Spherical platinum nanoparticles using cyanobacterium, *Plectonema boryanum*, were biosynthesized by Lengke et al. (2006). Riddin et al. (2009) reported the formation of platinum nanoparticles in which sulfate-reducing bacteria were used to understand the mechanism involved in platinum nanoparticle production. Prozorov et al. (2007) have used recombinant Mms6 protein for the formation of Fe_2O_4 nanocrystals. Microorganisms have also been suggested as nanofactories, for nanoparticle production (Mandal et al. 2006). Duran et al. *(*2005*)* highlighted the role of NADH and NADH-dependent nitrate reductase enzymes in metal nanoparticle fabrication. Several enzymes and their role in nanoparticles by microorganisms have been studied. Nangia et al. (2009) synthesized gold nanoparticles using a NADPH-dependent reductase enzyme. UV-vis absorption is used to understand the specific absorption of the prepared nanoparticle. Energy-dispersive X-ray (EDAX) analysis is used for studying the elemental composition of the synthesized nanoparticles. Electron microscopy techniques such as transmission electron microscopy (TEM) and scanning electron microscopy (SEM) are used to study the size and shape of the nanoparticles. Fourier-transform infrared (FTIR) analysis reveals the specific chemical bonds which are involved in binding to the nanoparticles. X-ray diffraction (XRD) analysis is used for understanding the crystal nature of the nanoparticles. The nature and specific composition of nanoparticles can be understood by photon correlation spectroscopy (PCS) and atomic force microscopy (AFM). *Thermomonospora* sp. produced gold nanoparticles extracellularly (Ahmad et al. 2003). Gericke and Pinches (2006) highlighted the role of pH, temperature, substrate concentration, and exposure time in synthesized nanoparticle size. pH was known to influence gold nanoparticle fabrication by *Rhodopseudomonas capsulata*. Baia et al. (2009a, b) used TEM analysis to know the size of nanoparticles synthesized by *Rhodopseudomonas palustris.* Gold nanoparticle synthesis using *Yarrowia lipolytica* occurred in a pH-dependent manner (Agnihotri et al. 2009). UV-vis spectroscopy, XRD, and SEM were used for characterization of gold nanocubes synthesized by *Bacillus licheniformis* (Kalishwaralal et al. 2008). The nanofabrication of nanoparticles using bacteria is presented in Table 4.2.

Table 4.2 Bacteria-mediated green fabrication of nanoparticles

Bacteria	Nanoparticle	References
Aeromonas sp. *SH10*	Silver	Rai et al. (2006)
Bacillus cereus	Silver	Sunkar and Nachiyar (2012)
Bacillus megaterium D01	Gold	Wen et al. (2009)
Bacillus subtilis 168	Gold	Beveridge and Murray (1980)
Bacillus subtilis	Silver	Saifuddin et al. (2009)
Clostridium thermoaceticum	Cadmium sulfide	Cunningham and Lundie (1993)
Corynebacterium sp. *SH09*	Silver	Zhang et al. (2005)
Desulfobacteraceae	Zinc sulfide	Labrenz et al. (2000)
Desulfovibrio desulfuricans	Palladium and selenium	Kessi et al. (1999), Yong et al. (2002)
Desulfovibrio vulgaris	Gold, uranium, and chromium	Kashefi et al. (2001)
Desulfovibrio magneticus strain RS-1	Magnetite	Pósfai et al. (2006)
Enterobacter cloacae	Silver and selenium	Kessi et al. (1999)
Escherichia coli MC4100	Gold	Deplanche and Macaskie (2008)
Geobacillus sp.	Gold	Correa-Llantén et al. (2013)
Klebsiella aerogenes	Cadmium sulfide	Holmes et al. (1995)
Klebsiella pneumonia	Silver	Shahverdi et al. (2007)
Lactobacillus strains	Gold	Nair and Pradeep (2002)
Magnetospirillum magnetotacticum	Magnetite	Philipse and Maas (2002)
Nocardiopsis sp. *MBRC-1*	Silver	Manivasagan et al. (2013)
Pseudomonas fluorescens	Gold	Rajasree and Suman (2012)
Pseudomonas putida NCIM 2650	Silver	Thamilselvi and Radha (2013)
Rhodopseudomonas palustris	Cadmium sulfide	Bai et al. (2009a, b)
Rhodobacter sphaeroides	Zinc sulfide	Bai et al. (2006)
Serratia nematodiphila	Silver	Malarkodi et al. (2013)
Shewanella algae	Platinum	Konishi et al. (2007a, b)

4.6 Fungus- and Alga-Based Synthesis

In fungus-mediated green synthesis, extracellular enzymes produced by them transform the metal ions to their counter metal oxide nanoparticles (Durán et al. 2011). Fungal production of nanoparticles is comparatively faster than bacterial synthesis as they produce more extracellular proteins. *Saccharomyces cerevisiae* was investigated for biofabrication of TiO_2 nanoparticles (Vainshtein et al. 2014). Zinc oxide nanoparticles from ZnNO3 were fabricated using *Aspergillus fumigatus* TFR-8 (Raliya et al. 2013). *Fusarium oxysporum* was reported to produce bismuth oxide (Jha and Prasad 2010). Algae are autotrophic and aquatic photosynthetic organisms which are unicellular and also multicellular in nature. Microalgal polysaccharides alginate, fucoidan, and laminarin are known to be involved in the formation of nanoparticles (Pinheiro et al. 2014; Lee and Lim 2014, 2015; Kimura et al. 2013). Govender et al. (2010) proposed that the hydrogenase enzyme from *Fusarium oxysporum* was involved in the formation of platinum nanoparticles. Yan et al. (2012) investigated yeasts for fabrication of zinc phosphate nanopowders with a size between 10 and 80 nm. The nanofabrication of nanoparticles using fungi is presented in Table 4.3.

4.7 Actinomycete-Based Nanoparticle Synthesis

Actinomycetes can also produce nanoparticles via extracellular or intracellular synthesis (Golinska et al. 2014). It was found that extracellular synthesis was the most preferred way of formation by these groups. *Rhodococcus* sp. could form nontoxic

Table 4.3 Green fabricated nanoparticles using fungi

Name of the organism/biomolecule	Nanoparticle synthesized	References
Candida glabrata	Cadmium sulfide	Krumov et al. (2007)
Schizosaccharomyces pombe	Cadmium sulfide	Dameron et al. (1989)
Schizosaccharomyces pombe	Cadmium sulfide	Kowshik et al. (2002)
Schizosaccharomyces pombe	Cadmium sulfide	Krumov et al. (2007)
Pichia jadinii (Candida utilis)	Gold	Pimprikar et al.(2009)
Fusarium oxysporum	Bismuth	Bansal et al. (2006)
Fusarium oxysporum	Cadmium	Kumar et al. (2007)
Fusarium oxysporum f. sp. *lycopersici*	Platinum	Govender et al. (2009)
Fusarium spp.	Zinc	Velmurugan et al. (2010)
Aspergillus	Mercury	Das et al. (2008)
Fungi	Zinc, magnesium, and titanium	Raliya and Tarafdar (2013)
Penicillium fellutanum	Silver	Kathiresan et al. (2009)
Penicillium strain J3	Silver	Maliszewska et al. (2009)
Cladosporium cladosporioides	Silver	Balaji et al. (2009)

gold nanoparticles with size between 5 and 15 nm (Abdeen et al. 2014). Cell wall enzymes were reported to be involved in intracellular reduction of gold and silver ions to form metallic nanoparticles (Balagurunathan et al. 2011; Chauhan et al. 2013; Prakasham et al. 2012; Sukanya et al. 2013). Karthik et al. (2014) investigated *Streptomyces* sp. LK-3 for extracellular synthesis of silver nanoparticles and the various factors which could induce the production of these nanoparticles. This study revealed that NADH-dependent nitrate reductase enzyme was essential for the formation of silver nanoparticles via an electron transfer mechanism. Similarly, for gold nanoparticles it was found that nitrate reductase enzymes played a pivotal role in the reduction process (Shah et al. 2012).

4.8 Viral Particles for Nanoparticle Synthesis

Viral particles are being used for the past decade for the synthesis of nanomaterials. Silicon dioxide and zinc sulfide, cadmium sulfide, and iron oxide nanoparticles along with quantum dots have been synthesized by Lee et al. (2002) and Mao et al. (2003). They are particularly attractive due to the presence of capsomeres which form the capsid protein. These capsid proteins contain many functional groups and form a good surface for the formation of metal ions (Makarov et al. 2014). Tobacco mosaic virus, a plant virus, has about 2130 capsid protein molecules which can act as attachment points for the deposition of nanoparticles (Aljabali et al. 2010; Gorzny et al. 2010; Kobayashi et al. 2012; Royston et al. 2008). They can also be used to produce three-dimensional vessels of nanoparticles for drug delivery in pharmaceutical industry (Zeng et al. 2013; Steinmetz et al. 2009). Tobacco mosaic virus was able to reduce silver and gold salts before adding plant extracts of *Nicotiana benthamiana* or *Hordeum vulgare* thus resulting in lower-sized nanoparticles (Love et al. 2014). Some other studies investigated viruses as a template for the formation of nanowires and nanotubes (Merzlyak and Lee 2006).

4.9 Biological Derivatives for Nanoparticle Synthesis

Biological derivatives like cellulose were used to fabricate TiO_2 nanoparticles between 5- and 10-nm size by Farag et al. (2016). *Cylindrotheca fusiformis* peptide R5 has been used for fabricating TiO_2 nanoparticles (Sewell and Wright 2006). Padalkar et al. (2012) using alpha synuclein protein reported the fabrication of titanium nanowires. Titanium nanotubes were synthesized using bacterial flagella by Li et al. (2012). The lignocellulose waste material-mediated synthesis of TiO_2 nanoparticles was reported by Ramimoghadam et al. (2014). Using different enzymes Chen et al. (2010) reported synthesis of TiO_2 nanoparticles. Glucose oxidase enzyme produced rutile TiO_2 nanoparticles, while the enzyme catalase formed anatase TiO_2 nanoparticles. Chen et al. (2009) used lysozyme for anatase TiO_2 nanoparticle production. Shankar et al. (2003) investigated the role of terpenoids in silver nanoparticle formation using extracts from geranium leaves. *Cinnamomum zeylanisum* (cinnamon) extracts with eugenol could reduce silver nitrate and aurium chloride to their respective nanoparticles (Singh et al. 2010). Flavonoids comprise of anthocyanins, chalcones, flavones, isoflavonoids, flavonols, and flavanones were also found to bioreduce metal ions into nanoparticles. *Ocimum basilicum* extracts containing flavonoids resulted in the fabrication of silver nanoparticles (Ahmad et al. 2010). Gold and silver nanoparticle synthesis with a size between 21 and 30 nm was done with apigenin glycoside from *Lawsonia inermis* (Kasthuri et al. 2009). Monosaccharides such as glucose can act as reducing agents, while keto group containing sugars can act as antioxidants. Maltose and lactose have also shown to reduce metallic nanoparticles to metals (Panigrahi et al. 2004). Glucose and fructose were reported to be involved in fabricating metal nanoparticles of various morphologies (Panigrahi et al. 2004). Magnolia vine extract could reduce gold salts to gold ions (Shankar et al. 2003). Gruen (1975) observed that lysine, methionine, cysteine, and arginine could bind to silver ions. Saikat et al. (2002) reported that aspartate can reduce gold salt, while valine and lysine could not do so.

Spectral studies indicated that nanoparticles formed using plants or plant extracts are bound to proteins (Zayed et al. 2012). Tan et al. (2010) established that tryptophan is the strongest reducer of gold ions, while histidine only binds to gold ions (Glusker et al. 1999). Tan et al. (2010) demonstrated that amino acid sequence determines the ability to bind or reduce metal ions. Glusker et al. (1999) reported that peptides with some amino acids were ineffective in reducing gold salts. Synthetic peptide GASLWWSEKL reduces metal ions to form small nanoparticles of size 10 nm, while SEKLWWGASL sequences form larger nanoparticles of 40 nm in size (Glusker et al. 1999). Biological nanoparticles can be derived from biomolecules having a size range of 10 nm–1 μm (Kumar and Lal 2014). Polysaccharides such as pectin and alginates, dextran, insulin, and chitosan are being primarily used for the fabrication of nanomaterials (Sahdev et al. 2014). Albumin, elastin, collagen, gelatin, zein, gliadin and ferritin, sericin, and fibroin have been so far investigated for the production of nanoparticle (Nitta and Numata 2013; Hazeri et al. 2012) and (Zhao et al. 2015). DNA and RNA can also be used for the production of nanomaterials where the size of nanoparticles can be controlled. They can also be tagged with fluorophores, oligonucleotides, and aptamers (Panigaj and Reiser 2016). Lipid-based nanoparticles are promising due to their longer shelf-life and stability and have good encapsulation capacity which can also be used for drug delivery (Tamjidi et al. 2013; Weiss et al. 2008; Willner et al. 2007).

4.10 Green Nanocatalysts

Catalysts will provide the sites for the reactants to be activated leading to the formation of the products. Nanocatalysts have a nanoscale dimension, which could be used for overcoming mass transfer limitation and reducing the cost of biofuel production. Nanoscale catalysts can be divided into various types based on carbon, graphite, and metal oxides. Nanocatalysts enhance the selectivity of the reactions at a lower temperature with higher recycling rates and less energy consumption. Hence, there are more effective alternatives when compared to conventional catalysts for the efficient production of biofuels. Wen et al. (2010) reported that the solid base nanocatalyst KF/CaO could be used for biodiesel production with higher yields (Chaturvedi et al. 2012). Wen et al. (2010) demonstrated that the nanocatalyst KF/CaO has increased catalytic activity with enhanced stability. Sidra et al. (2016) investigated the transesterification process using CaO-Al_2O_3 nanocatalyst for biodiesel formation using Jatropha oil. Sivakumar et al. (2013) used smoke-deposited nanocatalyst magnesium oxide for the production of biodiesel. Three to five times higher conversion ratios were seen due to enhancement of surface area than those with conventional catalyst earlier reported. Biodiesel produced from castor oil and *Pongamia pinnata* oil with methanol was studied using nano-iron catalyst by Mookan et al. (2014a, b). Gupta and Agarwal (2016) prepared and characterized CaO nanoparticles from calcium nitrate (CaO/CaN) and snail shell (CaO/SS) to estimate the performance of the catalysts in terms of biodiesel yield from soybean oil. They concluded that the nanocatalyst from snail shell showed good catalytic activity and enhanced stability for the transesterification reaction. Zhang and Misra (2007) prepared CaO catalyst which was used for the transesterification of soybean oil into biodiesel. Deng et al. (2011) designed a hydrotalcite Mg/Al (3:1) nanocatalyst for transesterification of Jatropha oil. Acid nanocatalysts can be used for low-quality oil feedstocks with high FFA content. Acid catalysts catalyze esterification of free fatty acids and also transesterification of triglycerides for biodiesel production simultaneously (Canakci and Van Gerpen 2001; Zhang et al. 2003). Wang et al. (2015) prepared a magnetic acid catalyst for biofuel production. Bifunctional nanocatalysts promote simultaneous esterification and transesterification in a single step. A bifunctional quintinte-3T nanocatalyst was prepared for transesterification and as esterification promoter by Kondamudi et al. (2011). Palladium and molybdenum phosphate nanoparticles on silica were used for phenol hydrodeoxygenation (Hess et al. 2017; Löfstedt et al. 2016; Xing et al. 2017). Enumula et al. (2017)) reported production of alkyl levulinates from lignocellulosic feedstocks using alumina/SBA-15 heterogeneous catalysts. Nanoparticle immobilization can maintain enzymatic activity efficiently at low cost and higher stability (Dinçer and Telefoncu 2007; Ho et al. 2008, Wu et al. 2005); Liao et al. 2008, (Cherian et al. 2015; Liao et al. 2010). Verma et al. (2013a, b) immobilized nanoparticles with β-glucosidase on functionalized magnetic nanoparticles for biofuel production. Nanomagnetic solid base catalyst can improve biodiesel production up to 95% (Hu et al. 2011). Ren et al. (2011) have devised a method for lipase immobilization on polydopamine-coated magnetic nanoparticles. Rhodium particles were entrapped in carbon nanotubes to enhance catalytic activity (Pan et al. 2007).

4.11 Bioenergy Applications of Nanoparticles

Developing clean and efficient energy technologies has become the need of the hour due to the ever-increasing energy demand. Nanotechnology application in biofuel production mainly focuses on breaking down the feedstock more efficiently, decreasing the transportation cost of feedstock, and improving the yield of biofuels. Biodiesel production improved when nanocatalysts were used from animal fats and vegetable oils were used as feedstock (Jennifer and Peter 2006). In biofuel production, economic viability can be achieved by enzyme immobilization in nanomaterials. In industry, covalent linking of an enzyme to a support is the most interesting method of immobilization which is based on activation of chemical groups to support its reaction with protein nucleophiles (Ahmad and Sardar 2015; Mateo et al. 2007). With the dwindling energy sources, the problem of energy crisis looming at large, all countries are looking for alternative sources of their energy needs. The European Union has a target to increase its biofuel energy source to 5.75% by 2010 and 20% by 2020 without any substantial breakthrough. A study has indicated that the world's fossil fuel reserves would be depleted by the year 2050. In India, the need for exploring alternate sources of energy has become the first agenda in terms of investment issues. Indians have primarily depended on conventional energy sources, such as fossil fuels such as coal and petroleum, which has caused rapid depletion of fossil fuels with significant effects on the environment. Bioenergy sources are eco-friendly and pose no serious environmental risks. There is an increasing interest on bioenergy research using nanomaterial as the potential to increase the yield. The application of nanomaterials can contribute to potential bioenergy research. Bioresources are renewable as they are self-prevalent and ubiquitous and leave no toxic end products or residues.

Energy resources are rapidly depleting, and we are facing an energy crisis due to which we are forced to look for alternative sources of energy. Biofuel is an alternative source of energy which is sustainable and environment-friendly. As of now the biodiesel and biogas production are being investigated for this purpose. For this purpose nanomaterials are also being explored to increase the production of biofuels. Various methods of nanoparticle production are being improvised for efficient and sustainable production of nanoparticles. Along with enzyme-based nanocatalysts are being used to enhance the production of biofuels. Due to increasing fuel demand and emission of greenhouse gases, biomass-based biofuels have been gaining intense attention. Farrell et al. (2006) have suggested the need to search for alternative renewable energy sources. Biomass can be divided into two types, namely, oleaginous feedstock and carbohydrates. Triglycerides make up the largest portion of oleaginous feedstock and have some of the highest hydrogen to carbon (H/C) ratios, combined with relatively low amounts of oxygen. Carbohydrates form the most of the biomass available for conversion which includes mono-, di-, and polysaccharides. Biofuels can reduce the global dependency and global warming due to petroleum (Kumar and Sharma 2014; Lee et al. 2014). The first-generation biofuels have some limitations (Naik et al. 2010); therefore, lignocellulosic materials are being investigated (Eggert and Greaker 2014; Patumsawad 2011).

Pseudomonas cepacia lipase bound to FeO nanoparticle was effective in the production of biofuel. Among the explored nanomaterials, ZrO_2 showed the highest conversion and selectivity for n-C17 alkanes. Ni/ZrO2 catalysts transformed microalgal oil to liquid alkanes. Kandel (2013) reported that aminopropyl-functionalized Ni-MSN catalysts increased octadecane yield. Ni-supported zeolites were investigated by Peng et al. (2012) for conversion of microalgal oil. Various metal oxides including SiO_2 and TiO_2 have been investigated for lignocellulosic biomass conversion apart from enzymatic degradation and ball milling method (Cherian et al. 2015; Hendriks and Zeeman 2009; Liao et al. 2010). Nanotechnology application in biofuel production mainly focuses on breaking down the feedstock more efficiently, decreasing the transportation cost of feedstock, and improving the yield of biofuels. Biodiesel production improved when nanocatalysts were used for animal fats and vegetable oils were used as feedstock (Jennifer and Peter 2006). Mahmood and Hussain (2010) used spent tea in a three-step reaction for formation of biofuels using nanomaterials with *Aspergillus niger*. Wen et al. (2010) produced biodiesel using KF/CaO nanocatalyst from tallow seed oil. Qiu et al. (2011) prepared nanocatalyst for biodiesel production from soybean oil and methanol. Immobilized cellulase on silica nanoparticles was used for efficient breakdown of lignocellulosic biomass (Lupoi and Smith 2011). Polyacrylonitrile nanofiber immobilized with polycaprolactone for biodiesel production was investigated by Sakai et al. (2010). Tran et al. (2012) investigated biodiesel production by immobilizing lipase on ferric silica nanocomposites. Immobilized enzyme nanocatalysts showed biodiesel production higher than 90 % with high methanol tolerance and reusability when compared to other types of nanocatalysts. Cho et al. (2012) immobilized cellulose onto gold nanoparticles for ethanol production. In biofuel production, economic viability can be achieved by enzyme immobilization in nanomaterials (Ahmad and Sardar 2015; Mateo et al. 2007). Immobilization of β-glucosidase and cellobiase enzymes on nanomaterials for improved bioethanol production was also reported by several workers (Ansari and Husain 2012; Datta et al. 2013; Verma et al. 2013a, b). Bioethanol production using β-glucosidase immobilized on nanomaterials was reported by other workers (Tsai and Meyer 2014; Verma et al. 2013a, b; Zhang et al. 2013). Verma et al. (2013a, b) immobilized β-glucosidase from *Aspergillus niger* on functionalized magnetic nanoparticles. Zheng et al. (2013) also reported similar results. *Trichoderma reesei* cellulase immobilized on silica was used for pretreatment of biomass (Das et al. 2011). Abraham et al. (2014) immobilized cellulase from *Trichoderma reesei* onto nanoparticles for hemp hurd biomass hydrolysis. Cherian et al. (2015) used MnO_2 nanoparticle-immobilized cellulase for bioethanol production. Nanocatalysts (Johnson et al. 2009) increased the conversion efficiency of lignocellulose. Carbon nanotubes coated with nano-size rhodium and manganese were used to produce ethanol from carbon monoxide and hydrogen (Pan et al. 2007). Rhodium- and manganese-loaded carbon tubes enhanced activation of carbon monoxide and improved ethanol production rate.

Laccases are extracellular enzymes produced by fungi which degrade various cellulosic material and produce sugars and phenolic products. Nanostructured enzymes have about 1000 times half-lives when compared to native enzymes and can also be reused. Laccase is used in the pretreatment of agrowastes for biofuel production. Laccase are immobilized with various nanomaterials to enhance biofuel production. Lipase enzymes are extracellular or intracellular in nature. Lipase from *Thermomyces lanuginosa* covalently bound to ferric oxide nanoparticle was investigated using sunflower oil (Xie and Ma 2010) and waste grease (Ngo et al. 2013) for the production of biodiesel. Magnetite nanoparticles are the most investigated due to their low toxicity and good biocompatibility (Lie et al. 2011). Casals et al. (2010) have used nano-iron oxide of diameter 7 nm to enhance methane production. Murahashi et al. (1992) have successfully employed the nano-iron powder as a catalyst to carry out the oxidation of cyclohexane at room temperature with a conversion efficiency of 11% and a selectivity of 70% for the formation of cyclohexanol and cyclohexanone. Kesavan et al. (1999) have reported a conversion of 40% and a selectivity of 80% using nanophased catalytic particles of iron and nickel from the catalytic mixtures of iron carbonyl, Fe(CO)5, and nickel carbonyls. Nanotubes (Martone et al. 2011) and nanoporous materials which possess high aspect ratios have made the immobilization process easier for enzyme loading, biocatalyst recovery, and reusability (Misson et al. 2015). Covalent binding of enzymes with nanostructured supports are being used for long-term enzyme stability (Ovsejevi et al. 2013). Hussain et al. (2011) have used nickel- and cobalt-based nanoparticles as catalysts for biofuel production, and they have also shown excellent improvement in terms of the biofuel yield from the process. Nickel nanoparticles when employed in combination with cobalt particles result in better catalysis (Guo et al. 1995). Spent tea is a type of solid waste. It is chiefly composed of milk and sugar. *Aspergillus niger* can biologically reduce the spent tea to give a mixture of liquid extracts, solid charcoal, and gaseous compounds (Mahmood and Hussain 2010). To improvise the current method, cobalt or cobalt oxide nanostructures were used (Son et al. 2002). Studies on nanocatalysts have reported the extensive and exhaustive use of cobalt nanoparticles for the purpose of gasification of biomass. The various factors which contributed to biomass gasification were investigated by Patra and Baek (2014). Lestari et al. (2008) have reported the active role of cobalt nanoparticles in the tea extract in their attempt to make biodiesel as esters. Kondamudi et al. (2008) have accessed the biodiesel formation and efficiency of the biofuel obtainable from tea as well as coffee. The chief advantages of incorporating cobalt nanoparticles as catalysts in this bioconversion stem from high reactivity and high surface areas which resulted in the decrease of the reaction temperature and the corresponding activation energy requirement. Suarez et al. (2009) have reported that the biodiesel produced had low sulfur and nitrogen content, which is more eco-friendly. Nanofarming technology has proved to be very effective by enabling the synthesis of biofuels with the advantage of reuse of biocatalysts. Nanobiocatalysts, which are highly selective mesoporous membranes supplemented with nanoparticle adsorption, eliminate the problem of algal biomass getting killed after the extraction process is over. Removal of water in harnessing the biofuels is important as it disturbs the ethanol extraction by formation of azeotropic mixtures. Abdelsalam et al. (2016) investigated fresh raw manure using nickel, magnetite, cobalt, and iron nanoparticles for improving biogas yields.

Similar enhancement was found by Wang et al. (2016) who utilized waste-activated sludge for methane production using nZVI and Fe2O3 nanoparticles. Inorganic nanocatalysts such as (CaO, MgO), hydrotalcites, zeolites, zirconia, and sulfated oxides are reported for the synthesis of biodiesel. Titanium dioxide (Gardy et al. 2017), strontium oxide (Liu et al. 2007), calcium oxide (Liu et al. 2008), magnesium oxide (Verziu et al. 2008), and metal oxide nanocatalysts have been developed with high catalytic performance for biodiesel production. Gardy et al. (2018) proposed that magnetic nanoparticle-based catalysts are more advantageous due to their easier separation and reuse apart from being cost effective. Yahya et al. (2016) prepared different mesoporous nanocatalysts, with enhanced catalytic activities, for biodiesel production. For biodiesel production carbon-based nanocatalysts (Guan et al. 2017), carbon nanofibers (Stellwagen et al. 2013), graphene oxide (Mahto et al. 2016), and biochar (Dehkhoda et al. 2010) were found to have potential applications for biodiesel production using a wide range of feedstocks. Enzyme immobilization during lipase-catalyzed biodiesel and cellulosic ethanol production processes was investigated by Kim et al. (2018). Trindade (2011) explained the benefits of nanostructures which include large surface area for high enzyme loading, higher enzymatic stability, enzyme reusability, and reduction of the operational costs of biofuel production. Trindade (2011) also proposed nanoparticles as fuel additives to boost fuel blend performance. Zhang et al. (2013) stated that silica nanoparticles, single-walled carbon nanotubes, metal oxide nanoparticles, and nano-clay have been studied for lipid accumulation, extraction, and transesterification processes in biofuel production. Basha and Anand (2011) have suggested utilization of alumina nanoparticles and carbon nanotubes to enhance the combustion characteristics of biodiesel-operated engines with less harmful emissions. Ganzoury and Allam (2015) have observed enhanced methane production using nano-iron oxide, nano-bottom ash, nano-zerovalent iron, nano-fly ash, and bioactive nano-metal oxides under anaerobic condition. Nickel and cobalt nanocatalyst was useful for gasification of waste animal fat (Mahmood et al. 2010). Biomass thermal decomposition with catalyst showed gases like CO, CO_2, CH_4, and soot (Davidian et al. 2006). Gates (2000) have used molecular nanocatalysts which have the advantages of improved activity and lifetime resistance to poisoning (Mahmood and Hussain 2010). Size and structure influencing the efficiency of cobalt, silicon, nickel, and magnesium nanoparticles were investigated by different researchers (Kesavan et al. 2001; Son et al. 2002; Hussain et al. 2011). Cobalt nanoparticles are inexpensive and give higher yields of products within shorter duration (Mahmood et al. 2010). For gasification purposes, nickel catalyst was widely used (Kimura et al. 2006; Hussain et al. 2011). The size and shape influence the melting point of nanoparticles. Reduction of aromatic and other carbonyl compounds can take place using nickel nanoparticles (Kidwai et al. 2006; Mahmood and Hussain 2010). Mahmood et al. (2010) have opined that Ni nanoparticles can be used as efficient catalyst for reduction of complex aromatic and carbonyl compounds. TiO2 particles were found to have unique properties, less toxicity, and more stability (Kidwai et al. 2006).

Table 4.4 Bioenergy applications of nanoparticles

Nanoparticles used for biofuel production	Biofuel	References
Ni-supported zeolites	Diesel-range alkanes	Peng et al. (2012).
Nanometal oxides of SiO2, TiO2	Biogas	Hendriks and Zeeman (2009), Liao et al. (2010)
KF/CaO nanocatalyst	Biodiesel	Wen et al. (2010).
PAN nanofiber immobilized with PCL	Biodiesel	Sakai et al. (2010)
Lipase immobilized on ferric silica nanocomposites	Biodiesel	Tran et al. (2012).
Gold nanoparticles	Ethanol	Cho et al. (2012)
β-Glucosidase and cellobiase enzymes immobilized on different nanomaterials	Bioethanol	Verma et al. (2013a, b)
β-Glucosidase immobilized on nanomaterials	Bioethanol	Zheng et al. (2013)
Magnetic nanoparticles	Sugars	Abraham et al. (2014)
Immobilized cellulase onto MnO_2 nanoparticles	Bioethanol	Cherian et al. (2015)
Cobalt nanoparticles	Biodiesel	Lestari et al. (2008)
Ni, Co, Fe nanoparticles	Methane	Abdelsalam et al. (2016)
Nano-iron oxide (Fe3O4 NPs)	Methane	Casals et al. (2010)
Nanocatalysts and sulfonic acid cobalt spinel ferrite magnetic nanoparticles	Biogas	Johnson et al. (2009)

Lestari et al. (2008) used cobalt for preparing biodiesel as esters. Kondamudi et al. (2008) also have reported the formation of biodiesel from tea and coffee. All applications in the area of bioenergy are illustrated in Table 4.4.

4.12 Prospective Applications of Green Synthesized Nanoparticles

Silver nanoparticles fabricated using *Pinus thunbergii* extracts exhibited good antibacterial activity against pathogens of *Xanthomonas oryzae*, *Burkholderia glumae*, *Pseudomonas syringae*, and *Bacillus thuringiensis* (Velmurugan et al. 2013). Green synthesized silver nanoparticles from *Tridax procumbens* (tridax daisy) extract showed similar antimicrobial activity to nanoparticles against bacteria *Escherichia coli*, *Shigella dysenteriae*, and *Vibrio cholerae* (Dhanalakshmi and Rajendran 2012). Silver nanoparticles synthesized in *Iresine herbstii* showed good cytotoxic activity against various tumor cell lines of HeLa cells, while silver nanoparticles from *Euphorbia nivulia* extracts were toxic to the A549 cell lines (Valodkar et al. 2012). *Nerium oleander* synthesized silver nanoparticles which demonstrated larvicidal activity (Suganya et al. 2013). Silver nanoparticles from *Cyamopsis tetragonoloba* (cluster bean) were used to make a biosensor to determine ammonia (Pandey et al. 2012). *Ocimum sanctum* extracts generated platinum nanoparticles which were shown to possess catalytic activity in hydrogen generation (Soundarrajan et al. 2012). *Sesbania drummondii* gold nanoparticles were shown to reduce aromatic nitro compounds such as toxic 4-nitrophenol to 2-amino-phenol. Green synthesized nanoparticles have a functionalized surface and have organic ligands, proteins, polysaccharides, and polyatomic alcohols which cannot be obtained using physical and chemical methods. This makes green synthesized nanoparticles as promising catalysts in organic synthesis (Sintubin et al. 2012). Table 4.5 shows the applications of green fabricated nanomaterials.

4.13 Advantages and Disadvantages of Green Synthesis

In green synthesis of nanoparticles, the precursors are the most expensive followed by the substrates which have to be procured (Kurakhmaeva et al. 2009). Functionalization step of the nanoparticle is eliminated when we use a bionanocatalsyt unlike in physiochemical synthesis (Zhang et al. 2011). Phytochemicals, pharmacological metabolites, and enzymes act as stabilizers in plant- and microbial-based synthesized nanoparticles. Many researchers have explained that green synthesized nanoparticles play an important role in the medical field for diagnostic applications (Gunalan et al. (2012a, b). Jayaseelana et al. (2012) concluded that green methods of nanoparticle synthesis show better antibacterial effects. Copper oxide nanoparticles fabricated using *Malva sylvestris* have antibacterial, antifungal, and antiparasitic activities (Gusseme et al. 2010; Sanghi and Verma 2009; Velayutham et al. 2012). *Lactobacillus fermentum* was investigated for biogenic silver production which was used for the removal of viruses from drinking water. Chen et al. (2005) have explored small interfering RNA (siRNA) with quantum dots for producing fluorescent nanoparticles. Elechiguerra et al. (2005) reported that nanoparticles could readily bind to HIV-1. Some researchers have clearly stated that nanoparticles may cause cell wall and membrane damage (Chaloupka et al. 2010; Gopinath et al. 2014). They can also be used for bioremediation (Njagi et al. 2011). Dhandapani et al. (2012) have opined that green nanomaterials can be used for bioremediation purposes. Nanoparticles can be exploited for cleaning hazardous waste sites and pollutants (Kalaiselvi et al. 2015; Sankar et al. 2014). *Azadirachta indica*-fabricated titanium dioxide nanoparticles were investigated for photocatalysis by Singh and Naraa (2013). Similarly, other workers have reported that iron NPs can be used for treating water and soil remediation (Mahdavi et al. 2013). Management of plant diseases has also been reported with nanoparticles rather than with pesticides (Khot et al. 2012). Bacteria and diatoms produced magnetite and siliceous material which were reported to be useful for optical coatings (Joerger et al. 1999). Biosynthesized TiO_2 by *Psidium guajava* showed significant antibacterial and antioxidant effect (Santhoshkumar et al. 2014). Khan et al. (2015) showed that the antimicrobial activity of CuO-NPs and ZnO-NPs against *E. coli*, *B. subtilis*, and *S. aureus* was higher when compared to chemically synthesized nanoparticles. The size and shape of the nanoparticles and the compounds involved in their synthesis are unknown (Oberdörster et al. 2006).

Table 4.5 Applications of green fabricated nanomaterials

Name of the organism	Name of the nanoparticle	Applications	References
Aloe barbadensis	Gold and silver	Cancer, optical coatings	Chandran et al. (2006)
Acalypha indica	Silver	Antibacterial activity against pathogens	Krishnaraj et al. (2010)
Apiin extracted from henna leaves	Silver and gold	Cancer and optical coatings	Kasthuri et al. (2009)
Azadirachta indica	Gold, silver, and silver-gold alloys	Bioremediation of toxic metals	Shankar et al. (2004)
Camellia sinensis	Gold and silver	Catalysts and sensors	Mondal et al. (2011)
Coriandrum sativum	Gold	Drug delivery, tissue imaging, photothermal therapy	Narayanan and Sakthivel (2008)
Cymbopogon flexuosus	Gold	IR-absorbing optical coatings	Shankar et al. (2005)
Garcinia mangostana	Silver	Antimicrobial activity	Veerasamy et al. (2010)
Gardenia jasminoides Ellis	Palladium	Nanocatalysts for *p*-nitrotoluene hydrogenation	Jia et al. (2009)
Syzygium aromaticum	Gold	Destruction of cancer cells	Raghunandan et al. (2010)
Medicago sativa	Iron oxide	Cancer, drug delivery	Herrera-Becerra et al. (2008)
Morus	Silver	Antimicrobial activity	Singh et al. (2017a, b)
Nelumbo nucifera	Silver	Larvicidal activity	Santhoshkumar et al. (2011)
Ocimum sanctum	Gold and silver	Biolabeling, biosensor	Philip and Unni (2011)
Pear fruit extract	Gold	Catalysis, biosensing	Ghodake et al. (2010)
Sedum alfredii	Zinc oxide	Nanoelectronics	Qu et al. (2011)
Bacillus mycoides	Titanium	Green solar cells and antibacterial activity	Aenishanslins et al. (2014)
Bacillus subtilis	Titanium	Suppress aquatic biofilm growth	Dhandapani et al. (2012).
Bacteria strains NS2 and NS6	Lead sulfide	Bioremediation	Singh and Naraa (2013)
Cassia alata	Copper	Medicinal value	Jayalakshmi and Yogamoorthi (2014)
Stigmaphyllon littorale	Silver	Antibacterial activity	Kudle et al. (2013a, b, c)
Tectona grandis	Silver	Antibacterial activity	Nalvothula et al. (2014)
Phototrophic bacteria	Silver	Antibacterial activity	Manisha et al. (2014a, b).

(continued)

Table 4.5 (continued)

Name of the organism	Name of the nanoparticle	Applications	References
Allmania nodiflora	Silver	Antimicrobial activities	Kudle et al. (2013a, b, c).
Ixora coccinea	Silver	Antitumor and antimicrobial activity	Nalvolthula et al. (2015a, b).
Justicia adhatoda	Silver	Antitumor activity on human epithelioid carcinoma cells	Kudle et al. (2014a, b)
Acacia senegal	Silver	Antibacterial activity	Kudle et al. (2014a, b)
Citrus limon	Silver	Antitumor activity	Nalvolthula et al. (2015a, b).
Coriandrum sativum	Silver	Antimicrobial activity	Manisha et al. (2014a, b).
Sterculia foetida	Silver	Antimicrobial activity	Kudle et al. (2013a, b, c)
Toddy palm	Copper and silver	Antibacterial activity	Deshpande et al. (2018)
Fruit pulp of Palmyra fruit	Silver and copper	Antibacterial, antioxidant activity	Chitturi et al. (2018)
Toddy palm	Silver and nickel	Antibacterial, antioxidant activity	Padala et al. (2019)
Cassia auriculata	Zinc oxide	Stabilizing and reducing agent for nanoparticle synthesis	Ramesh et al. (2014)
Catharanthus roseus	Titanium	Effective against *Hippobosca maculata* and *Bovicola ovis*	Velayutham et al. (2012)
Euphorbia condylocarpa	Palladium, iron	Recyclable catalyst	Nasrollahzdeha et al. (2015)
Gloriosa superba	Copper	Antibacterial activity	Naikaa et al. (2015)
Camellia sinensis	Zinc oxide	Effective against *Klebsiella pneumoniae*	Senthilkumar and Sivakumar (2014)
Gum karaya	Copper	Antimicrobial activity	Padil et al. (2013)
Humicola sp.	Copper	Medicinal applications	Khan and Ahmad (2013)
Malva sylvestris	Copper	Antimicrobial activity	Awwad et al. (2015)
Phyllanthus amarus	Copper	Antimicrobial activity	Acharyulu et al. (2014)
Oryza sativa	Titanium	Photocatalysts	Ramimoghadam et al. (2014)
Raphanus sativus	Silver	Antioxidant and antibacterial	Koyyati et al. (2013)
Amaranthus viridis	Silver	Antibacterial activity	Koyyati et al. (2014)
Pleurotus florida	Gold	Anticancer activity	Bhat et al. (2013)

(continued)

Table 4.5 (continued)

Name of the organism	Name of the nanoparticle	Applications	References
Acalypha indica	Silver, gold	Cytotoxic activity	Krishnaraj et al. (2014)
Rosa damascena	Silver	Anticancer activity	Venkatesan et al. (2014)
Eucalyptus chapmaniana	Silver	Antimicrobial and cytotoxic activity	Sulaiman et al. (2013)
Dracocephalum kotschyi	Gold	Anticancer, antibacterial	Dorosti and Jamshidi (2016)
Inonotus obliquus	Gold	Antibacterial, antioxidant, and cytotoxic activities	Lee et al. (2015)
Quercetin and gallic acid	Silver and selenium	Antitumor and antimicrobial activity	Mittal et al. (2014a, b)
Gymnema sylvestre	Silver and gold	Antioxidant and anticancer activities	Nakkala et al. (2015)
Syzygium aromaticum	Silver	Anticancer activity	Venugopal et al. (2017).
Cissus quadrangularis	Silver	Antimicrobial and anticancer potentials	Renugadevi et al. (2012).
Ganoderma neo-japonicum Imazeki	Silver	Antitumor activity	Gurunathan et al. (2013)
Dioscorea bulbifera	Platinum and palladium	Anticancer and antioxidant activities	Ghosh et al. (2015)
Pleurotus ostreatus	Silver	Anticandidal and anticancer activities	Yehia and Al-Sheikh (2014)
Gracilaria edulis	Silver and zinc oxide	Anticancer activity	Priyadharshini et al. (2014)
Borago officinalis	Silver	Anticancer and antibacterial activities	Singh et al. (2017a, b)
Corallina officinalis	Gold	Cytotoxic activity	Kassas and El-Sheekh (2014)
Dimocarpus longan	Silver	Antibacterial activity and anticancer	He et al. (2016)

4.14 Future Directions and Conclusions

Nanotechnology can overcome many technical hurdles due to their unique properties and small size compared to macrosize particles. Carbon and graphene are the major nanomaterial sources that are being widely explored. Only few kinds of nanomaterials such as nanotubes and nanofibers, nanoplates, and nanocrystals are being studied. Multi-walled nanotubes, nanocomposites, and other kinds of nanomaterials also need to be studied to enhance the effectivity and efficiency of the process. Recyclable, highly active, and selective heterogeneous nanocatalysts should be synthesized using biological approaches. Faster methods of synthesis, kinetics of nanoparticle formation, process parameter optimization, and nondestructive characterization which is nonexpensive should be understood for the production of nanomaterials. Catalysts normally employed in the synthesis of chemicals are expensive and have several disadvantages. Green synthesized nanocatalysts with intermediate characteristics between homogeneous and heterogeneous systems can be used effectively to overcome these disadvantages. Green-based syntheses of base nanocatalysts, acid nanocatalysts, and bifunctional nanocatalysts which are advanced biocatalysts have many applications for biofuel as well as high-density biofuel production. The potential prospects of this technology are still in a nascent stage and need to be explored with biological materials so that they can be effectively used for many applications. But due to their extremely small size, they are toxic to the environment. The risks should be properly investigated before going for optimization and large-scale production and subsequent applications of these particles.

Acknowledgement I would also like to show my gratitude to my teachers Prof. S.M. Reddy, Emeritus Professor, Kakatiya University, and Prof. S. Girisham, Kakatiya University, for sharing their pearls of wisdom with me during the writing of this chapter. I am immensely grateful to Prof. Katta Narasimha Reddy, former Vice-Chancellor, Mahatma Gandhi University, for encouraging me to write this chapter.

References

Abdeen S, Geo S, Sukanya S, Praseetha PK, Dhanya RP (2014) Biosynthesis of Silver nanoparticles from Actinomycetes for therapeutic applications. Int J Nano Dimens 5:155–162

Abdelsalam E, Samer M, Attia YA, Abdel-Hadi MA, Hassan HE, Badr Y (2016) Comparison of nanoparticles effects on biogas and methane production from anaerobic digestion of cattle dung slurry. Renew Energy 87:592–598

Abraham RE, Verma ML, Barrow CJ, Puri M (2014) Suitability of magnetic nanoparticle immobilised cellulases in enhancing enzymatic saccharification of pretreated hemp biomass. Biotechnol Biofuels 7(1):90

Acharyulu NPS, Dubey RS, Swaminadham V, Kalyani RL, Kollu P, Pammi SVN (2014) Green synthesis of CuOnanoparticles using Phyllanthus amarus leaf extract and their antibacterial activity against multidrug resistancebacteria. Int J Eng Res Technol 3:639–641

Aenishanslins NAO, Saona LA, Duran-Toro VM, Monras JP, Bravo DM, Donoso JMP (2014) Use of titanium dioxidenanoparticles biosynthesized by Bacillus mycoides inquantum dot sensitized solar cells. Microb Cell Factories 13:90

Agnihotri M, Joshi S, Kumar AR, Zinjarde S, Kulkarni S (2009) Biosynthesis of gold nanoparticles by the tropical marine yeast *Yarrowia lipolytica* NCIM 3589. Mater Lett 3:1231–1234. https://doi.org/10.1016/j.matlet.2009.02.042

Ahmad R, Sardar M (2015) Enzyme immobilization: an overview on nanoparticles as immobilization matrix. Biochem Anal Biochem 4:2

Ahmad A, Senapati S, Khan MI, Kumar R, Sastry M (2003) Extracellular biosynthesis of monodisperse gold nanoparticles by a novel extremophilic actinomycete, Thermomonospora sp. Langmuir 19(8):3550–3553

Ahmad N, Sharma S, Alam MK, Singh VN, Shamsi SF, Mehta BR, Fatma A (2010) Colloids Surf B Biointerfaces 81:81–86

Ahmed S, Annu SAC, Ikram S (2017) A review on biogenic synthesis of ZnO nanoparticles using plant extracts and microbes: a prospect towards green chemistry. J Photochem Photobiol B 166:272–284

Aljabali AAA, Barclay JE, Lomonossoff GP, Evans DJ (2010) Virus templated metallic nanoparticles. Nanoscale 2:2596–2600

Ankamwar B (2010) Biosynthesis of gold nanoparticles (green-gold) using leaf extract of Terminalia catappa. E-J Chem 7:1334–1339

Ansari SA, Husain Q (2012) Potential applications of enzymes immobilized on/in nano materials: a review. Biotechnol Adv 30(3):512–523

Armendariz V, Herrera I, Peralta-Videa JR et al (2004) Size controlled goldnanoparticle formation by *Avena sativa* biomass: use of plants innanobiotechnology. J Nanoparticle Res 6:377–382

Asha A, Sivaranjani T, Thirunavukkarasu P, Asha S (2016) Green synthesis of silver nanoparticle from different plants – a review. Int J Pure Appl Biosci 4(2):118–124

Awwad AM, Albiss BA, Salem NM (2015) Antibacterial activity of synthesized copper oxide nanoparticles using Malva sylvestris leaf extract. SMU Med J 2:91–101

Baesman SM, Bullen TD, Dewald J, Zhang D, Curran S, Islam FS et al (2007) Formation of tellurium nanocrystals during anaerobic growth of bacteria that use Te oxyanions as respiratory electron acceptors. Appl Environ Microbiol 73(7):2135–2143

Bai H, Zhang Z, Gong J (2006) Biological synthesis of semiconductor zinc sulfide nanoparticles by immobilized Rhodobacter sphaeroides. Biotechnol Lett 28(14):1135–1139

Bai HJ, Zhang ZM, Guo Y, Yang GE (2009a) Biosynthesis of cadmium sulfide nanoparticles by photosynthetic bacteria Rhodopseudomonas palustris. Colloids Surf B: Biointerfaces 70(1):142–146

Bai H, Zhang Z, Guo Y, Jia W (2009b) Biological synthesis of size-controlled cadmium sulfide nanoparticles usingimmobilized Rhodobacter sphaeroides. Nanoscale Res Lett 4:717–723

Balagurunathan R, Radhakrishnan M, Rajendran RB, Velmurugan D (2011) Biosynthesis of gold nanoparticles by actinomycete Streptomyces viridogens strain HM10. J Biochem Biophys 48:331–335

Balaji DS, Basavaraja S, Deshpande R, Mahesh DB, Prabhakar BK, Venkataraman A (2009) Extracellular biosynthesis of functionalized silver nanoparticles by strains of Cladosporium cladosporioides fungus. Colloids Surf B Biointerfaces 68:88–92

Bansal V, Poddar P, Ahmad A, Sastry M (2006) Room-temperature biosynthesis of ferroelectric barium titanate nanoparticles. J Am Chem Soc 128:11958–11963

Bao S-J, Lei C, Xu M-W, Cai C-J, Jia D-Z (2012) Environment-friendly biomimetic synthesis of TiO_2 nanomaterials for photocatalytic application. Nanotechnology 23(20):205601. https://doi.org/10.1088/0957-4484/23/20/205601

Bar H, Bhui DK, Sahoo GP et al (2009) Green synthesis of silver nanoparticles using latex of *Jatropha curcas*. Colloids Surf A Physicochem Eng Asp. https://doi.org/10.1016/j.colsurfa.2009.02.008

Basha JS, Anand RB (2011) Role of nanoadditive blended biodiesel emulsion fuel on the working characteristics of a diesel engine. J Renew Sust Energy 3(2):023106

Beveridge TJ, Murray RGE (1980) Sites of metal deposition in the cell wall of Bacillus subtilis. J Bacteriol 141(2):876–887

Bhat R, Sharanabasava VG, Deshpande R, Shetti U, Sanjeev G, Venkataraman A (2013) Photo-bio-synthesis of irregular shaped functionalized gold nanoparticles using edible mushroom Pleurotus florida and its anticancer evaluation. J Photochem Photobiol B: Biol 125:63–69

Canakci M, Van Gerpen J (2001) Tran Asae 44:1429–1436

Casals E, Pfaller T, Duschl A, Oostingh GJ, Puntes V (2010) Time evolution of the nanoparticle protein corona. ACS Nano 4:3623–3632

Chaloupka K, Malam Y, Seifalian AM (2010) Nanosilver as anew generation nanoproduct in biomedical applications. Trends Biotechnol 28:580–588

Chandran SP, Chaudhary M, Pasricha R et al (2006) Synthesis of gold nanotrianglesand silver nanoparticles using aloe vera plant extract. Biotechnol Prog 22(2):577–583

Chaturvedi S, Dave PN, Shah NK (2012) Applications of nano-catalyst in new era. J Saudi Chem Soc 16:307–325

Chauhan R, Kumar A, Abraham J (2013) A biological approach to the synthesis of Silver nanoparticles with streptomyces sp JAR1 and its antimicrobial activity. Sci Pharm 81:607–621. 148

Chen AA, Derfus AM, Khetani SR, Bhatia SN (2005) Quantum dots to monitor RNAi delivery and improve gene silencing. Nucleic Acids Res 33:1–8
Chen J, Saggar JK, Corey P, Thompson LU (2009) Flaxseed and pure secoisolariciresinol diglucoside, but not flaxseed hull, reduce human breast tumor growth (MCF-7) in athymic mice. J Nutr 139(11):2061–2066
Chen G, Li M, Li F, Sun S, Xia D (2010) Protein-mediated synthesis of nanostructured titania with different polymorphs at room temperature. Adv Mater 22(11):1258–1262
Cherian E, Dharmendirakumar M, Baskar G (2015) Immobilization of cellulase onto MnO2 nanoparticles for bioethanol production by enhanced hydrolysis of agricultural waste. Chin J Catal 36(8):1223–1229
Chitturi KL, Garimella S, Marapaka AK, Kudle KR, Merugu R (2018) Single pot green synthesis, characterization, antitumor antibacterial, antioxidant activity of bimetallic silver and copper nanoparticles using fruit pulp of palmyra fruit. J Bionanosci 12(2):284–289
Cho EJ et al (2012) Co-immobilization of three cellulases on Audoped magnetic silica nanoparticles for the degradation of cellulose. Chem Commun 48:886–888
Correa-Llantén DN, Muñoz-Ibacache SA, Castro ME, Muñoz PA, Blamey JM (2013) Gold nanoparticles synthesized by Geobacillus sp. strain ID17 a thermophilic bacterium isolated from Deception Island, Antarctica. Microb Cell Fact 12(1):75
Cunningham DP, Lundie LL Jr (1993) Precipitation of cadmium by Clostridium thermoaceticum. Appl Environ Microbiol 59(1):7–14
Dameron CT, Reese RN, Mehra RK, Kortan AR, Carroll PJ, Steigerwald ML, Brus LE, Winge DR (1989) Biosynthesis of cadmium sulphide quantum semiconductor crystallites. Nature 338:596–597
Das S, Das A, Guha A (2008) Adsorption behavior of mercury on functionalized Aspergillus versicolor mycelia: Atomic force microscopic study. Langmuir 25:360–366
Das RK, Gogoi N, Bora U (2011) Green synthesis of gold nanoparticles using Nyctanthes arbortristis flower extract. Bioprocess Biosyst Eng 34:615–619
Datta A, Patra C, Bharadwaj H, Kaur S, Dimri N et al (2017) Green synthesis of zinc oxide nanoparticles using parthenium hysterophorus leaf extract and evaluation of their antibacterial properties. J Biotechnol Biomater 7:271. https://doi.org/10.4172/2155-952X.1000271
Datta S, Christena LR, Rajaram YRS (2013) Enzyme immobilization: an overview on techniques and support materials. 3. Biotech 3(1):1–9
Davidian T, Guihaume N, Iojoiu E, Provendier H, Mirodatos C (2006) Hydrogen production from crude pyrolysis oil by a sequential catalytic process. Appl Catal B Environ 73(1–2):116–127
Dehkhoda AM, West AH, Ellis N (2010) Biochar based solid acid catalyst for biodiesel production. Appl Catal A General 382(2):197–204
Deng X, Fang Z, Liu Y, Yu CL (2011) Production of biodiesel from Jatropha oil catalyzed by nanosized solid basic catalyst. Energy 36:777–784
Deplanche K, Macaskie LE (2008) Biorecovery of gold by Escherichia coli and Desulfovibrio desulfuricans. Biotechnol Bioeng 99(5):1055–1064
Deplanche K, Caldelari I, Mikheenko IP, Sargent F, Macaskie LE (2010) Involvement of hydrogenases in the formation of highly catalytic Pd (0) nanoparticles by bioreduction of Pd (II) using *Escherichia coli* mutant strains. Microbiology 156(9):2630–2640
Deshpande PK, Upputuri N, Gothalwal R, Merugu R (2018) In vitro dissolution studies and enhanced antimicrobial activity of some antibiotics loaded with copper and silver nanoparticles in polyvinyl alcohol films. J Bionanosci 12(4):569–573
Dhanalakshmi T, Rajendran S (2012) Arch Appl Sci Res 4:1289–1293
Dhandapani P, Maruthamuthu S, Rajagopal G (2012) Bio-mediated synthesis of TiO2 nanoparticles and its photocatalytic effect on aquatic biofilm. J Photochem Photobiol B110:43–49
Dinçer A, Telefoncu A (2007) Improving the stability of cellulase by immobilization on modified polyvinyl alcohol coated chitosan beads. J Mol Catal B: Enzym 45(1–2):10–14
Dobrucka R (2017) Synthesis of titanium dioxide nanoparticles using echinacea purpurea herba. Iran J Pharm Res 16:756–762

Dorosti N, Jamshidi F (2016) Plant-mediated gold nanoparticles by Dracocephalum kotschyi as anticholinesterase agent: synthesis, characterization, and evaluation of anticancer and antibacterial activity. J Appl Biomed 14(3):235–245
Dubey M, Bhadauria S, Kushwah BS (2009) Green synthesis of nanosilver particles from extract of *Eucalyptus hybrida* (Safeda) leaf. Dig J Nanomater Biostruct 4:537–543
Duran N, Marcato PD, Alves O, Souza GI, Esposito E (2005) J Nanobiotechnol. 3:8
Durán N, Marcato PD, Durán M, Yadav A, Gade A, Rai M (2011) Mechanistic aspects in the biogenic synthesis of extracellular metal nanoparticles by peptides, bacteria, fungi, and plants. Appl Microbiol Biotechnol *90*(5):1609–1624
Eggert H, Greaker M (2014) Promoting second generation biofuels: does the first generation pave the road? Energies 7(7):4430–4445
Elechiguerra JL, Burt JL, Morones JR, Camacho-Bragado A, Gao X, Lara HH et al (2005) Interaction of silver nanoparticles with HIV-1. J Nanobiotechnol 3:1–10
Engelmann W, Aldrovandi A, Berger Filho AG (2013) Prospects for the regulation of nanotechnology applied to food and biofuels. Vigilância Sanitária em Debate 1(4):110–121
Enumula SS, Koppadi KS, Babu Gurram VR, Burri DR, Rao Kamaraju SR (2017) Sust Energy Fuels 1:644–651
El-Kassas HY, El-Sheekh MM (2014) Cytotoxic activity of biosynthesized gold nanoparticles with an extract of the red seaweed *Corallina officinalis* on the MCF-7 human breast cancer cell line. Asian Pac J Cancer Prev 15(15):4311–4317
Farag AA, El-shafei A, Ibrahim HM, Asker MS (2016) Influence of bacterial cellulose for synthesis and application of titanium dioxide nanoparticles compared with sol-gel method. Int J Life Sci Res 4(1):69–75
Farrell AE, Plevin RJ, Turner BT, Jones AD, O'hare M, Kammen DM (2006) Ethanol can contribute to energy and environmental goals. Science 311(5760):506–508
Feynman in 1959: In December of 1959, Richard Feynman gave a talk called "There's Plenty of Room at the Bottom" at an annual meeting of the American Physical Society at Caltech
Fukuda H, Kondo A, Noda H (2001) Biodiesel fuel production by transesterification of oils. J Biosci Bioeng 92(5):405–416
Ganzoury MA, Allam NK (2015) Impact of nanotechnology on biogas production: a mini-review. Renew Sust Energy Rev 50:1392–1404
Gardea-Torresdey JL, Parsons J, Gomez E, Peralta-Videa J, Troiani H, Santiago P, Yacaman M (2002) Nano Lett 2:397–401
Gardy J, Hassanpour A, Lai X, Ahmed MH, Rehan M (2017) Biodiesel production from used cooking oil using a novel surface functionalised TiO2 nano-catalyst. Appl Catal B Environ 207:297–310
Gardy J, Osatiashtiani A, Céspedes O, Hassanpour A, Lai X, Lee AF, Wilson K, Rehan M (2018) A magnetically separable SO4/Fe-AlTiO2 solid acid catalyst for biodiesel production from waste cooking oil. Appl Catal B Environ 234:268–278
Gates BC (2000) Supported metal cluster catalysts. J Mol Catal A Chem 163:55–65
Gericke M, Pinches A (2006) Biological synthesis of metal nanoparticles. Hydrometallurgy 83(1):132–140
Ghidan AY, Al-Antary TM, Awwad A (2016) Green synthesis of copper oxide nanoparticles using *Punica granatum* peels extract: effect on green peach Aphid. Environ Nanotechnol Monit Manag 6:95–98
Ghodake G, Deshpande N, Lee Y, Jin E (2010) Pear fruit extract-assisted room-temperature biosynthesis of gold nanoplates. Colloids Surf B Biointerfaces 75:584–589
Ghosh S, Nitnavare R, Dewle A, Tomar GB, Chippalkatti R, More P et al (2015) Novel platinum–palladium bimetallic nanoparticles synthesized by Dioscorea bulbifera: anticancer and antioxidant activities. Int J Nanomed 10:7477
Glusker J, Katz A, Bock C (1999) Rigaku J 16(2):8–16. 44
Golinska P, Wypij M, Ingle AP, Gupta I, Dahm H, Rai M (2014) Biogenic synthesis of metal nanoparticlesfrom actinomycetes: Biomedical applications and cytotoxicity. Appl Microbiol Biotechnol 98:8083–8097

Gopinath K, Shanmugam VK, Gowri S, Senthilkumar V, Kumaresan S, Arumugam A (2014) Antibacterial activityof ruthenium nanoparticles synthesized using Gloriosasuperba L. leaf extract. J Nanostruct Chem 4:83

Gorzny ML, Walton AS, Evans SD (2010) Synthesis of high-surface-area platinum nanotubes using a viraltemplate. Adv Funct Mater 20:1295–1300

Govender Y, Riddin T, Gericke M, Whiteley CG (2009) Bioreduction of platinum salts into nanoparticles: a mechanistic perspective. Biotechnol Lett 31:95–100

Govender Y, Riddin TL, Gericke M, Whiteley CG (2010) On the enzymatic formation of platinum nanoparticles. J Nanopart Res 12(1):261–271

Gruen LC (1975) Biochim Biophys Acta 386:270–274

Gunalan S, Sivaraj R, Rajendran V (2012a) Green synthesized ZnO nanoparticles against bacterial and fungal pathogens. Prog Nat Sci 22:693–700

Gunalan S, Sivaraj R, Venckatesh R (2012b) Aloe barbadensismiller mediated green synthesis of mono-disperse copperoxide nanoparticles optical properties. Spectrochim Acta Part A 97:1140–1144

Guan Q, Li Y, Chen Y, Shi Y, Gu J, Li B et al (2017) Sulfonated multi-walled carbon nanotubes for biodiesel production through triglycerides transesterification. RSC Adv 7(12):7250–7258

Guo T, Nikolaev P, Thess A, Colbert DT, Smalley RE (1995) Catalytic growth of single-walled nano tubes by laser vaporization. Chem Phys Lett 243(1–2):49–54

Gupta J, Agarwal M (2016) Preparation and characterization of CaO nanoparticle for biodiesel production. 2nd International Conference on Emerging Technologies. American Institute of Physics, Jaipur

Gurunathan S, Raman J, Malek SNA, John PA, Vikineswary S (2013) Green synthesis of silver nanoparticles using Ganoderma neo-japonicum Imazeki: a potential cytotoxic agent against breast cancer cells. Int J Nanomed 8:4399

Gusseme DB, Sintubin L, Baert L, Thibo E, Hennebel T, Vermeulen G, Uyttendaele M, Verstraete W, Boon N (2010) Biogenic silver for disinfection of water contaminated with viruses. Appl Environ Microbiol 76:1082–1087

Harris AT, Bali R (2008) J Nanoparticle Res 10:691–695

Hassan MHJ, Abul Kalam M (2013) An overview of biofuel as a renewable energy source: development and challenges. 5th BSME International Conference on Thermal Engineering. Proc Eng 56:39–53

Haverkamp RG, Marshall AT (2009) The mechanism of metal nanoparticleformation in plants: limits on accumulation. J Nanoparticle Res 11:1453–1463

Hazeri N, Tavanai H, Moradi AR (2012) Production and properties of electrosprayed sericin nanopowder. Sci Technol Adv Mater 13:035010. (7pp)

He Y, Du Z, Ma S, Cheng S, Jiang S, Liu Y et al (2016) Biosynthesis, antibacterial activity and anticancer effects against prostate cancer (PC-3) cells of silver nanoparticles using Dimocarpus Longan Lour. peel extract. Nanoscale Res Lett 11(1):300–4317

Hendriks ATWM, Zeeman G (2009) Pretreatments to enhance the digestibility of lignocellulosic biomass. Bioresour Technol 100(1):10–18

Herrera-Becerra R, Zorrilla C, Rius JL, Ascencio JA (2008) Electron microscopy characterization of biosynthesized iron oxide nanoparticles. Appl Phys A Mater Sci Process 91:241–246

Hess SK, Schunck NS, Goldbach V, Ewe D, Kroth PG, Mecking S (2017) J Am Chem Soc 139:13487–13491

Ho M, Mao X, Gu L, Li P (2008) Facile route to enzyme immobilization: coreshell nanoenzyme particles consisting of well-defined poly(methyl methacrylate) cores and cellulase shells. Langmuir 24:11036–11042

Holmes JD, Smith PR, Evans-Gowing R, Richardson DJ, Russell DA, Sodeau JR (1995) Energy-dispersive X-ray analysis of the extracellular cadmium sulfide crystallites of Klebsiella aerogenes. Arch Microbiol 163(2):143–147

Hu S, Guan Y, Wang Y, Han H (2011) Nano-magnetic catalyst KF/CaO–Fe3O4 for biodiesel production. Appl Energy 88(8):2685–2690

Huang Q, Li D, Sun Y et al (2007) Biosynthesis of silver and gold nanoparticles by novel sundried *Cinnamomum camphora* leaf. Nanotechnology 1:1
Hunter WJ, Manter DK (2008) Bio-reduction of selenite to elemental red selenium by Tetrathiobacter kashmirensis. Curr Microbiol 57:83–88
Hussain ST, Ali SA, Bano A, Mahmood T (2011) Use of nanotechnology for the production of biofuels from butchery waste. International Journal of Physical Sciences 6(31):7271–7279
Imran M, Rani A (2015) Recent advances in the synthesis and stabilization of nickel and nickeloxide nanoparticles: a green adeptness. Int J Anal Chem 2:1–14
Jahn MK, Haderlein SB, Meckenstock RU (2005) Anaerobic degradation of benzene, toluene, ethylbenzene, and o-xylene in sediment-free iron-reducing enrichment cultures. Appl Environ Microbiol 71(6):3355–3358
Jain, Pradeep (2004) Silver nanoparticle-coated polyurethane foam as antibacterial water filter. Biotechnol Bioeng 90(1):59–63
Jalill RDA, Nuaman RS, Abd AN (2016) Biological synthesis of titanium dioxide nanoparticles by curcuma longa plant extract and study its biological properties. World Sci News 49(2):204–222
Jayalakshmi, Yogamoorthi A (2014) Green synthesis of copper oxide nanoparticles using aqueous extract of flowers of Cassia alata and particles characterisation. Int J Nanomat Biostruct 4:66–71
Jayaseelana C, Rahumana AA, Kirthi AV, Marimuthua S, Santhoshkumara T, Bagavana A et al (2012) Novel microbial route to synthesize ZnO nanoparticles using Aeromonashydrophila and their activity against pathogenic bacteria and fungi. Spectrochim Acta Part A 90:78–84
Jennifer K, Peter VH (2006) Nanotechnology in agriculture and food production. Woodrow Wilson International Center, Washington, DC
Jha AK, Prasad K (2010) Green synthesis of silver nanoparticles using cycas leaf. Int J Green Nanotechnol Phys Chem 1:110–117
Jia L, Zhang Q, Li Q, Song H (2009) The biosynthesis of palladium nanoparticles by antioxidants in *Gardenia jasminoides* Ellis: long lifetime nanocatalysts for *p*-nitrotoluene hydrogenation. Nanotechnology 20:385601
Joerger R, Klaus T, Olsson E, Granqvist CG (1999) Spectrally selective solar absorber coatings prepared by a biomimetic technique. Proc Soc Photo-Opt Instrum Eng 3789:2–7
Johnson RL, Johnson GOB, Nurmi JT, Tratnyek PG (2009) Natural organic matter enhanced mobility of nano zerovalent iron. Environ Sci Technol 43(14):5455–5460
Kalaiselvi A, Roopan SM, Madhumitha G, Ramalingam C, Elango G (2015) Synthesis and characterization of palladium nanoparticles using Catharanthus roseus leaf extract and its application in the photo-catalytic degradation. Spectrochim Acta Part A 135:116–119
Kalishwaralal K, Deepak V, Ramkumarpandian S, Nellaiah H, Sangiliyandi G (2008) Extracellular biosynthesis of silver nanoparticles by the culture supernatant of Bacillus licheniformis. Mater Lett 62(29):4411–4413
Kandel K (2013) Multitasking mesoporous nanomaterials for biorefinery applications Graduate Theses and Dissertations. 13214.
Karthik L, Kumar G, Vishnu-Kirthi A, Rahuman AA, Rao VB (2014) Streptomyces sp. LK3 mediated synthesis of silver nanoparticles and its biomedical application. Bioprocess Biosyst Eng 37:261–267
Kashefi K, Lovley DR (2000) Reduction of Fe (III), Mn (IV), and toxic metals at 100 C by Pyrobaculum islandicum. Appl Environ Microbiol 66(3):1050–1056
Kashefi K, Tor JM, Nevin KP, Lovley DR (2001) Reductive precipitation of gold by dissimilatory Fe(III)-reducing bacteria and archaea. Appl Environ Microbiol 67(7):3275–3279
Kasthuri J, Veerapandian S, Rajendiran N (2009) Biological synthesis of silver and gold nanoparticles using apiin as reducing agent. Colloids Surf B Biointerfaces 68:55–60
Kathiresan K, Manivannan S, Nabeel M, Dhivya B (2009) Studies on silver nanoparticles synthesized by a marine fungus, Penicillium fellutanum isolated from coastal mangrove sediment. Colloids Surf B Biointerfaces 71:133–137
Kesavan V, Sivanand PS, Chandrasekaran S, Koltypin Y, Gedanken A (1999) Angew Chem Int Ed 38:3521

Kesavan V, Dhar D, Koltypin Y, Perkas N, Palchik O, Gedanken A, Chandrasekaran S (2001) Nanostructured amorphous metals, alloys, and metal oxides as new catalysts for oxidation. Pure Appl Chem 73(1):85–91

Kessi J, Ramuz M, Wehrli E, Spycher M, Bachofen R (1999) Reduction of selenite and detoxification of elemental selenium by the phototrophic bacterium Rhodospirillum rubrum. Appl Environ Microbiol 65(11):4734–4740

Khan SA, Ahmad A (2013) Fungus mediated synthesis of biomedically important cerium oxide nanoparticles. Mater Res Bull 48:4134–4138

Khan YA, Singh BR, Ullah R, Shoeb M, Naqvi AH, Abidi SM (2015) Anthelmintic effect of biocompatible zinc oxide nanoparticles (ZnO NPs) on Gigantocotyle explanatum, a neglected parasite of Indian water buffalo. PLoS One 10:e0133086

Khot LR, Sankaran S, Maja JM, Ehsani R, Schuster EW (2012) Applications of nanomaterials in agricultural production and crop protection: a review. Crop Prot 35:64–70

Kidwai M, Bansal V, Saxena A, Shankar R, Mozumdar S (2006) Ni nano particles: an efficient green catalyst for chemo selective reduction of aldehydes. Tetrahed Lett 47(25):4161–4165

Kim KH, Lee OK, Lee EY (2018) Nano-immobilized biocatalysts for biodiesel production from renewable and sustainable resources. Catalysts 8(2):68

Kimura T, Miyazawa T, Nishikawa J, Kado S, Okumura K, Miyao T, Naito S, Kunimori K, Tomishige K (2006) Development of Ni catalysts for tar removal by steam gasification of biomass. Appl Catal B-Environ 68:160–170

Kimura R, Rokkaku T, Takeda S, Senba M, Mori N (2013) Cytotoxic effects of fucoidan nanoparticles against osteosarcoma. Mar Drugs 11(11):4267–4278

Klaus-Joerger T, Joerger R, Olsson E, Granqvist CG (2001) Bacteria as workers in the living factory: metal-accumulating bacteria and their potential for materials science. Trends Biotechnol 19(1):15–20

Kobayashi M, Tomita S, Sawada K, Shiba K, Yanagi H, Yamashita I, Uraoka Y (2012) Chiral metamolecules consisting of gold nanoparticles and genetically engineered tobacco mosaic virus. Opt Express 20:24856–24863

Kondamudi N, Mohapatra SK, Misra M (2008) Spent coffee grounds as a versatile source of green energy. J Agric Food Chem 56:11757–11760

Kondamudi N, Mohapatra SK, Misra M (2011) Appl Catal A-General 393:36–43

Konishi Y, Tsukiyama T, Tachimi T, Saitoh N, Nomura T, Nagamine S (2007a) Microbial deposition of gold nanoparticles by the metal-reducing bacterium Shewanella algae. Electrochim Acta 53(1):186–192

Konishi Y, Ohno K, Saitoh N et al (2007b) Bioreductive deposition of platinum nanoparticles on the bacterium Shewanella algae. J Biotechnol 128(3):648–653

Kowshik M, Deshmukh N, Vogel W, Urban J, Kulkarni SK, Paknikar KM (2002) Microbial synthesis of semiconductor CdS nanoparticles, their characterization, and their use in the fabrication of an ideal diode. Biotechnol Bioeng 78:583–588

Koyyati R, Nagati V, Merugu R, Manthurpadigya P (2013) Biological synthesis of silver nanoparticles using Raphanus sativus var. longipinnatus leaf extract and evaluation of their antioxidant and antibacterial activity. Int J Med Pharma Sci 3(4):89–100

Koyyati R, babu Nagati V, Nalvothula R, Merugu R, Kudle KR, Marx P, Padigya PRM (2014) Antibacterial activity of silver nanoparticles synthesized using Amaranthus viridis twig extract. Int J Res Pharm Sci 5(1):32–39

Krishnaraj C, Jagan EG, Rajasekar S et al (2010) Synthesis of silver nanoparticles using *Acalypha indica* leaf extracts and its antibacterial activity against water borne pathogens. Colloids Surf B Biointerfaces 1:1

Krishnaraj C, Muthukumaran P, Ramachandran R, Balakumaran MD, Kalaichelvan PT (2014) Acalypha indica Linn: biogenic synthesis of silver and gold nanoparticles and their cytotoxic effects against MDA-MB-231, human breast cancer cells. Biotechnol Rep 4:42–49

Krumov N, Oder S, Perner-Nochta I, Angelov A, Posten C (2007) Accumulation of CdS nanoparticles by yeasts in a fed-batch bioprocess. J Biotechnol 132:481–486

Kudle KR, Donda MR, Merugu R, Prashanthi Y, Rudra MP (2013a) Microwave assisted green synthesis of silver nanoparticles using Stigmaphyllon littorale leaves their characterization and anti-microbial activity. Int J Nanomater Biostruct 3(1):13–16

Kudle KR, Donda MR, Rudra MP, Merugu R (2013b) Synthesis of silver nanoparticles using the medicinal plant Allmania nodiflora and evaluation of its anti microbial activities. Int J Res Pharm Sci 4(4):504

Kudle KR, Donda MR, Merugu R, Prashanthi Y, Kudle MR, Rudra MP (2013c) Green synthesis of silver nanoparticles using water soluble gum of Sterculia foetida and evaluation of its antimicrobial activity. Int J Res Pharm Sci 4(4):563–568

Kudle KR, Donda MR, Merugu R, Prashanthi Y, Rudra MP (2014a) Investigation on the cytotoxicity of green synthesis and characterization of silver nanoparticles using Justicia adhatoda Leaves on human epitheloid carcinoma cells and evaluation of their antibacterial activity. Int J Drug Dev Res 6(1):0975–9344

Kudle KR, Kudle MR, Prashanthi Y, Merugu R, Rudra MP (2014b) Silver nanoparticles synthesis, stabilization and characterization by different concentrations of Acacia senegal (l.) willd. Extract and Evaluation of their antibacterial activity. Int J Pharmacogn Phytochem Res 6(2):210–212

Kumar R, Lal S (2014) Synthesis of organic nanoparticles and their applications in drug delivery and food nanotechnology: A Review. J Nanomater Mol Nanotechnol 3:4

Kumar M, Sharma MP (2014) Potential assessment of microalgal oils for biodiesel production: a review. J Mater Environ Sci 3:757–766

Kumar SA, Ansary AA, Ahmad A, Khan MI (2007) Extracellular biosynthesis of CdSe quantum dots by the fungus, Fusarium oxysporum. J Biomed Nanotechnol 3:190–194

Kurakhmaeva KB, Djindjikhashvili IA, Petrov VE, Balabanyan VU, Voronina TA, Trofimov SS, Kreuter J, Gelperina S, Begley D, Alyautdin RN (2009) Brain targeting of nerve growth factor using poly (butyl cyanoacrylate) nanoparticles. J Drug Target 17(8):564–574

Labrenz M, Druschel GK, Thomsen-Ebert T et al (2000) Formation of sphalerite (ZnS) deposits in natural biofilms of sulfate-reducing bacteria. Science 290(5497):1744–1747

Lee EJ, Lim KH (2014) Polyelectrolyte complexes of chitosan self-assembled with fucoidan: an optimum condition to prepare their nanoparticles and their characteristics. Korean J Chem Eng 31:664–675

Lee EJ, Lim KH (2015) Formation of chitosan-fucoidan nanoparticles and their electrostatic interactions: quantitative analysis. J Biosci Bioeng 121:73–83

Lee SW, Mao C, Flynn C, Belcher AM (2002) Ordering of quantum dots using genetically engineeredviruses. Science 296:892–895. 128

Lee AF, Bennett JA, Manayil JC, Wilson K (2014) Heterogeneous catalysis for sustainable biodiesel production via esterification and transesterification. Chem Soc Rev 43(22):7887–7916

Lee KD, Nagajyothi PC, Sreekanth TVM, Park S (2015) Eco-friendly synthesis of gold nanoparticles (AuNPs) using Inonotus obliquus and their antibacterial, antioxidant and cytotoxic activities. J Ind Eng Chem 26:67–72

Lengke MF, Ravel B, Fleet ME, Wanger G, Gordon RA, Southam G (2006) Mechanisms of gold bioaccumulation by filamentous cyanobacteria from gold (III)–chloride complex. Environ Sci Technol 40:6304–6309

Lestari S, Simakova I, Tokarev A, Maki-Arvela P, Eranen K, Murzin DY (2008) Synthesis of biodiesel via deoxygenation of stearic acid over supported Pd/C catalyst. Catal Lett 122:247–251

Li N, Huang J, Dufresne A (2012) Preparation, properties and applications of polysaccharide nanocrystals in advanced functional nanomaterials: a review. Nanoscale 4:3274–3294

Liangwei D, Huang M, Xu Q, Zhang J (2015) Direct electrochemistry and Electrocatalysis of Hemoglobin immobilized on eggshell membrane modified glassy carbon electrode. Asian J Chem 27(1)

Liao J, Mangold MA, Grunder S, Mayor M, Schönenberger C, Calame M (2008) Interlinking au nanoparticles in 2D arrays via conjugated dithiolated molecules. New J Phys 10(6):065019

Liao H, Chen D, Yuan L, Zheng M, Zhu Y, Liu X (2010) Immobilized cellulase by polyvinyl alcohol/Fe2O3 magnetic nanoparticle to degrade microcrystalline cellulose. Carbohydr Polym 82(3):600–604

Lie Y, Jia S, Wu Q, Ran J, Zhang W, Wu S (2011) Studies of Fe3O4-chitosan nanoparticles prepared by co-precipitation under the magnetic field for lipase immobilization. Catal Commun 12:717–720

Liu X, He H, Wang Y, Zhu S (2007) Transesterification of soybean oil to biodiesel using SrO as a solid base catalyst. Catal Commun 8(7):1107–1111

Liu X, He H, Wang Y, Zhu S, Piao X (2008) Transesterification of soybean oil to biodiesel using CaO as a solid base catalyst. Fuel 87(2):216–221

Lo HM, Chiu HY, Lo SW, Lo FC (2012) Effects of micro-nano and non micro-nano MSWI ashes addition on MSW anaerobic digestion. Bioresour Technol 114:90–94

Löfstedt J, Dahlstrand C, Orebom A, Meuzelaar G, Sawadjoon S, Galkin MV, Agback P, Wimby M, Corresa E, Mathieu Y, Sauvanaud L, Eriksson S, Corma A, Samec JSM (2016) ChemSusChem 9:1392

Love AJ, Makarov VV, Yaminsky IV, Kalinina NO, Taliansky ME (2014) The use of tobacco mosaic virus and cowpea mosaic virus for the production of novel metal nanomaterials. Virology 449:133–139

Lupoi and Smith (2011) Biotechnol Bioeng 108:2835–2843

Lioyd JR, Yong P, Macaskie LE (1998) Enzymatic recovery of elemental palladium by using sulfate-reducing bacteria. Appl Environ Microbiol 64:4607–4609

Maensiri S, Laokul P, Klinkaewnarong J, Phokha S, Promarak V, Seraphin S (2008) Indium oxide (In2O3). Indium oxide nanoparticles using Aloe vera plant extract: synthesis and optical properties. J. Optoelectron Adv Mater 10:161–165

Mahdavi M, Namvar F, Ahmad MB, Mohamad R (2013) Green Biosynthesis and characterization of magnetic iron oxide (Fe3O4) nanoparticles using seaweed Sargassum muticumaqueous extract. Molecules 18:5954–5964

Mahmood T, Hussain ST (2010) Nanobiotechnology for the production of biofuels from spent tea. Afr J Biotechnol 9(6):858–868

Mahmood T, Hussain ST, Malik SA (2010) New nanomaterial and the process for the production of biofuel from metal hyper accumulator water hyacinth. Afr J Biotechnol 9(16):2381–2391

Mahto TK, Jain R, Chandra S, Roy D, Mahto V, Sahu SK (2016) Single step synthesis of sulfonic group bearing graphene oxide: a promising carbo-nano material for biodiesel production. J Environ Chem Eng 4(3):2933–2940

Makarov VV, Love AJ, Sinitsyna OV, Makarova SS, Yaminsky IV, Taliansky ME, Kalinina NO (2014) Green nanotechnologies: Synthesis of metal nanoparticles using plants. Acta Naturae 6:35–44

Malarkodi C, Rajeshkumar S, Paulkumar K, Vanaja M, Jobitha GDG, Annadurai G (2013) Bactericidal activity of bio mediated silver nanoparticles synthesized by Serratia nematodiphila. Drug Invent Today 5(2):119–125

Maliszewska I, Szewczyk K, Waszak K (2009) Biological synthesis of silver nanoparticles. J Phys Conf Ser 146. https://doi.org/10.1088/1742-6596/146/1/012025

Manceau A, Nagy KL, Marcus MA, Lanson M, Geoffroy N, Jacquet T, Kirpichtchikova T (2008) Environ Sci Technol 42(5):1766–1772

Mandal D, Bolander ME, Mukhopadhyay D, Sarkar G, Mukherjee P (2006) The use of microorganisms for the formation of metal nanoparticles and their application. Appl. Microbiol. Biotechnol 69:485–492

Manisha DR, Ramchander M, Prashanthi Y, Pratap MPR (2014a) Phototrophic bacteria mediated synthesis, characterisation and antibacterial activity of silver nanoparticles. Nanosci Nanotechnol Int J 4(2):20–24

Manisha DR, Merugu R, Vijaybabu AR, Pratap Rudra MP (2014b) Microwave assisted biogenic synthesis of silver nanoparticles using dried seed extract of Coriandrum sativum, characterization and antimicrobial activity. Int J Chem Tech Res 6:3957–3961

Manivasagan P, Venkatesan J, Senthilkumar K, Sivakumar K, Kim S (2013) Biosynthesis, antimicrobial and cytotoxic effect of silver nanoparticles using a novel Nocardiopsis sp. MBRC-1. BioMed Research International 2013:9. Article ID 287638, 9 pages
Mao C, Flynn CE, Hayhurst A, Sweeney R, Qi J, Georgiou G, Iverson B, Belcher AM (2003) Viralassembly of oriented quantum dot nanowires. Proc Natl Acad Sci USA 100:6946–6951
Marimuthu S, Rahuman AA, Jayaseelan C, Kirthi AV, Santhoshkumar T, Velayutham K, Bagavan A, Kamaraj C, Elango G, Iyappan M (2013) Acaricidal activity of synthesized titanium dioxide nanoparticles using calotropis gigantea against *Rhipicephalus microplus* and *Haemaphysalis bispinosa*. Asian Pac J Trop Med 6(9):682–688
Martone G, Faiella V, Antonucci M, Giordano MZ (2011) The effect of the aspect ratio of carbon nanotubes on their effective reinforcement modulus in an epoxy matrix. Compos Scie Technol 71(8):1117–1123., Elsevier
Mateo C, Palomo JM, Fernandez-Lorente G, Guisan JM, FernandezLafuente R (2007) Improvement of enzyme activity, stability and selectivity via immobilization techniques. Enzyme Microb Technol 40:1451–1463
Merzlyak A, Lee SW (2006) Phage as template for hybrid materials and mediators for nanomaterials synthesis. Curr Opin Chem Biol 10:246–252
Misson M, Zhang H, Jin B (2015) Nanobiocatalyst advancements and bioprocessing applications. J R Soc Interface 12:20140891
Mittal AK, Kaler A, Banerjee UC (2012) Free radical scavenging and antioxidant activity ofsilvernanoparticles synthesized from flower extract of Rhododendron dauricum. Nano Biomed Eng 4:118–124
Mittal AK, Bhaumik J, Kumar S, Banerjee UC (2014a) Biosynthesis of silver nanoparticles: elucidation of prospective mechanism and therapeutic potential. J Colloid Interface Sci 415:39–47
Mittal AK, Kumar S, Banerjee UC (2014b) Quercetin and gallic acid mediated synthesis of bimetallic (silver and selenium) nanoparticles and their antitumor and antimicrobial potential. J Colloid Interface Sci 431:194–199
Mochochoko T, Oluwafemi OS, Jumbam DN, Songca SP (2013) Green synthesisof silver nanoparticles using cellulose extracted from an aquaticweed; water hyacinth. Carbohydr Polym 98:290–294
Mondal S, Roy N, Laskar RA et al (2011) Biogenic synthesis of Ag, Au andbimetallic Au/Ag alloy nanoparticles using aqueous extract of mahogany (*Swietenia mahogani* JACQ.) leaves. Colloids Surf B Biointerfaces 82:497–504
Mookan RSM, Senthilkumar HV, Ravichandran B et al (2014a) Transesterification of castor oil using nano-sized iron catalyst for the production of biodiesel. J Chem Pharm Sci 2:108–112. 97
Mookan RKA, Sundaresan M, Velan P (2014b) Biodiesel production from pongamia pinnata oil using synthesized iron nanocatalyst. Int J Chem Tech Res 6:4511–4516
Mude N, Ingle A, Gade A, Rai M (2009) Synthesis of silver nanoparticles usingcallus extract of *Carica papaya*—a first report. J Plant Biochem Biotechnol 18:83–86
Murahashi ST, Oda Y, Naota T (1992) J Am Chem Soc 114:7913
Nadeem M, Abbasi BH, Younas M, Ahmad W, Khan T (2017) A review of the green syntheses and anti-microbial applications of gold nanoparticles. Green Chem Lett Rev 10(4):216–227
Naik SN, Goud VV, Rout PK, Dalai AK (2010) Production of first and second generation biofuels: a comprehensive review. Renew Sust Energy Rev 14(2):578–597
Naikaa HR, Lingarajua K, Manjunathb K, Kumar D, Nagarajuc SD, Nagabhushanae H (2015) Green synthesis of CuO nanoparticles using Gloriosa superba L. extract and their antibacterial activity. J Taibah Univ Sci 9:7–12
Nair B, Pradeep T (2002) Coalescence of nanoclusters and formation of submicron crystallites assisted by Lactobacillus strains. Cryst Growth Des 2(4):293–298. View at Publisher ·View at Google Scholar · View at Scopus
Nakkala JR, Mata R, Bhagat E, Sadras SR (2015) Green synthesis of silver and gold nanoparticles from Gymnema sylvestre leaf extract: study of antioxidant and anticancer activities. J Nanopart Res 17(3):151

Nalvolthula R, Merugu R, Rudra MP (2015a) Phytochemical analysis, synthesis, antitumor and antimicrobial activity of silver nanoparticles using flower extracts of Ixora coccinea. Int J Chem Tech Res 7:2374. 80.504-511

Nalvolthula R, Merugu R, Alwala J, Rudra MP (2015b) Cytotoxicity of biologically synthesized silver nanoparticles from Citrus lemon against some cell lines. Int J Pharm Tech Res 8:691–695

Nalvothula R, Nagati VB, Koyyati R, Merugu R, Padigya PRM (2014) Biogenic synthesis of silver nanoparticles using Tectona grandis leaf extract and evaluation of their antibacterial potential. Int J Chem Tech Res 6(1):293–298

Nangia Y, Wangoo N, Goyal N, Shekhawat G, Suri CR (2009) A novel bacterial isolate Stenotrophomonas maltophilia as living factory for synthesis of gold nanoparticles. Microb Cell Factories 8(1):39

Narayanan KB, Sakthivel N (2008) Coriander leaf mediated biosynthesis of goldnanoparticles. Mater Lett 62:4588–4590

Nasrollahzadeha M, Sajadib SM, Vartoonia AR, Khalajc M (2015) Green synthesis of Pd/Fe3O4 nanoparticles using Euphorbia condylocarpa M. bieb root extract and theircatalytic applications as magnetically recoverable and stable recyclable catalysts for the phosphine-free Sonogashira and Suzuki coupling reactions. J Mol Catal A 396:31–39

Nestor ARV, Mendieta VS, Lopez MAC, Espinosa RMG, Lopez MAC (2008) J.a.a. AlatorreSolventless synthesis and optical properties of au and ag nanoparticles using Camiellia sinensis extract. Mater Lett 62:3103–3105

Ngo TPN, Li A, Tiew KW, Li Z (2013) Efficient transformation of grease to biodiesel using highly active and easily recyclable magnetic nanobiocatalyst aggregates. Bioresour Technol 145:233–239

Nitta SK, Numata K (2013) Biopolymer-based nanoparticles for drug/gene delivery and tissue engineering. Int J Mol Sci 14:1629–1654

Njagi HH, Stafford L, Genuino H, Galindo HM, Collins JB et al (2011) Biosynthesis of iron and silver nanoparticles atroom temperature using aqueous sorghum bran extracts Eric C. Langmuir 27:264–271

Oberdörster E, Zhu S, Blickley TM, McClellan-Green P, Haasch ML (2006) Ecotoxicology of carbon-based engineered nanoparticles: effects of fullerene (C60) on aquatic organisms. Carbon 44(6):1112–1120

Ovsejevi K, Manta C, Batista-Viera F (2013) Reversible covalent immobilization of enzymes via disulfide bonds. Methods Mol Biol 1051:89–116

Padala G, Nageshwari B, Kudle KR, Dathar V, Merugu R (2019) Toddy mediated synthesis of Ag-Ni bimetallic nanoparticles and their biological evaluation. IJGHC, December 2018 –February 2019. Sec A: Green Chem 8(1):000–000. https://doi.org/10.24214/IJGHC/GC/8/1/09908

Padalkar S, Schroeder K, Won Y, Jang H, Stanciu L (2012) Biotemplated silica and titania nanowires: synthesis, characterization and potential applications. J Nanosci Nanotechnol 12(1):227–235

Padil V, Thekkae V, Černík M (2013) Green synthesis of copper oxide nanoparticles using gum karaya as a biotemplate and their antibacterial application. Int J Nanomedicine 8:889

Pan X, Fan Z, Chen W, Ding Y, Luo H, Bao X (2007) Enhanced ethanol production inside carbon-nanotube reactors containing catalytic particles. Nat Mater 6(7):507

Pandey G, Madhuri S, Mandloi AK (2012) PI Arch 12:1–4

Panigaj M, Reiser J (2016) Aptamer guided delivery of nucleic acid-based nanoparticles. DNA RNA Nanotechnol 2:42–52

Panigrahi S, Kundu S, Ghosh S, Nath S, Pal T (2004) J Nanopart Res 6(4):411–414. 39

Pantidos N, Horsfall LE (2014) Biological synthesis of metallic nanoparticles by bacteria, fungi and plants. J Nanomed Nanotechnol 5(5):1. https://doi.org/10.4172/2157-7439.1000233

Patra JK, Baek K-H (2014) Green nanobiotechnology: factors affecting synthesis and characterization techniques. J Nanomater 2014:12. Article ID 417305, 12 pages

Patumsawad S (2011) 2nd generation biofuels: technical challenge and R&D opportunity in Thailand. J Sustain Energy Environ (Special Issue) 1:47–50

Peng B, Yao Y, Zhao C, Lercher J a (2012) Towards quantitative conversion of microalgae oil to diesel-range alkanes with bifunctional catalysts. Ang Chem 124(9):2072–2075
Philip D (2011) Mangifera indica leaf-assisted biosynthesis of well-dispersed silver nanoparticles. Spectrochim Acta A 78:327–331
Philip D, Unni C (2011) Extracellular biosynthesis of gold and silver nanoparticlesusing Krishna tulsi (*Ocimum sanctum*) leaf. Phys E Low Dimens Syst Nanostruct 43:1318–1322
Philipse P, Maas D (2002) Magnetic colloids from magnetotactic bacteria: chain formation and colloidal stability. Langmuir 18(25):9977–9984. View at Publisher · View
Pimprikar PS, Joshi SS, Kumar AR, Zinjarde SS, Kulkarni SK (2009) Influence of biomass and gold salt concentration on nanoparticle synthesis by the tropical marine yeast Yarrowia lipolytica NCIM 3589. Colloids Surf B Biointerfaces 74:309–316
Pinheiro AC, Bourbon AI, Cerqueira MA, Maricato É, Nunes C, Coimbra MA, Vicente AA (2014) Chitosan/fucoidan multilayer nanocapsules as a vehicle for controlled release of bioactive compounds. Carbohydr Polym 115:1–9
Poinern GEJ, Shah M, Chapman P, Fawcett D (2013a) Green biosynthesis of silver nanocubes using the leaf extracts from Eucalyptus macrocarpa. Nano Bull 2:1–7
Poinern GEJ, Le X, Chapman P, Fawcett D (2013b) Green biosynthesis of gold nanometre scale plate using the leaf extracts from an indigenous Australian plant Eucalyptus macrocarpa. Gold Bull 46:165–173
Pósfai M, Moskowitz BM, Arató B et al (2006) Properties of intracellular magnetite crystals produced by Desulfovibrio magneticus strain RS-1. Earth Planet Sci Lett 249(3–4):444–455
Prakasham RS, Buddana SK, Yannam SK, Guntuku GS (2012) Characterization of silver nanoparticles synthesized by using marine isolate Streptomyces albidoflavus. J Microbiol Biotechnol 22:614–621
Prathna TC, Chandrasekaran N, Raichur AM, Mukherjee A (2011) Biomimetic synthesis of silver nanoparticles by *Citrus limon* (lemon) aqueous extract and theoretical prediction of particle size. Colloids Surf B Biointerfaces 82:152–159
Priyadharshini RI, Prasannaraj G, Geetha N, Venkatachalam P (2014) Microwave-mediated extracellular synthesis of metallic silver and zinc oxide nanoparticles using macro-algae (Gracilaria edulis) extracts and its anticancer activity against human PC3 cell lines. Appl Biochem Biotechnol 174(8):2777–2790
Prozorov T, Mallapragada SK, Narasimhan B, Wang L, Palo P, Nilsen-Hamilton M et al (2007) Protein-mediated synthesis of uniform superparamagnetic magnetite nanocrystals. Adv Funct Mater 17(6):951–957
Pum D, Sleytr UB (1999) Trends Biotechnol 17:8
Qiu F, Li Y, Yang D, Li X, Sun P (2011) Heterogeneous solid base nanocatalyst: preparation, characterization and application in biodiesel production. Bioresour Technol 102(5):4150–4156
Qu J, Luo C, Hou J (2011) Synthesis of ZnO nanoparticles from Zn-hyperaccumulator (*Sedum alfredii* Hance) plants. IET Micro Nano Lett 6:174–176
Raghunandan D, Basavaraja S, Mahesh B, Balaji S, Manjunath SY, Venkataraman A (2009) Biosynthesisof stable polyshaped gold nanoparticles from microwave-exposed aqueous extracellular anti-malignant Guava (Psidium guajava) leaf extract. Nanobiotechnology 5:34–41
Raghunandan D, Bedre MD, Basavaraja S et al (2010) Rapid biosynthesisof irregular shaped gold nanoparticles from macerated aqueousextracellular dried clove buds (*Syzygium aromaticum*) solution. Colloids Surf B Biointerfaces 79:235–240
Rai M, Yadav A (2008) A Gade Current trends in phytosynthesis of metal nanoparticles. Crit Rev Biotechnol 28:277–284
Rai A, Singh A, Ahmad A, Sastry M (2006) Role of halide ions and temperature on the morphology of biologically synthesized gold nanotriangles. Langmuir 22(2):736–741
Rajasree SR, Suman TY (2012) Extracellular biosynthesis of gold nanoparticles using a gram negative bacterium Pseudomonas fluorescens. Asian Pac J Trop Dis 2(2):S796–S799
Rajiv P, Rajeshwari S, Venckatesh R (2013) Bio-Fabrication of zinc oxide nanoparticles using leaf extract of Parthenium hysterophorus L. and its size-dependent antifungal activity against plant fungal pathogens. Spectrochim Acta Part A: Mol Biomol Spectrosc 112:384–387

Raliya R, Tarafdar JC (2013) ZnO nanoparticle biosynthesis and its effect on phosphorous-mobilizing enzyme secretion and gum contents in clusterbean (Cyamopsis tetragonoloba L.). Agric Res 2(1):48–57

Ramesh P, Rajendran A, Shisundaram MM (2014) Green synthesis of zinc oxide nanoparticles using flower extractcassia. J Nanosci Nanotechnol 2:41–45

Ramezani F, Ramezani M, Talebi S (2010) Mechanistic aspects of biosynthesis of nanoparticles by several microbes. Nano 10(12–14):1–7

Ramimoghadam D, Bagheri S, Bee S, Hamid A (2014) Biotemplated synthesis of anatase titanium dioxidenanoparticles via lignocellulosic waste material. Bio Med Res Int 2014:205636

Ren Y, Rivera JG, He L, Kulkarni H, Lee DK, Messersmith PB (2011) Facile, high efficiency immobilization of lipase enzyme on magnetic iron oxide nanoparticles via a biomimetic coating. BMC Biotechnol 11(1):63

Renugadevi K, Inbakandan D, Bavanilatha M, Poornima V (2012) Cissus quadrangularis assisted biosynthesis of silver nanoparticles with antimicrobial and anticancer potentials. Int J Pharm Biol Sci 3(3):437–445

Riddin TL, Govender Y, Gericke M, Whiteley CG (2009) Two different hydrogenase enzymes from sulphate reducing bacteria are responsible for the bioreductive mechanism of platinum into nanoparticles. Enzym Microb Technol 45:267–273

Roden EE, Lovley DR (1993) Dissimilatory Fe(III) reduction by the marine microorganism Desulfuromonas acetoxidans. Appl Environ Microbiol 59(3):734–742

Roopan SM, Bharathi A, Prabhakarn A, Rahuman AA, Velayutham K, Rajakumar G, Padmaja R, Lekshmi M, Madhumitha G (2012) Efficient phyto-synthesis and structural characterization of rutile TiO2 nanoparticles using annona squamosa peel extract. Spectrochim Acta Part A 98:86–90

Royston E, Ghosh A, Kofinas P, Harris MT, Culver JN (2008) Self-assembly of virus-structured high surface area nanomaterials and their application as battery electrodes. Langmuir 24:906–912. 185

Sadhasivam S, Shanmugam P, Yun K (2010) Biosynthesis of silver nanoparticles by Streptomyces hygroscopicus and antimicrobial activity against medically important pathogenic microorganisms. Colloids Surf B: Biointerfaces 81(1):358–362

Sahdev P, Ochyl LJ, Moon JJ (2014) Biomaterials for nanoparticle vaccine delivery systems. Pharma Res 31:2563–2582

Saifuddin N, Wong CW, Yasumira AAN (2009) Rapid biosynthesis of silver nanoparticles using culture supernatant of bacteria with microwave irradiation. E-J Chem 6(1):61–70

Saikat Mandal P, Selvakannan R, Phadtare S, Renu P, Sastry M (2002) J Chem Sci 114:513–520. 42

Sakai S, Liu Y, Yamaguchi T, Watanabe R, Kawabe M, Kawakami K (2010) Production of butyl-biodiesel using lipase physically-adsorbed onto electrospun polyacrylonitrile fibers. Bioresour Technol 101(19):7344–7349

Sangeetha G, Rajeshwari S, Venckatesh R (2011) Green synthesis of zinc oxide nanoparticles by aloe barbadensis miller leaf extract: structure and optical properties. Mater Res Bull 46(12):2560–2566

Sanghi R, Verma P (2009) A facile green extracellular biosynthesisof CdS nanoparticles by immobilized fungus. Chem Eng J 155:886–891

Sankar R, Rizwana K, Shivashangari KS, Ravikumar V (2014) Ultra-rapid photocatalytic activity of zadirachta indica engineered colloidal titanium dioxide nanoparticles. Appl Nanosci 5:731–736

Santhoshkumar T, Rahuman AA, Rajakumar G et al (2011) Synthesis of silvernanoparticles using *Nelumbo nucifera* leaf extract and its larvicidal activityagainst malaria and filariasis vectors. Parasitol Res 108:693–702

Santhoshkumar T, Rahuman AA, Jayaseelan C, Rajakumar G, Marimuthu S, Kirthi AV, Kim SK (2014) Green synthesis of titanium dioxide nanoparticles using Psidium guajava extract and its antibacterial and antioxidant properties. Asian Pac J Trop Med 7(12):968–976

Sathishkumar J, Narendhirakannan RT (2011) Dig J Nanomater Biostruct 6:961

Sathishkumar M, Sneha K, Yun YS (2009) Palladium nanocrystals synthesis using Curcuma longa tuber extract. Int J Mater Sci 4:11–17
Sathishkumar M, Krishnamurthy S, Yun YS (2010) Bioresour Technol 101:7958
Satyanarayana RA, Chen C, Jean J et al (2010) Biological synthesis of gold and silver nanoparticles mediated by the bacteria Bacillus subtilis. J Nanosci Nanotechnol 10(10):6567–6574
Satyavani K, Ramanathan T, Gurudeekan S (2011) Green synthesis of silver nanoparticles using stem dried callus extract of bitter apple (Citrullus colocynthis). Dig J Nanomater Biostruct 6:1019–1024
Senthilkumar SR, Sivakumar T (2014) Green tea (Camellia sinensis) mediated synthesis of zinc oxide (ZnO) nanoparticles and studies on their antimicrobial activities. Int J Pharm Sci 6:461–465
Sewell SL, Wright DW (2006) Biomimetic synthesis of titanium dioxide utilizing the R5 peptide derived from cylindrotheca F Usiformis. Chem Mater 18(13):3108–3113
Shah R, Oza G, Pandey S, Sharon M (2012) Biogenic fabrication of gold nanoparticles using Halomonas salina. J Microbiol Biotechnol Res 2:485–492
Shahverdi R, Minaeian S, Shahverdi HR, Jamalifar H, Nohi A (2007) Rapid synthesis of silver nanoparticles using culture supernatants of Enterobacteria: a novel biological approach. Process Biochem 42(5):919–923. View at Publisher · View at Google Scholar
Shankar SS, Ahmad A, Pasricha R, Sastry M (2003) Bioreduction of chloroaurateions by geranium leaves and its endophytic fungus yields goldnanoparticles of different shapes. J Mater Chem 13:1822
Shankar SS, Rai A, Ahmad A, Sastry M (2004) Rapid synthesis of Au, Ag, andbimetallic Au core Ag shell nanoparticles using Neem (*Azadirachtaindica*) leaf broth. J Colloid Interface Sci 1:1
Shankar SS, Rai A, Ahmad A, Sastry M (2005) Controlling the optical propertiesof lemongrass extract synthesized gold nanotriangles and potentialapplication in infrared-absorbing optical coatings. Chem Mater 17:566–572
Sidra HSG, Mahmood T, Nawaz U et al (2016) Biodiesel production by using CaOAl2O3 nano catalyst. Int J Eng Res Sci 2:2395–6992
Sindhura KS, Prasad TNVKV, Selvam PP, Hussain OM (2013) Synthesis, characterization and evaluation of effect of phytogenic zinc nanoparticles on soil exo-enzymes. Appl Nanosci 1:1–9
Singh P (2016) Biosynthesis of titanium dioxide nanoparticles and their antibacterial property. World Acad Sci Eng Technol Int J Chem Mol Nucl Mater Metall Eng 10(2):275–278
Singh N, Naraa S (2013) Biological synthesis and characterization of lead sulfide nanoparticles using bacterial isolatesfrom heavy metal rich sites. Int J Agric Food Sci Technol 4:16–23
Singh BR, Singh O (2012) In: Khan DS (ed) Global trends of fossil fuel reserves and climate change in the 21st century, fossil fuel and the environment. https://doi.org/10.5772/38655. ISBN: 978-953-51-0277-9
Singh A, Talat M, Singh D, Srivastava ON (2010) J Nanopart Res 12:1667–1675
Singh P, Kim Y, Zhang D, Yang D (2016) Biological synthesis of nanoparticles from plants and microorganisms. Trends Biotechnol 34:588–599
Singh J, Singh N, Rathi A et al (2017a) Facile approach to synthesize andcharacterization of silver nanoparticles by using mulberry leaves extractin aqueous medium and its application in antimicrobial activity. J Nanostruct 7:134–140
Singh H, Du J, Yi TH (2017b) Green and rapid synthesis of silver nanoparticles using Borago officinalis leaf extract: anticancer and antibacterial activities. Artif Cells, Nanomed Biotechnol 45(7):1310–1316
Sintubin L, Verstraete W, Boon N (2012) Biotechnol Bioeng 109:2422–2436
Sivakumar P, Sankaranarayanan S, Renganathan S et al (2013) Studies on sono-chemical biodiesel production using smoke deposited nano mgo catalyst. Bull Chem React Eng Cataly 8:89–96
Sivaranjani V, Philominathan P (2016) Synthesize of titanium dioxide nanoparticles using moringa oleifera leaves and evaluation of wound healing activity. Wound Med 12:1–5
Son S, Lee S, Chung Y, Kim S, Hyeon T (2002) The first intramolecular Pauson–Khand reaction in water using aqueous colloidal cobalt nano particles as catalysts. Organ Lett 4:277–279

Song JY, Kwon EY, Kim BS (2010) Biological synthesis of platinum nanoparticles using Diopyros kaki leaf extract. Bioprocess Biosyst Eng 33:159–164

Soni N, Prakash S (2011) Factors affecting the geometry of silver nanoparticles synthesis in Chrysosporium tropicum and Fusarium oxysporum. Am J Nanotechnol 2(1):112–121

Soundarrajan C, Sankari A, Dhandapani P, Maruthamuthu S, Ravichandran S, Sozhan G, Palaniswamy N (2012) Bioprocess Biosyst Eng 35:827–833

Spring H, Schleifer KH (1995) Appl Micribiol 18:147

Steinmetz NF, Shah SN, Barclay JE, Rallapalli G, Lomonossoff GP, Evans DJ (2009) Virus-templated silica nanoparticles. Small 5(7):813–816

Stellwagen DR, van der Klis F, van Es DS, de Jong KP, Bitter JH (2013) Functionalized carbon nanofibers as solid-acid catalysts for transesterification. ChemSusChem 6(9):1668–1672

Suarez PAZ, Moser BR, Sharma BK, Erhan SZ (2009) Comparing the lubricity of biofuels obtained from pyrolysis and alcoholysis of soybean oil and their blends with petroleum diesel. Fuel 88:1143–1147

Suganya A, Murugan K, Kovendan K, Mahesh Kumar P, Hwang JS (2013) Parasitol Res 112:1385–1397

Sukanya MK, Saju KA, Praseetha PK, Sakthivel G (2013) Therapeutic potential of biologically reduced silver nanoparticles from Actinomycete cultures. J Nanosci 2013:940719

Sulaiman GM, Mohammed WH, Marzoog TR, Al-Amiery AAA, Kadhum AAH, Mohamad AB (2013) Green synthesis, antimicrobial and cytotoxic effects of silver nanoparticles using Eucalyptus chapmaniana leaves extract. Asian Pac J Trop Biomed 3(1):58–63

Sundrarajan M, Bama K, Bhavani M, Jegatheeswaran S, Ambika S, Sangili A, Nithya P, Sumathi R (2017) Obtaining titanium dioxide nanoparticles with spherical shape and antimicrobial properties using M. citrifolia leaves extract by hydrothermal method. J Photochem Photobiol B 171:117–124. https://doi.org/10.1016/j.jphotobiol.2017.05.003

Sunkar S, Nachiyar CV (2012) Biogenesis of antibacterial silver nanoparticles using the endophytic bacterium Bacillus cereus isolated from Garcinia xanthochymus. Asian Pac J Trop Biomed 2(12):953–959

Sutradhar P, Saha M (2016) Green synthesis of zinc oxide nanoparticles using tomato (Lycopersiconesculentum) extract and its photovoltaic application. J Exp Nanosci 11(5):314–327

Tamjidi F, Shahedi M, Varshosaz J, Nasirpour A (2013) Nanostructured lipid carriers (NLC): a potential delivery system for bioactive food molecules. Innov Food Sci Emerg Technol 19:29–43

Tan YN, Lee JY, Wang DI (2010) J Am Chem Soc 132(16):5677–5686

Thamilselvi V, Radha KV (2013) Synthesis of silver nanoparticles from Pseudomonas putida NCIM 2650 in silver nitrate supplemented growth medium and optimization using response surface methodology. Dig J Nanomater Biostruct 8(3):1101–1111

Tran DT, Chen CL, Chang JS (2012) Immobilization of Burkholderia sp. lipase on a ferric silica nanocomposite for biodiesel production. J Biotechnol 158(3):112–119

Trindade SC (2011) Nanotech biofuels and fuel additives. In: Dos Santos Bernardes MA (ed) Biofuel's engineering process technology. IntechOpen, Rijeka, pp 103–114

Tsai and Meyer (2014) Enzymatic cellulose hydrolysis: enzyme reusability and visualization of β-glucosidase immobilized in calcium alginate. Molecules 19:19390–19406

Vainshtein M, Belova N, Kulakovskaya T, Suzina N, Sorokin V (2014) Synthesis of magnetosensitive iron-containing nanoparticles by yeasts. J Ind Microbiol Biotechnol 41(4):657–663

Valodkar M, Rathore PS, Jadeja RN, Thounaojam M, Devkar RV, Thakore S (2012) J Hazard Mater 201:244–249

Veerasamy R, Xin TZ, Gunasagaran S et al (2010) Biosynthesis of silver nanoparticles using mangosteen leaf extract and evaluation of their antimicrobial activities. J Saudi Chem Soc 15:113–120

Velayutham K, Rahuman AA, Rajakumar G, Santhoshkumar T, Marimuthu S, Jayaseelan C, Bagavan A, Kirthi AV, Kamaraj C, Zahir AA (2012) Evaluation of catharanthus roseus leaf extract-mediated biosynthesis of titanium dioxide nanoparticles Against Hippobosca

Maculata and Bovicola Ovis. Parasitol Res 111(6):2329–2337. https://doi.org/10.1007/s00436-011-2676-x
Velmurugan P, Shim J, You Y, Choi S, Kamala-Kannan S, Lee KJ, Kim HJ, Oh BT (2010) Removal of zinc by live, dead, and dried biomass of Fusarium spp. isolated from the abandoned-metal mine in South Korea and its perspective of producing nanocrystals. J Hazard Mater 182:317–324
Velmurugan P, Lee SM, Iydroose M, Lee KJ, Oh BT (2013) Appl Microbiol Biotechnol 7:361–368
Venkatesan B, Subramanian V, Tumala A, Vellaichamy E (2014) Rapid synthesis of biocompatible silver nanoparticles using aqueous extract of Rosa damascena petals and evaluation of their anticancer activity. Asian Pac J Trop Med 7:S294–S300
Venugopal K, Rather HA, Rajagopal K, Shanthi MP, Sheriff K, Illiyas M et al (2017) Synthesis of silver nanoparticles (Ag NPs) for anticancer activities (MCF 7 breast and A549 lung cell lines) of the crude extract of Syzygium aromaticum. J Photochem Photobiol B: Biol 167:282–289
Verma ML, Barrow CJ, Puri M (2013a) Nanobiotechnology as a novel paradigm for enzyme immobilisation and stabilisation with potential applications in biodiesel production. Appl Microbiol Biotechnol 97(1):23–39
Verma ML, Chaudhary R, Tsuzuki T, Barrow CJ, Puri M (2013b) Immobilization of ß-glucosidase on a magnetic nanoparticle improves thermostability: application in cellobiose hydrolysis. Bioresour Technol 135:2–6
Verziu M, Cojocaru B, Hu J, Richards R, Ciuculescu C, Filip P, Parvulescu VI (2008) Sunflower and rapeseed oil transesterification to biodiesel over different nanocrystalline MgO catalysts. Green Chem 10(4):373–381
Wang HW, Covarrubias J, Prock H, Wu XR, Wang DH, Bossmann SH (2015) J Phys Chem C 119:26020–26028
Wang T, Zhang D, Dai L, Chen Y, Dai X (2016) Effects of metal nanoparticles on methane production from waste-activated sludge and microorganism community shift in anaerobic granular sludge. Sci Rep 6:25857
Wei X, Luo M, Li W et al (2012) Synthesis of silver nanoparticles by solar irradiation of cell-free Bacillus amyloliquefaciens extracts and AgNO3. Bioresour Technol 103(1):273–278
Weiss J, Decker EA, McClements DJ, Kristbergsson K, Helgason T, Awad T (2008) Solid lipid nanoparticles as delivery systems for bioactive food components. Food Biophys 3:146–154
Wen L, Lin Z, Gu P et al (2009) Extracellular biosynthesis of monodispersed gold nanoparticles by a SAM capping route. J Nanopart Res 11(2):279–288
Wen L, Wang Y, Lu D et al (2010) Preparation of KF/CaO nanocatalyst and its application in biodiesel production from Chinese tallow seed oil. Fuel 89:2267–2271
Willner I, Willner B, Katz E (2007) Biomolecule–nanoparticle hybrid systems for bioelectronic applications. Bioelectrochemistry 70:2–11
Wu J, Xiao YZ, Yu HQ (2005) Degradation of lignin in pulp mill wastewaters by white-rot fungi on biofilm. Bioresour Technol 96(12):1357–1363
Xie W, Ma N (2010) Enzymatic transesterification of soybean oil by using immobilized lipase on magnetic nano-particles. Biomass Bioenergy 34(6):890–896
Xing SY, Lv PM, Yuan HR, Yang LM, Wang ZM, Yuan ZH, Chen Y (2017) Green Chem 19:4157–4168
Yadav V, Sharma N, Prakash R, Raina KK, Bharadwaj LM, Prakash NT (2008) Generation of selenium containing nano-structures by soil bacterium, *Pseudomonas aeruginosa*. Biotechnology 7(2):299–304
Yahya NY, Ngadi N, Jusoh M, Halim NAA (2016) Characterization and parametric study of mesoporous calcium titanate catalyst for transesterification of waste cooking oil into biodiesel. Energy Convers Manag 129:275–283. (Guan et al., 2017),
Yan G, Huang Y, Bu Q, Lv L, Deng P, Zhou J et al (2012) Zinc oxide nanoparticles cause nephrotoxicity and kidney metabolism alterations in rats. J Environ Sci Health A Tox Hazard Subst Environ Eng 47:577–588. https://doi.org/10.1080/10934529.2012.650576

Yehia RS, Al-Sheikh H (2014) Biosynthesis and characterization of silver nanoparticles produced by Pleurotus ostreatus and their anticandidal and anticancer activities. World J Microbiol Biotechnol 30(11):2797–2803

Yong P, Rowson NA, Farr JPG, Harris IR, Macaskie LE (2002) Bioreduction and biocrystallization of palladium by Desulfovibrio desulfuricans NCIMB 8307. Biotechnol Bioeng 80(4):369–379

Zare B, Faramarzi MA, Sepehrizadeh Z, Shakibaie M, Rezaie S, Shahverdi AR (2012) Biosynthesis and recovery of rod-shaped tellurium nanoparticles and their bactericidal activities. Mater Res Bull 47(11):3719–3725

Zayed MF, Eisa WH, Shabaka AA (2012) Spectrochim Acta A Mol Biomol Spectrosc 98:423–428

Zeng C, Vangveravong S, McDunn JE, Hawkins WG, Mach RH (2013) Sigma-2 receptor ligand as a novel method for delivering a SMAC mimetic drug for treating ovarian cancer. Br J Cancer 109(9):2368

Zhang J, Misra RDK (2007) Magnetic drug-targeting carrier encapsulated with thermosensitive smart polymer: core–shell nanoparticle carrier and drug release response. Acta Biomater 3:838

Zhang Y, Dube MA, McLean DD, Kates M (2003) Bioresour Technol 90:229–240

Zhang HR, Li QB, Lu YH et al (2005) Biosorption and bioreduction of diamine silver complex by Corynebacterium. J Chem Technol Biotechnol 80(3):285–290

Zhang XD, Wu D, Shen X, Liu PX, Yang N, Zhao B, Fan FY (2011) Size-dependent in vivo toxicity of PEG-coated gold nanoparticles. Int J Nanomed 6:2071–2081

Zhang X, Yan S, Tyagi R, Surampalli R (2013) Biodiesel production from heterotrophic microalgae through transesterification and nanotechnology application in the production. Renew Sust Energy Rev 26:2y

Zhao P, Feng X, Huang D, Yang G, Astruc D (2015) Basic concepts and recent advances in nitrophenol reduction by gold-and other transition metal nanoparticles. Coord Chem Rev 287:114–136

Zheng P, Wang J, Lu C, Xu Y, Sun Z (2013) Immobilized β-glucosidase on magnetic chitosan microspheres for hydrolysis of straw cellulose. Process Biochem 48(4):683–687

Nanomaterials: Types, Synthesis and Characterization

5

T. C. Mokhena, M. J. John, M. A. Sibeko, V. C. Agbakoba, M. J. Mochane, A. Mtibe, T. H. Mokhothu, T. S. Motsoeneng, M. M. Phiri, M. J. Phiri, P. S. Hlangothi, and T. G. Mofokeng

Abstract

Nanoparticles are generally defined as particles having one or more dimensions of sizes ranging from 1 to 100 nm. Nanoparticles can be classified into organic, inorganic and carbon-based materials. In comparison with conventional micro-size particles, nanoparticles show enhanced properties, such as high reactivity, strength, surface area, sensitivity and stability due to their nanosize. Various preparation methods, viz. physical, chemical and mechanical, have been employed to synthesize different nanoparticles. This chapter presents an overview on nanoparticles and their types, properties, synthesis methods and application in bioconversion of biomass into biofuels.

T. C. Mokhena (✉) · M. J. John · V. C. Agbakoba
Department of Chemistry, Nelson Mandela University, Port Elizabeth, South Africa

CSIR Materials Science and Manufacturing, Polymers and Composites, Port Elizabeth, South Africa

M. A. Sibeko · M. M. Phiri · M. J. Phiri · P. S. Hlangothi
Department of Chemistry, Nelson Mandela University, Port Elizabeth, South Africa

M. J. Mochane
Department of Life Sciences, Central University of Technology Free State, Bloemfontein, South Africa

A. Mtibe · T. S. Motsoeneng
CSIR Materials Science and Manufacturing, Polymers and Composites, Port Elizabeth, South Africa

T. H. Mokhothu
Department of Chemistry, Durban University of Technology, Durban, South Africa

T. G. Mofokeng
DST/CSIR National Centre for Nanostructured Materials, Council for Scientific and Industrial Research, Pretoria, South Africa

M. Srivastava et al. (eds.), *Nanomaterials in Biofuels Research*, Clean Energy Production Technologies, https://doi.org/10.1007/978-981-13-9333-4_5

Keywords
Nanoparticles · Synthesis · Characterization · Properties · Biofuel application

5.1 Introduction

Nanotechnology has been a topical subject in the academia and industrial communities for the past years (Zhang et al. 2013; Romero and Moya 2012; Margulis-Goshen and Magdassi 2012; Moreno-Vega et al. 2012). This technology offers a new dimension to a science of working and manipulating materials having a size of 100 nm or less at least in one dimension (Zhang et al. 2013). With respect to their nanosize, these materials are recognized by their unique properties, viz. physicochemical and mechanical properties such as reactivity, tenacity, elasticity, strength and excellent electrical and thermal conductivities. These nanomaterials can easily be employed in a wide range of fields including crop production, cosmetics, drug delivery, photonic crystals, analysis, food, coatings, paints, bioremediation, catalysis and material science. The nanomaterials' size, shape and morphology play a significant role on their properties; therefore, more research has been dedicated in finding means to fine-tune these properties (Romero and Moya 2012; López-Serrano et al. 2014). For example, gold (Au), platinum (Pt), silver (Ag) and palladium (Pd) nanoparticles all with a size of 20 nm have characteristic colours, i.e. wine red, yellowish-grey, black and dark black colours, respectively (Khan et al. 2017). The overall shapes can be 0D, 1D, 2D or 3D.

Nanotechnology can simply be defined as the science of designing, synthesizing and characterizing nanomaterials as well as their applications. Since the introduction of the term "nanotechnology", there has been a lot of reports based on the classification, preparation and characterization of nanomaterials. Different synthesizing routes have been developed in order to control the size, structure and morphology of the resulting nanomaterials. These include chemical vapour deposition (CVD), sol-gel technique and mechanochemical processes. The preparation route depends on the resulting nanoparticle type and properties and thus intended application. Inorganic nanoparticles such metal oxides are often produced via wet chemistry, sol-gel, chemical microemulsion, hydrothermal, solvothermal, microwave-assisted combustion, sonochemical and direct precipitation, whereas carbon-based nanomaterials are usually produced through techniques such as arc discharge (AD), laser ablation and chemical vapour deposition (CVD). In the case of graphene nanosheets, techniques such as chemical, mechanical and thermal exfoliation are used. Despite the success that has been made on the production of various nanomaterials, the use of green methods in order to produce these materials has been one of the major research activities for the past decade (Shao et al. 2018; Logeswari et al. 2015). In this case, the plant extract or natural material is either used as reducing or stabilizing agent for nanoparticle production. Chitosan and alginate were recently reported as a suitable stabilizing and reducing agent in synthesizing silver nanoparticles (AgNPs) (Shao et al. 2018; Logeswari et al. 2015; Mokhena and Luyt 2017a; Mokhena and Luyt 2017b).

Nanotechnology application in biofuel production provides very intriguing solutions to some challenges faced in this field. Nanomaterials can be used as favourable carriers to immobilize some catalyst in order to facilitate their recovery from liquid phase by filtration or centrifugation (Zhang et al. 2013). For instance, it was demonstrated that modified nanosphere silica realizes the extraction from alive microalgae without harm which avoids recultivation by sending the microalgae back for lipid accumulation (Lin et al. 2009). Moreover, nanoparticles can be used directly as heterogeneous catalysts in order to improve the biofuel conversion yields and resultant fuel quality (Wen et al. 2010). Nanoparticles such as CaO, Al_2O_3 and MgO are already being recognized as heterogeneous catalysts with conversion rate of more than 99% with less amount of oil (i.e. ˂1%) (Zhang et al. 2013; Wen et al. 2010; Venkat Reddy et al. 2006). There has been quite a lot of publications, reviews and book chapters covering the use of nanoparticles in order to facilitate the biofuel production and resultant fuel quality (Zhang et al. 2013; Lee et al. 2015). This chapter presents an overview on nanoparticles and their types, properties, synthesis methods and application in bioconversion of biomass into biofuels.

5.2 Classification

Nanoparticles can be classified either by source (natural or anthropogenic); chemical composition (organic or inorganic); synthetic route (biogenic, geogenic, anthropogenic and atmospheric); their shape, size or structure; or their applications (Romero and Moya 2012; Margulis-Goshen and Magdassi 2012; Khan et al. 2017; Cui et al. 2018). In this chapter, we reviewed the nanoparticles based on their chemical composition as discussed below. In general, the fabrication/preparation of the nanoparticles can either be "bottom-up" or "top-down" approach as schematically presented in Fig. 5.1. Bottom-up approach as reflected by its name involves building up NPs from single molecule or atom, e.g. sedimentation and reduction techniques. In this context, precursor or source is used as simple substance to synthesize NPs either by sedimentation or reduction. Examples include sol-gel, biosynthesis, spinning and pyrolysis. On the other hand, top-down approach involves breaking down bulk materials into smaller units which are further converted into suitable NPs. Examples include milling, CVD, physical vapour deposition and decomposition processes.

5.2.1 Organic Nanoparticles

Organic nanoparticles feature unique properties such as biodegradability, biocompatibility and high stability in biological fluids (Romero and Moya 2012; Margulis-Goshen and Magdassi 2012; Moreno-Vega et al. 2012). These particles are available in nature as protein aggregates, lipid bodies, milk emulsions or more complex organized structures such as viruses, etc. Organic nanoparticles have been used as a part of different industrial products, such as food (e.g. creams, chocolate and cakes) and

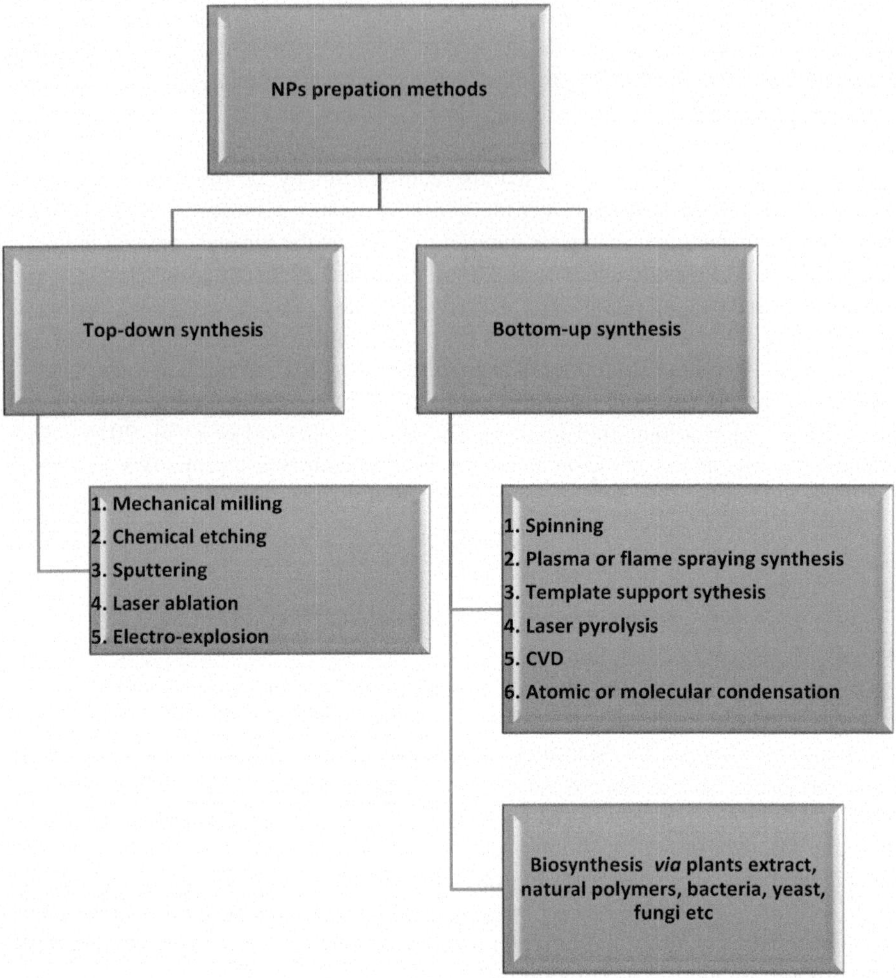

Fig. 5.1 Schematic presentation of synthetic routes for NPs

cosmetics (as nanoemulsions), and in pharmaceutical formulations (viz. liposome vectors, polymersomes and polymer-protein and polymer-drug conjugates). Organic nanoparticles are often fabricated by both "top-down" and "bottom-up" approaches as discussed in the next sections (Romero and Moya 2012; Margulis-Goshen and Magdassi 2012; Cui et al. 2018).

5.2.1.1 Synthesis of Organic Nanoparticles

The "top-down" approach such as milling, lithography and microfluidics is often used to physically reduce the size of bulk material into nanomaterials (Romero and Moya 2012; Rolland et al. 2005). Milling is the most used process especially for drug NPs made from poor water-soluble compounds. The limitations of the milling process include unavoidable contamination of the produced NPs and their wide size distribution. Elsewhere, it was demonstrated that a wet milling process can be employed to produce particles having sizes of 230 nm from poor water-soluble drug crystals (Sigfridsson et al. 2011). The obtained NPs were chemically stable for 10 months in both room temperature and when refrigerated and physically stable (particle size) for 10 months under refrigeration and 3 years at room temperature. Microfluidic and lithography are often employed in order to gain more control over the size and the shape of the resulting NPs. A very general top-down approach, particle replication in nonwetting templates (PRINT), was used to fabricate monodisperse particles with simultaneous control over structure (i.e. shape, size and composition) and function (i.e. cargo, surface structure) by Rolland et al. (Rolland et al. 2005). They prepared monodispersed NPs of poly(ethylene glycol diacrylate), triacrylate resin, poly(lactic acid) and pyrrole having sizes below 200 nm. Considering the compatibility of PRINT with other synthetic schemes (free radical polymerization, metal-catalysed high-temperature reaction, oxidative coupling using strong acids), it demonstrated its chemical flexibility and tolerance which is important for production of NPs for various applications.

Bottom-up approach is based on building NPs through physicochemical processes, starting from single molecule or atoms (Romero and Moya 2012). It involves single atom or molecule combining via synthetic chemistry and self-organization to produce various organic nanoparticles, such as micelles, vesicles and liposomes, polymersomes, polymer conjugates, dendrimers, capsules and polymeric NPs (Romero and Moya 2012).

5.2.2 Inorganic Nanoparticles

5.2.2.1 Metal Oxide and Metallic Nanoparticles

Metal and metal oxide nanoparticles include silver, silver oxide, copper oxide, magnesium oxide, ZnO, Fe_3O_4 and many more (Jamdagni et al. 2018). Metal oxide nanoparticles feature unique characteristics such as antifungal, antimicrobial, optical absorption, large surface-to-volume ratio, semiconducting and chemical-sensing properties. In addition, their non-toxicity and biocompatibility characteristics offer them opportunity to be applied in different fields such as biomedical, food packaging, biofuel conversion and agriculture. Several researchers have investigated the small-scale production of metal and metal oxide NPs using chemical, green, physical and a combination of these methods with some showing a great potential for large-scale production (Jamdagni et al. 2018) (Table 5.1).

5.2.2.1.1 Synthesis of Metal and Metal Oxide Nanoparticles

Several methods have been applied to synthesize metal and metal oxide nanoparticles, such as physical, chemical, enzymatic and biological methods(Ahmed et al. 2017). Physical methods include plasma arcing, ball milling, thermal evaporation, spray pyrolysis, ultrathin films, pulsed laser desorption, lithographic techniques, sputter deposition, layer-by-layer growth and diffusion flame synthesis. Chemical

Table 5.1 Selected studies on the synthesis of nanoparticles and their properties

Type	Precursor	Route	Techniques	Highlights	References
AgNPs	Silver nitrate ($AgNO_3$)	Biosynthesis using chitosan as stabilizing and reducing agent	UV-vis, TEM and FTIR	The optimum preparation time for AgNPs was 12 h heating at 90 °C to obtain NPs of sizes of 5–20 nm	Mokhena and Luyt (2017a, b)
ZnNPs	Zinc acetate dehydrate ($Zn(CH_3COO)_2{\cdot}2H_2O$)	Biosynthesis using extract from *Nyctanthes arbor-tritis*	UV-vis, FTIR, XRD, DLS and TEM	NPs had size range of 12–32 nm	Jamdagni et al. (2018)
ZnNPs	$ZnCl_2$	Hydrothermal method	UV-vis, XRD, SEM, TEM, SAED	ZnO hollow spheres	Yu and Yu (2008)
TiO_2	Tetrabutyl orthotitanate ($(Ti(OC_4H_9)_4)$)	Hydrothermal method	XRD, XPS, TEM, BET	Mesoporous fluorinated anatase TiO_2 powders having primary particles of 11 ± 2 nm	Yu et al. (2009)
ZnNPs	Zinc acetate	Spray pyrolysis	XRD	Polycrystalline films having hexagonal wurtzite-type crystal structure with grain size of 20 nm	Ashour et al. (2006)
MgO	Magnesium nitrate ($mg(NO_3)_2.6H_2O$)	Spray pyrolysis	XRD, TEM, UV-vis	Crystallite size of 9 nm was obtained	Nemade (2014)
ZnNPs	ZnO powder	Chemical vapour deposition	SEM, TEM, SAED, XRD	Nanowires with diameter of 5 nm and length of 50 nm	Xiang et al. (2007)

methods include electrodeposition, sol-gel process, chemical solution deposition, chemical vapour deposition, Langmuir-Blodgett method, soft chemical method, hydrolysis, co-precipitation and wet chemical methods. One of the most widely used chemical syntheses is co-precipitation (Dobrucka and Długaszewska 2016). In this case, the metal oxide precursor is mixed with stabilizing agent to form nanoparticles. Chemical synthesis is disadvantageous because it requires hard labour and is time consuming and costly (Dobrucka and Długaszewska 2016). Moreover, the chemical used is toxic and may contaminate the resulting NPs and produces large amount of secondary wastes. In solvothermal method, two steps are involved: (i) NPs are synthesized from chemical method and (ii) then calcined at high temperatures. In this method, it is possible to control the size, shape and crystallinity of the metal oxide or metal nanoparticles by changing the experimental conditions such as temperature, reaction time, solvent type, surfactant type and precursor type.

Hydrothermal synthesis has also been utilized to fabricate metal or metal oxide nanoparticles (Yu and Yu 2008; Yu et al. 2009). Similar to solvothermal method, it involves high temperatures and pressures (in autoclave), but the solution is non-aqueous (Yu and Yu 2008). Yu and Yu (2008) synthesized hollow spheres of ZnO nanoparticles by hydrothermal approach from a mixture of sugar and $ZnCl_2$. This mixture was treated at 180 °C for 24 h and then calcined in air at 500 °C for 4 h. The average diameter of as-prepared microsphere was 10 μm (Fig. 5.2a) which drastically decreased to about 800 nm after calcination (Fig. 5.2b). The hollow spheres had shell thickness of about 60 nm (Fig. 5.2c) which were polycrystalline (inset in

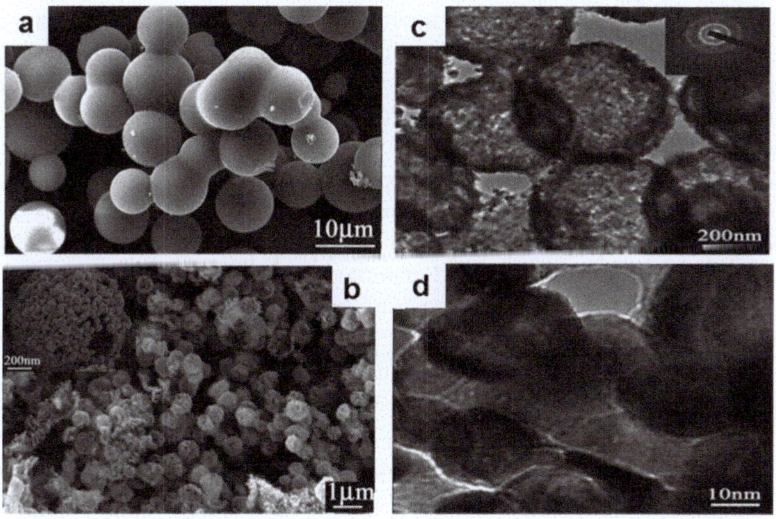

Fig. 5.2 SEM images of the ZnO hollow spheres prepared before calcination (**a**) and after calcination (**b**) at 500 °C for 4 h. TEM image (**c**), high magnification. TEM image (**d**) and SAED pattern (inset in **c**) of the ZnO hollow spheres prepared and calcined at 500 °C for 4 h. (Reprinted with permission from Yu and Yu 2008 with permission from American Chemical Society)

Fig. 5.2a). It was observed that the hollow spheres were composed of randomly aggregated nanocrystal particles with sizes of about 18 nm.

Biosynthesis is a new area of nanotechnology in which natural materials are used to synthesize different nanoparticles (Jamdagni et al. 2018; Ahmed et al. 2017; Dobrucka and Długaszewska 2016; Matinise et al. 2017). These materials can be used as either stabilizing or reducing agent or both. The economic and environmental benefit associated with this technique opens the door for future NP production especially to overcome the challenges associated with chemical and physical methods. Natural materials such as gelatine, alginate, chitosan and many more have been studied for the synthesis of metal oxide nanoparticles as depicted in Table 5.1. In general, the natural polymer or plant extracts are mixed with precursor and then heated under certain temperature and for a given period to obtain NPs (Jamdagni et al. 2018; Matinise et al. 2017). The extracts or natural polymers may act as either reducing agent or stabilizing agent or both.

Another strategy to synthesize NPs involves sol-gel technique (Tamilselvi et al. 2013). This process is divided into two reaction steps (considered as chemical method), namely, hydrolysis and condensation. Hydrolysis is when hydroxyl groups (—OH) are produced, whereas condensation involves polycondensation of the formed hydroxyl groups and residual alkoxy groups to form a 3D network (Mtibe et al. 2018). These steps solely depend on the presence of catalyst, type and concentration as well as the ratio of water and alkoxide. The sol-gel can further be classified into (i) hydrolytic route which involves water as ligand and solvent and (ii) nonhydrolytic route which involves ethers, alcohols and ketones as oxygen donors (Mtibe et al. 2018). The sol-gel technique offers the processor the opportunity to control the properties of the resulting nanoparticles through large number of parameters such as pH, temperature, hydrolysis and condensation rate of the metal oxide precursor, the nature and concentration of anions, the method of mixing and the mixing rate (Mtibe et al. 2018). The advantage of this technique includes a control on particle-size distribution, choice of precursor, high yield and low overall production cost (Tamilselvi et al. 2013).

Chemical vapour deposition (CVD) is known as a parent to a family of processes used to coat almost any metallic or ceramic compound, including elements, metals and their alloys and intermetallic compounds (Guo et al. 2011; Carlsson and Martin 2010; Wang 2011). It involves a solid material being deposited from a vapour by a chemical reaction occurring in the vicinity of a normally heated substrate. The resulting solid materials include thin films, powder or single crystal. It is recognized that by varying the experimental conditions, substrate material, substrate temperature and composition of the reaction gas mixture, total pressure gas flows' different materials having a wide range of physical, tribological and chemical properties can be produced (Guo et al. 2011; Carlsson and Martin 2010; Wang 2011). The CVD process includes chemical vapour deposition, metal-organic chemical vapour deposition, low-pressure chemical vapour deposition, laser chemical vapour deposition, photochemical vapour deposition, chemical vapour infiltration, chemical beam epitaxy, plasma-assisted chemical vapour deposition and plasma-enhanced chemical vapour deposition. The most used methods are thermal chemical vapour deposition, plasma-enhanced chemical vapour deposition and laser chemical vapour deposition (Carlsson and Martin 2010). This technique has been employed in the production of oxides, nitride and carbide nanopowders.

Spray pyrolysis involves the particle formation through atomization of liquid feed in a spray nozzle, combustion of the precursor, evaporation and decomposition of the metal precursor molecules and then nucleation of the particle clusters and growth of particles by condensation and coagulation (Nemade 2014; Meland et al. 2006; Patil 1999). Flame pyrolysis can be utilized to produce metal oxide nanoparticles with large surface through flame conditions and the burner design. The properties of the resulting NPs can further be modified by changing the conditions during the synthesis. For instance, the rutile phase appeared when the preparation temperature was increased and surpassed anatase when temperature was above 1000 °C; however, below 800 °C rutile was not observed (Seo et al. 2006). In addition, the average sizes of the crystallite increased with an increase in preparation temperature. This method is capable of yielding high-purity homogeneous samples, and up-scaling is possible (Meland et al. 2006).

5.2.3 Carbon-Based

5.2.3.1 Graphene

Graphene is recognized as a monolayer of graphite consisting of a honeycomb network of sp^2-hybridized carbon atoms (Goriparti et al. 2014; Kucinskis et al. 2013). The carbons are bonded into 2D sheets having nanometre thickness (single-atom thickness). It has distinctive properties which include high electrical conductivity, good mechanical properties, large surface area and high values of charge mobility, hence receiving unprecedented interest from various fields such as chemical, physical, biological and engineering sciences (Goriparti et al. 2014; Kucinskis et al. 2013). Graphene can be fabricated using various methods, such as micromechanical exfoliation of highly oriented pyrolytic graphite with or without previous processing surface, epitaxial growth, chemical vapour deposition (CVD) and reduction of graphene oxide (GO) (Goriparti et al. 2014; Kucinskis et al. 2013). The most preferable route is the reduction of GO because this process is the most suitable for large-scale production.

5.2.3.1.1 Synthesis of Graphene

Natural graphite also known as expandable graphite is one of the carbon-based material used to produce graphene by various exfoliation processes such as mechanical, chemical or electrochemical and thermal exfoliation techniques. These techniques can be used alone or in combination in order to produce graphene of high-quality and fine-tuned morphology or structure. Mechanical exfoliation or micromechanical cleavage is the most used method to obtain graphene layers. This is the first method which was reported for graphene layers' production using scotch/adhesive tape by repeated peeling of graphite crystal (Novoselov et al. 2004). Another mechanical technique involves rubbing multilayer graphene against a flat surface to obtain a single graphene layer (Mtibe et al. 2018; Bonaccorso et al. 2012). The use of atomic force microscopy (AFM) tip by gluing graphite on it and scratching on Si substrate also offers the production of 2D graphene layers (Bonaccorso et al. 2012). This method, however, offers minimal control over separation and number of resulting graphene layers; hence, it is still difficult for most of these mechanical techniques to be realized as scalable and cost-effective (Mtibe et al. 2018; Bonaccorso et al. 2012).

Chemical exfoliation involves the use of strong oxidants and acids to oxidize graphite through Hummers, Staudenmaier and Brodie methods or by electrochemical oxidative exfoliation of bulk natural graphite (expandable graphite) to yield expanded graphite (EG) or graphite oxide (GO) (Mtibe et al. 2018; Cooper et al. 2014; Zhong and Swager 2012). The oxygen functionality on GO facilitates the dispersion of the sheets in aqueous and organic solvents, however at the expense of electronic conductivity of graphene. The disadvantage of chemical process is the use of harsh chemicals, which are not environment friendly and not easy to handle, and the long production periods associated with this process (Mtibe et al. 2018; Cooper et al. 2014; Zhong and Swager 2012). Thermal exfoliation involves the exposure of expandable graphite to rapid heat, and thus, forcing the graphite sheets apart results in an enormous increase in volume and high surface area material with low bulk density. The chemical and thermal processes can be used together to afford maximum exfoliation of the expandable graphite.

X-ray diffraction (XRD), Raman spectroscopy, transmission electron microscopy (TEM) and scanning electron microscopy (SEM) are usually used to establish the extent of exfoliation or the properties of the resulting graphene layer. Among these techniques, TEM is the most reliable technique to elucidate the properties of resulting material. For instance, the difference between the products obtained from graphite electrode using ionic liquid-assisted electrochemical exfoliation was reported by Lu et al. (2009). They studied time dependence of products exfoliated from graphite in three different stages in the electrochemical exfoliation (Fig. 5.3) using high-resolution TEM. Each stage resulted in different products as reflected by colour change, and their structure is analysed by TEM. Stage I led to hexagonal-shaped water-soluble and fluorescent carbon nanocrystals sized 8–10 nm (Fig. 5.4a). The nanocrystals had a lattice spacing of 0.21–0.25 nm comparable to the (100) facet of graphite (Fig. 5.4b). In stage II, fluorescent nanoribbons of 10 nm × (60 ± 20) nm size with flat rectangular edges with some graphene were obtained. The nanoribbons had a lattice distance of 0.34 nm that corresponds to the (002) plane of graphite (Fig. 5.4c, d), while ultrathin graphene sheets (200 nm × 500 nm) were obtained as shown in Fig. 5.4e, f.

5.2.3.2 Carbon Nanotubes (CNTs)

Carbon nanotube features unique characteristics such as high electrical and thermal conductivities and extremely large surface because of their 1D tubular structure (Goriparti et al. 2014; Liu et al. 2012; Mubarak et al. 2014; Fang et al. 2016). CNTs are recognized as cylinders consisting of rolled-up graphene sheets with hollow central core having diameters on nanometer scale and end caps with a hemisphere

Fig. 5.3 Time evolution of IL electrolyte and highly oriented pyrolytic graphite (HOPG) anode during exfoliation in 60 wt % water/[BMIm][BF_4] electrolyte. Stages I, II and III are shown correspondingly in panels b, c and d. Heavily expanded HOPG is obtained in panel f. (Copied from Lu et al. 2009 with permission from American Chemical Society)

of fullerene structure. They are categorized into single-walled CNTs (SWCNTs) and multi-walled CNTs (MWCNTs) based on the number of graphene layers. As reflected by their names, SWCNTs consist of a single graphene layer, while MWCNTs consist of two or more graphene layers with van der Waals forces between these layers (Goriparti et al. 2014; Liu et al. 2012). In general, CNTs are often synthesized using various techniques, such as arc discharge (Mubarak et al. 2014; Fang et al. 2016; Zhao et al. 2006), laser ablation (Choi et al. 2016; Bota et al. 2015; Vander Wal et al. 2003), gas-phase catalytic growth from carbon monoxide and chemical vapour deposition (CVD) from hydrocarbons. Among these methods, gas-phase methods such as CVD have a potential for large-scale production at fairly low cost. Chemical bonding of CNTs consists of entirely sp^2 carbon-carbon bonds which are stronger than the sp^3 bonds found in diamond, thus providing CNTs with such remarkable mechanical properties (Young's modulus of 1.2 TPa and tensile strength of 50–200 GPa) (Goriparti et al. 2014; Liu et al. 2012).

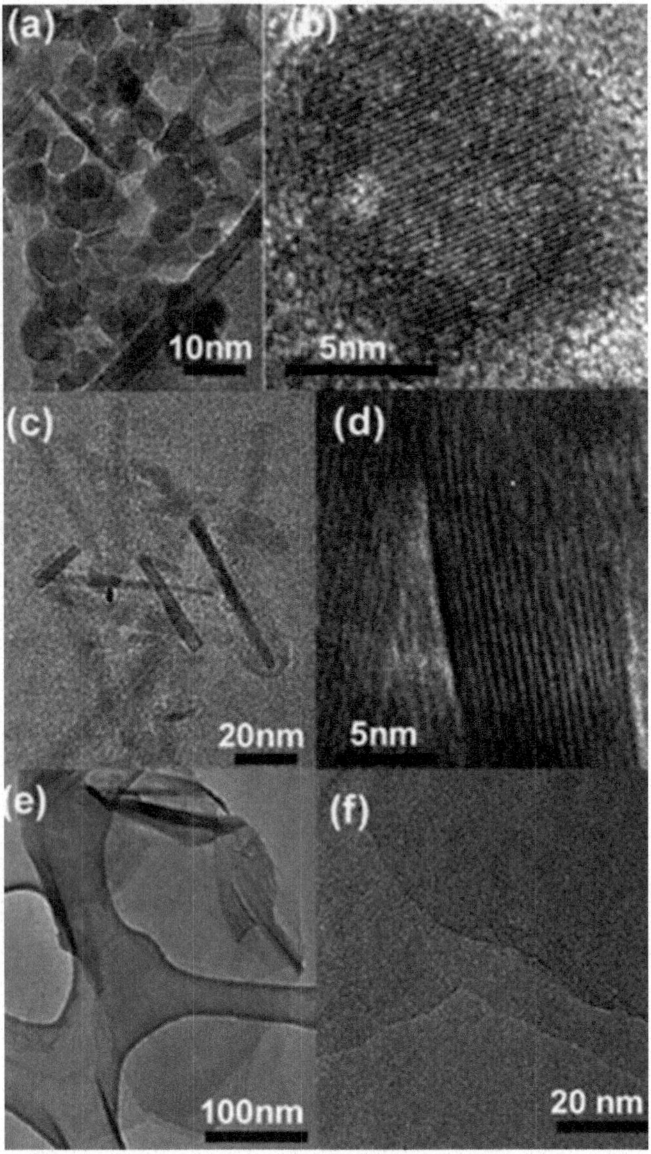

Fig. 5.4 TEM images of carbon nanoparticles (**a**, **b**), carbon nanoribbons (**c**, **d**) and graphene sheets (**e**, **f**) produced in the one-pot electrochemical exfoliation (Reprinted from Lu et al. 2009 with permission from American Chemical Society)

5.2.3.3 Synthesis of CNTs

Different types of CNTs can be produced using various techniques which include electric AD, laser ablation and CVD. These techniques require carbon source, catalyst and energy to produce CNTs. In the case of electric AD method, electric discharge is used as energy source with electric current ranging from 40 to 125 A and potential difference ranging from 20 to 50 V (Fang et al. 2016; Zhao et al. 2006; Kim and Kim 2006; Berkmans et al. 2015; Maria and Mieno 2015; Su et al. 2011, 2014). The graphite electrodes are placed close to each other (1 mm) under inert atmosphere (i.e. at a pressure of 500 Torr) (Mtibe et al. 2018). Moreover, the metal catalyst such as cobalt, nickel, yttrium or iron and high temperatures (2000–3000 °C) are required in order to evaporate the carbons from the electrodes (Mubarak et al. 2014; Fang et al. 2016; Zhao et al. 2012). The evaporated carbon is then condensed in the cathode to form CNTs with diameters of 0.6–1.2 nm. The processors can vary experimental parameters such as metal catalyst concentration, electrode gap, current flow rate, temperature distribution gas type and gas pressure in order to afford the production of either MWCNTs or SWNTs (Mubarak et al. 2014; Fang et al. 2016; Zhao et al. 2006; Kim and Kim 2006; Berkmans et al. 2015; Maria and Mieno 2015; Su et al. 2011). By varying the ratio of catalyst (yttrium and nickel) and carbon source concentration, different carbon-based materials were obtained (Berkmans et al. 2015). It was revealed that a combination of SWCNT having a diameter of ~5.7 nm and single-walled carbon nanohorn (SWCNH) bundles sized about 25 nm can be produced using arc discharge technique in an open air atmosphere which makes this method cheaper and effective for hybrid carbon materials. However, low yields and the quality of CNTs produced via this method limit its industrial production (Mubarak et al. 2014; Berkmans et al. 2015; Su et al. 2011, 2014). The effect of the ratio of gases, viz. carbon monoxide (CO) and helium, and pressure was investigated by Su et al. (Su et al. 2011). It was reported that impurities increased with an increase in the concentration of CO in the gas mixture; however, the gas pressure $>$ 4 kPa can result in smaller diameters. Laser ablation uses the same principle as AD but uses laser pulses as source rather than heat (Choi et al. 2016; Bota et al. 2015; Vander Wal et al. 2003; Chrzanowska et al. 2015). It is recognized that the yield and quality of the resulting CNTs depend on the laser power, catalyst and temperature (Bota et al. 2015; Chrzanowska et al. 2015). In a similar manner, this technique suffers from low yield and low quality of resulting material and requires high level of purification process (Bota et al. 2015; Chrzanowska et al. 2015).

CVD merits special interest due to its capability for large-scale production of high-quality and well-oriented CNTs (Ji et al. 2017; Marchand et al. 2013; Meysami et al. 2013a). In this case, the carbon sources are the hydrocarbons methane, ethylene and benzene and metal catalysts Ni, Co and Fe. Furthermore, high temperature is also required for the production of CNTs. In this regard, the hydrocarbons' catalytic decomposition on monolayer metal catalyst is carried out at temperatures 700–900 °C to form CNTs. The quality of CNTs, yield and structure can be improved by using a single or more metal catalyst. Quite a lot of studies have been carried out to modify CVD in order to improve growth rate, quality and yield as well as to produce different types of CNTs. These modified systems include aerosol-assisted chemical vapour deposition (AACVD) (Marchand et al. 2013; Meysami et al. 2013a, b, 2015), plasma-enhanced chemical vapour deposition (PECVD) (Ji et al. 2017), hydrogen-free spray pyrolysis CVD (Ionescu et al. 2011), catalytic CVD (Almkhelfe et al. 2017; Maruyama et al. 2016; Miura et al. 2018), low-temperature CVD (Maruyama et al. 2002) and alcohol combustion CVD (Hou et al. 2017).

5.3 Characterization

Chemical composition, structure, size and shape play significant role on the properties of resulting NPs; thus, several analytical techniques were utilized to characterize these properties (Hoo et al. 2008).

5.3.1 Size Determination

NPs are often characterized based on their size, morphology and surface charge using various microscopic techniques, such as scanning electron microscopy (SEM), transmission electron microscopy (TEM), scanning tunnelling microscopy (STM) and atomic force microscopy (AFM) (Hoo et al. 2008; Binnig and Rohrer 1983). TEM and SEM are essential for providing information based on the NPs' surface, crystal structure, elemental composition, size and shape. It is recognized that the treatment of the sample before the analysis (coating, drying, staining, etc.) and vacuum conditions in the chamber may result in imaging artefacts. Field emission scanning electron microscopy (FESEM) can overcome these limitations by imaging samples which are not pretreated; however, in most cases, the samples are still treated. On the other hand, AFM provides information based on the size, morphology and surface texture and roughness (Hoo et al. 2008). However, AFM may overestimate the dimensions of the NPs especially when the geometry of the tip is larger than the NPs themselves. In the case of STM, atoms and molecules made up of NPs can be identified by alternating the environment that the specimen is observed, such as vacuum, liquid environment or gaseous environment (Binnig and Rohrer 1983).

Light scattering techniques (e.g. dynamic light scattering (DLS)) are mostly employed to establish the NPs' size (Hoo et al. 2008; Glatter 2018). Dynamic light scattering (DLS), also known as photon correlation spectroscopy, is the most used technique to determine the size of the particle in colloidal suspensions. The suspended particles undergo random Brownian motion; hence, the light scattered off the particles in suspension renders calculation of particle size. The calculation is obtained from analysing fluctuation intensities of the light scattered by the particles during irradiation by a laser beam (Glatter 2018; Ramos 2017).

X-ray-based methods include X-ray absorption (XAS), X-ray fluorescence (XRF), X-ray photoelectron spectroscopy (XPS), energy-dispersive X-ray spectroscopy (EDS), small-angle X-ray scattering (SAXS) and X-ray diffraction (XRD) (Balasubramanian et al. 2014; Dao et al. 2015). These techniques are often used to provide information about surface properties and coatings, crystallographic structure, bonding environment or elemental composition. EDS and XPS are usually combined with either SEM or TEM to elemental assessment and quantitative analysis (Dao et al. 2015). X-ray absorption is used to characterize unoccupied electronic states, the chemical composition and the bonding environment in NPs (Balasubramanian et al. 2014; Dao et al. 2015; Mottana 2014). XRD is used to establish the crystal phase and crystallinity of the NPs (Balasubramanian et al. 2014). The comparison between patterns of the as-synthesized nanoparticles with the standard (JCPDS data) gives information based on the structure and properties of the resulting NPs (Balasubramanian et al. 2014). Spectroscopic techniques are used in NPs with plasmon resonance by collective oscillations of their conduction band electrons in response to electromagnetic waves (Dutta and Ganguly 2012; Begum et al. 2018; Okitsu 2013). UV-visible is utilized to characterize the properties of metal NPs especially the size and shape as well as surface property in the case of functionalized NPs (Begum et al. 2018; Okitsu 2013). This results from plasmonic NPs absorbing radiations of visible to near-infrared region (NIR) depending on their size and shape. This character could be related to collective oscillation of surface electrons of nanoparticles and known as surface plasmon resonance (SPR) (Begum et al. 2018; Okitsu 2013). Because of the SPR property of the nanoparticles, dispersed plasmonic NPs in suspension give one or more peaks that can be used to extract valuable information with regard to the shape, size and size distribution of the nanoparticles (Begum et al. 2018; Okitsu 2013).

Nanoparticle tracking analysis (NTA) is one of commonly used systems for particles of sizes below ~20–1000 nm, with the lower detection limit being dependent on the refractive index of the NPs (van der Pol et al. 2014; Zhou et al. 2015; James and Driskell 2013). It uses a highly sensitive CCD video camera to capture scattered light from each NPs and track the Brownian motion of individual NPs to yield independent size measurements (James and Driskell 2013). Yet another technique is hyperspectral imaging which relies on scattering the obliquely incident visible and near-infrared light in an enhanced dark field. It provides information based on spatial distribution and spectral characteristics depending on NP type, at a sensitivity of a single nanoparticle (size ˂10 nm) (Pena et al. 2016; Badireddy et al. 2012).

5.3.2 Quantification

NP quantification is another important step to understand the fate, behaviour and occurrence of NPS in different environments. Different sensitive analytical techniques have been used to detect and characterize nanoparticles as discussed in the next section.

ICP spectrometry is commonly employed during NP synthesis because it is capable of determining the total element concentration of the colloidal solution (Scheffer et al. 2008). Nowadays, it has been further extended to the size characterization of NPs. ICP mass spectroscopy (ICP-MS) was recently used to establish the size distribution of nanoparticles. In this regard, the determination of discrete ion clouds generated from the atomization of a single nanoparticle in the plasma causes an intensity signal at the detector associated with the size of the particle (Scheffer et al. 2008; Degueldre et al. 2004; Degueldre and Favarger 2004). Single-particle ICP (SP-ICP) is one of the analytical techniques used to determine NP concentration, size and size distribution. It measures the intensity signal produced after ionization of an individual particle rather than a continuous flow of ions reaching the plasma (de la Calle et al. 2017). ICP optical emission spectroscopy (ICP-EOS) has been also utilized for NP analysis (Scheffer et al. 2008; Fischer et al. 2007).

Liquid chromatography mass spectroscopy (LC-MS) is commonly used to analyse fullerenes in environmental matrices (Chen et al. 2008a). Since this technique relies on the rate at which the NPs elute from the stationery phase over a mobile phase gradient, the extraction of C60 has been exploited by its solubility in toluene (Chen et al. 2008a).

Fourier transform infrared (FTIR) and Raman spectroscopy are also used for identification of NPs. The most important range for NPs is the fingerprint region which provides signature information of the sample. Surface-enhanced Raman spectroscopy (SERS) is considered the new innovative method due to its signal-enhanced capability via SPR phenomenon (Muehlethaler et al. 2015; Ma et al. 2011).

5.4 Applications of the Nanoparticles in Biofuels

Nanomaterials can be employed in biodiesel production from biomass as substrate for lipid accumulation, lipid extraction and transesterification process as either catalyst or catalyst support. Due to their high surface area, NPs have been applied as immobilizing agent for lipids; thus, they can be easily recovered from liquid phase by either filtration or centrifugation. Since organic solvents such as hexane and methanol were mainly used in lipid extraction due to their strong affinity towards lipids, immobilization of organic solvent-like chemical onto NPs provides easy recovery and extraction of the lipids. Elsewhere, it was demonstrated that the modified nanosilica afforded the extraction of alive algae which can be sent back for further lipid accumulation again (Lin et al. 2009). Some studies also revealed that NPs enhance the growth of microalgae which improves lipid accumulation (Gao et al. 2010). However, selection of the NPs has to be carefully considered in order to avoid any negative effect on the microbes (Jin et al. 2007; Magrez et al. 2006; Williams et al. 2006).

Transesterification is one of the most reliable and simple technologies applied in biodiesel production (Lee et al. 2015). In this context, various oils from different sources such as animal, plant or oleaginous microorganisms react with alcohol to produce fatty acid methyl esters (FAMEs or biodiesel) (Lee et al. 2015). There are two conditions in which the reaction takes place, viz. either severe high temperatures and pressures or mild conditions at the presence of catalyst. In the latter, acids (H_2SO_4 and HCl) and base (NaOH and KOH) are often employed as catalyst for biodiesel production in the lab and industrial scale, respectively (Xu et al. 2006; Vicente et al. 2009; Ullah et al. 2009). The corrosiveness of the acids and soap formation for base catalyst opened door for enzymatic biocatalyst lipase which is environmentally friendly and efficient (Vicente et al. 2009; Ullah et al. 2009). However, the cost associated with the use of this enzymatic biocatalyst has been recognized as the major limitation that hurdles its success (Du et al. 2004; Pizarro and Park 2003; Noureddini et al. 2005). Different propositions have been made to reduce the overall cost, such as reducing lipase production cost, enhancing its efficiency and reusing it for more than once. The latter has been considered as a feasible route to reduce the overall cost. In this regard, lipase has been immobilized on different carriers, which include fibre cloth, acrylic resin, silica gel, hydrotalcite and micro- and macroporous materials (Noureddini et al. 2005; Yang et al. 2006; Bai et al. 2006). In recent years, nanomaterials received tremendous interest due to their unique attributes such as large surface-to-volume ratio which can provide enormous surface area for lipase immobilization. Different nanomaterials were reported to be employed for lipase immobilization and had the capability of retaining the enzyme and improving its efficiency, selectivity and stability as summarized in Table 5.2.

It was demonstrated that the activity of the lipase immobilized onto nanosized silica was 93% after 7 months of storage as compared to 40% activity for free lipase (Kwon et al. 2007). Elsewhere, the activity was further enhanced by immobilizing lipase onto functionalized zirconia (~214%) and remained high as initially after reusing for eight times (Chen et al. 2008b). Recent study by Li et al. (Li et al. 2018) also demonstrated that other nanoparticles such as carbon nitride can also be utilized for lipase immobilization. Carbon nitride (C_3N_4) as one of the types of two-dimensional nanomaterials has the capability of immobilizing lipase to improve its efficiency. It was reported that nanosheets for immobilization of lipase exhibited satisfactory enzyme loading (44.76 mg/g), pH flexibility, thermostability (after 180 min at 50 °C, 67% of the initial activity remained) and recyclability (after 10 runs, 72% of the initial activity). The concern with regard to immobilization of biocatalyst onto nanomaterials is the recovery of these materials which can end up being cumbersome. In some studies, centrifugation has been used as a recovery technique; however, it consumes a lot of energy and time. In order to easily recover both nanomaterials and immobilized biocatalyst, the use of magnetic nanomaterials has been a major subject in research in the past decades (Tsang et al. 2006; Konwarh et al. 2009; Dyal et al. 2003; Solanki and Gupta 2011; Amirkhani et al. 2016; Netto et al. 2009; Xie and Ma 2009). The magnetic nanoparticles can be easily recovered by using external magnet, thus improving biocatalyst recovery and reusability. However, the aggregation and the nature of chemical activity of magnetic nanoparticles have been the major drawbacks (Amirkhani et al. 2016). One of the proposed methods in order to solve these problems involves coating the magnetic NPs with inorganic layers such as carbon and silica (Amirkhani et al. 2016; Lu et al. 2007; Horak et al. 2007). Moghaddas and co-workers synthesized magnetic silica aerogel support by iron oxide nanoparticles and sodium silicate precursors in a sol-gel process followed by chemical modification and ambient pressure drying (Amirkhani et al. 2016). A maximum adsorption capacity of lipase through physical adsorption was about 81.9 mg/g.

Apart from enzymatic catalyst, heterogeneous catalysts (viz. calcined Li-CaO, Mg-Al hydrotalcites, calcium oxides, magnesia-rich magnesium aluminate spinel, Mg/Zr), which are solid acid or base, have been investigated in biofuel production (Wen et al. 2010; Venkat Reddy et al. 2006; Wang et al. 2009; Sree et al. 2009). Among these, nanocatalyst renders high catalytic efficiency and eases separation from products. Biofuel production through transesterification using nanocatalyst

Table 5.2 Selected studies on the immobilization of lipase

Lipase source	NP type	Synthesizing route and highlights	Activity remaining (%)	Reusability	References
Candida rugosa	MWCNTs	MWCNTs were commercially procured having diameters of 30 ± 5 nm	97	–	Shah et al. (2007)
Candida rugosa	Fe_3O_4	Co-precipitation method to obtain NPs having sizes of 10 ± 2 nm	80	4	Solanki and Gupta (2011)
Candida rugosa	ZrO_2	Chemical method followed by calcination to obtain ZrO_2 nanoparticles having diameter of ~20 nm	214	8	Chen et al. (2008b)
Candida rugosa	Carbon nitride (C_3N_4)	Thermal exfoliation (oxidation) to obtain 2D graphene-like nanosheets	~100	10	Li et al. (2018)
Candida rugosa	γ-Fe_3O_4	Sonication of Fe(CO) in decalin followed by annealing to obtain NPs of average size of 20 ± 10 nm	˂100	–	Dyal et al. (2003)
Candida antarctica	Fe_3O_4	Co-precipitation resulted in nearly cubic particles of sizes ranging from 5 to 10 nm	200	4	Netto et al. (2009)
Thermomyces lanuginosus	SiO_2	Sol-gel process resulted in mesoporous structure having a pore size of 15 nm satisfying lipase-chitosan conjugation	93	–	Kwon et al. (2007)
Candida antarctica	Fe_3O_4	Co-precipitation to produce almost spherical NPs having diameters of ~11.2 nm	70	4	Xie and Ma (2009)

from various oils, which include plant oils and waste oils from selected studies, is summarized in Table 5.3. It is recognized that the required nanocatalyst is 30% of that of conventional catalyst to obtain similar reaction conditions. In addition, the reaction was found to be not dependent on moisture and the free fatty acid content (Venkat Reddy et al. 2006; Boz et al. 2009).

Table 5.3 Selected studies on heterogeneous nanocatalysts

Catalyst	Size and synthesizing route	Oil type	Catalyst:oil ratio (%w/w)	Reaction time (hours)	Yield (%)	References
MWCNTs	Commercially procured and impregnated with KOH and calcined	Canola	12	4	94.2	Omraei et al. (2013)
CaO-NaY-Fe3O4	Grind and then calcined resulted in NPs of average particle size of 200–500 nm with surface area of 359.42 m^2/g	Canola	–	4	95.37	Firouzjaee and Taghizadeh (2017)
CaO-AuNPs	Impregnation by deposition of AuNPs onto CaO	Sunflower	–	3	94	Bet-Moushoul et al. (2016)
TiO2-ZnO	Combustion resulted in spherical (34.2 nm) particles	Palm	–	5	92	Madhuvilakku and Piraman (2013)
CaO	Commercially procured NPs with diameters of ˃160 nm	Soybean	16	6	93.5	Luz Martínez et al. (2010)
CaO	Commercially procured with size ranging from 20 to 40 nm	Poultry oil	0.6	12	99	Venkat Reddy et al. (2006)
$Cs_2Mg(CO_3)_2$	Co-precipitation to yield NPs with crystallite size of 17.5 ± 5.3 nm and surface area of 70 m^2/g	Butter	–	3	100	(Montero et al. 2010)
KF/Al_2O_3	Commercial procured	Soybean	3	8	99.84	Boz et al. (2009)
KF/CaO-MgO	Co-precipitation resulted in mesoporous structure with pore size of about 34 nm and the grain size in the range of 100–300 nm	Rapeseed	3	3	95	Wang et al. (2009)
KF/CaO-Fe3O4	Co-precipitation followed by impregnation resulted in NPs having average diameter of 50 nm and surface area of 20.8 m^2/g	Stillingia	4	3	95	Hu et al. (2011)
$K_2O/\gamma\text{-}Al_2O_3$	Grinding calcination method to afford spherical NPs of about 50 nm	Rapeseed	3	3	94	Han and Guan (2009)

(continued)

Table 5.3 (continued)

Catalyst	Size and synthesizing route	Oil type	Catalyst:oil ratio (%w/w)	Reaction time (hours)	Yield (%)	References
Li-CaO	Impregnation method afforded a mixture of rod-shaped and hexagonal particles having sizes ranging from 50 to 70 nm	Karanja and jatropha	5	1	100	Kaur and Ali (2011)
MgO	Hydrolysis followed by supercritical solvent removal and then dehydrated to form monocrystalline particles of 1–3 nm with lattice spacing of 2.1 Å and surface area of 435 m^2/g	Sunflower and rapeseed	1.5	6	90	Verziu et al. (2008)
MgO	Decomposition of $mg(OH)_2$	Palm	0.5	4	51.3	Yacob et al. (2009)
MgO	Commercially procured having average of 60 nm	Soybean	2	17	99	Wang and Yang (2007)

5.5 Conclusion and Future Remarks

In this chapter, different fabrication methods for various nanomaterials and their characterization as well as classification were presented. It can be concluded that nanoparticles received tremendous interest towards biofuel applications due to their inherited properties, such as particle size, larger surface area, tuneable porosity, crystallinity, shape, compositions and many more. A wide variety of methods, such as microscopic, X-ray-based, light-scattering, spectroscopic and other methods, have been employed to establish these properties because they are of significance with regard to their application. From microscopic viewpoint, TEM is the most reliable technique to provide information on the size distribution and shape of nanoparticles, whereas other quantification methods can also be used to complement TEM especially towards the intended application.

Besides the fact that the utilization of the nanomaterials for advancement of biofuel production is being in its infancy, there has been progress on applying nanomaterials instead of organic solvent with regard to safety and health issues. In addition, nanomaterials can enhance lipid production (by providing stress on biomass) and lipid extraction (even without harming microalgae) and can be used as biocatalyst carriers and/or as heterogeneous catalyst in oil transesterification to biodiesel (Zhang et al. 2013). Further studies are, however, required to further our understanding with regard to cost-effectiveness recycling and impact of these NPs on the environment in order to avoid possible negative consequences on human and other organism.

Acknowledgements The authors would like to thank the financial support from the National Research Funding (NRF) and Department of Science and Technology (DST) Biorefinery Program in South Africa.

References

Ahmed S, Annu CSA, Ikram S (2017) A review on biogenic synthesis of ZnO nanoparticles using plant extracts and microbes: a prospect towards green chemistry. J Photochem Photobiol B Biol 166:272–281. https://doi.org/10.1016/j.jphotobiol.2016.12.011

Almkhelfe H, Li X, Rao R, Amama PB (2017) Catalytic CVD growth of millimeter-tall single-wall carbon nanotube carpets using industrial gaseous waste as a feedstock. Carbon 116:181–190

Amirkhani L, Moghaddas J, Jafarizadeh-Malmiri H (2016) *Candida rugosa* lipase immobilization on magnetic silica aerogel nanodispersion. RSC Adv 6(15):12676–12687

Ashour A, Kaid MA, El-Sayed NZ, Ibrahim AA (2006) Physical properties of ZnO thin films deposited by spray pyrolysis technique. Appl Surf Sci 252(22):7844–7848. https://doi.org/10.1016/j.apsusc.2005.09.048

Badireddy AR, Wiesner MR, Liu J (2012) Detection, characterization, and abundance of engineered nanoparticles in complex waters by hyperspectral imagery with enhanced Darkfield microscopy. Environ Sci Technol 46(18):10081–10088. https://doi.org/10.1021/es204140s

Bai Y-X, Li Y-F, Yang Y, Yi L-X (2006) Covalent immobilization of triacylglycerol lipase onto functionalized novel mesoporous silica supports. J Biotechnol 125(4):574–582

Balasubramanian C, Joseph B, Gupta P, Saini NL, Mukherjee S, Di Gioacchino D, Marcelli A (2014) X-ray absorption spectroscopy characterization of iron-oxide nanoparticles synthesized

by high temperature plasma processing. J Electron Spectrosc Relat Phenom 196:125–129. https://doi.org/10.1016/j.elspec.2014.02.011
Begum R, Farooqi ZH, Naseem K, Ali F, Batool M, Xiao J, Irfan A (2018) Applications of UV/Vis spectroscopy in characterization and catalytic activity of noble metal nanoparticles fabricated in responsive polymer microgels: a review. Crit Rev Anal Chem 48(6):503–516
Berkmans J, Jagannatham M, Reddy R, Haridoss P (2015) Synthesis of thin bundled single walled carbon nanotubes and nanohorn hybrids by arc discharge technique in open air atmosphere. Diam Relat Mater 55:12–15
Bet-Moushoul E, Farhadi K, Mansourpanah Y, Nikbakht AM, Molaei R, Forough M (2016) Application of CaO-based/Au nanoparticles as heterogeneous nanocatalysts in biodiesel production. Fuel 164:119–127. https://doi.org/10.1016/j.fuel.2015.09.067
Binnig G, Rohrer H (1983) Scanning tunneling microscopy. Surf Sci 126(1–3):236–244
Bonaccorso F, Lombardo A, Hasan T, Sun Z, Colombo L, Ferrari AC (2012) Production and processing of graphene and 2d crystals. Mater Today 15(12):564–589. https://doi.org/10.1016/S1369-7021(13)70014-2
Bota P, Dorobantu D, Boerasu I, Bojin D, Enachescu M (2015) New laser ablation chamber for producing carbon nanomaterials using excimer laser. Mater Res Innov 19(1):33–39
Boz N, Degirmenbasi N, Kalyon DM (2009) Conversion of biomass to fuel: transesterification of vegetable oil to biodiesel using KF loaded nano-γ-Al_2O_3 as catalyst. Appl Catal B Environ 89(3–4):590–596
Carlsson J-O, Martin PM (2010) Chapter 7. Chemical vapor deposition. In: Martin PM (ed) Handbook of deposition technologies for films and coatings, 3rd edn. William Andrew Publishing, Boston, pp 314–363. https://doi.org/10.1016/B978-0-8155-2031-3.00007-7
Chen Z, Westerhoff P, Herckes P (2008a) Quantification of C60 fullerene concentrations in water. Environ Toxicol Chem 27(9):1852–1859
Chen YZ, Yang CT, Ching CB, Xu R (2008b) Immobilization of lipases on hydrophobilized zirconia nanoparticles: highly enantioselective and reusable biocatalysts. Langmuir 24(16):8877–8884
Choi S, Byeon C, Park D, Jeong M (2016) Polarization-selective alignment of a carbon nanotube film by using femtosecond laser ablation. J Korean Phys Soc 68(2):210–214
Chrzanowska J, Hoffman J, Małolepszy A, Mazurkiewicz M, Kowalewski TA, Szymanski Z, Stobinski L (2015) Synthesis of carbon nanotubes by the laser ablation method: effect of laser wavelength. Phys Status Solidi B 252(8):1860–1867
Cooper AJ, Wilson NR, Kinloch IA, Dryfe RA (2014) Single stage electrochemical exfoliation method for the production of few-layer graphene via intercalation of tetraalkylammonium cations. Carbon 66:340–350
Cui J, Björnmalm M, Ju Y, Caruso F (2018) Nanoengineering of poly(ethylene glycol) particles for stealth and targeting. Langmuir 34(37):10817–10827. https://doi.org/10.1021/acs.langmuir.8b02117
Dao ATN, Mott DM, Maenosono S (2015) Characterization of metallic nanoparticles based on the abundant usages of X-ray techniques. In: Handbook of nanoparticles, pp 1–24
de la Calle I, Menta M, Klein M, Séby F (2017) Screening of TiO_2 and Au nanoparticles in cosmetics and determination of elemental impurities by multiple techniques (DLS, SP-ICP-MS, ICP-MS and ICP-OES). Talanta 171:291–306. https://doi.org/10.1016/j.talanta.2017.05.002
Degueldre C, Favarger P-Y (2004) Thorium colloid analysis by single particle inductively coupled plasma-mass spectrometry. Talanta 62(5):1051–1054
Degueldre C, Favarger P-Y, Bitea C (2004) Zirconia colloid analysis by single particle inductively coupled plasma-mass spectrometry. Anal Chim Acta 518(1–2):137–142
Dobrucka R, Długaszewska J (2016) Biosynthesis and antibacterial activity of ZnO nanoparticles using Trifolium pratense flower extract. Saudi J Biol Sci 23(4):517–523. https://doi.org/10.1016/j.sjbs.2015.05.016
Du W, Xu Y, Liu D, Zeng J (2004) Comparative study on lipase-catalyzed transformation of soybean oil for biodiesel production with different acyl acceptors. J Mol Catal B Enzym 30(3–4):125–129

Dutta S, Ganguly BN (2012) Characterization of ZnO nanoparticles grown in presence of folic acid template. J Nanobiotechnol 10(1):29

Dyal A, Loos K, Noto M, Chang SW, Spagnoli C, Shafi KV, Ulman A, Cowman M, Gross RA (2003) Activity of *Candida rugosa* lipase immobilized on γ-Fe_2O_3 magnetic nanoparticles. J Am Chem Soc 125(7):1684–1685

Fang X, Shashurin A, Teel G, Keidar M (2016) Determining synthesis region of the single wall carbon nanotubes in arc plasma volume. Carbon 107:273–280

Firouzjaee MH, Taghizadeh M (2017) Optimization of process variables for biodiesel production using the nanomagnetic catalyst CaO/NaY-Fe_3O_4. Chem Eng Technol 40(6):1140–1148

Fischer HC, Fournier-Bidoz S, Chan W, Pang K (2007) Quantitative detection of engineered nanoparticles in tissues and organs: an investigation of efficacy and linear dynamic ranges using ICP-AES. NanoBiotechnology 3(1):46–54

Gao C, Zhai Y, Ding Y, Wu Q (2010) Application of sweet sorghum for biodiesel production by heterotrophic microalga Chlorella protothecoides. Appl Energy 87(3):756–761

Glatter O (2018) Chapter 11. Dynamic Light Scattering (DLS). In: Glatter O (ed) Scattering methods and their application in colloid and Interface science. Elsevier, Amsterdam, pp 223–263. https://doi.org/10.1016/B978-0-12-813580-8.00011-0

Goriparti S, Miele E, De Angelis F, Di Fabrizio E, Proietti Zaccaria R, Capiglia C (2014) Review on recent progress of nanostructured anode materials for Li-ion batteries. J Power Sources 257:421–443. https://doi.org/10.1016/j.jpowsour.2013.11.103

Guo J, Li J, Kou H (2011) Chapter 19. Chemical preparation of advanced ceramic materials. In: Xu R, Pang W, Huo Q (eds) Modern inorganic synthetic chemistry. Elsevier, Amsterdam, pp 429–454. https://doi.org/10.1016/B978-0-444-53599-3.10019-8

Han H, Guan Y (2009) Synthesis of biodiesel from rapeseed oil using K_2O/γ-Al_2O_3 as nano-solid-base catalyst. Wuhan Univ J Nat Sci 14(1):75–79

Hoo CM, Starostin N, West P, Mecartney ML (2008) A comparison of atomic force microscopy (AFM) and dynamic light scattering (DLS) methods to characterize nanoparticle size distributions. J Nanopart Res 10(1):89–96

Horak D, Babič M, Mackova H, Beneš MJ (2007) Preparation and properties of magnetic nano- and microsized particles for biological and environmental separations. J Sep Sci 30(11):1751–1772

Hou B, Wu C, Inoue T, Chiashi S, Xiang R, Maruyama S (2017) Extended alcohol catalytic chemical vapor deposition for efficient growth of single-walled carbon nanotubes thinner than (6,5). Carbon 119:502–510. https://doi.org/10.1016/j.carbon.2017.04.045

Hu S, Guan Y, Wang Y, Han H (2011) Nano-magnetic catalyst KF/CaO–Fe_3O_4 for biodiesel production. Appl Energy 88(8):2685–2690

Ionescu MI, Zhang Y, Li R, Sun X, Abou-Rachid H, Lussier L-S (2011) Hydrogen-free spray pyrolysis chemical vapor deposition method for the carbon nanotube growth: parametric studies. Appl Surf Sci 257(15):6843–6849

Jamdagni P, Khatri P, Rana JS (2018) Green synthesis of zinc oxide nanoparticles using flower extract of Nyctanthes arbor-tristis and their antifungal activity. J King Saud Univ Sci 30(2):168–175. https://doi.org/10.1016/j.jksus.2016.10.002

James AE, Driskell JD (2013) Monitoring gold nanoparticle conjugation and analysis of biomolecular binding with nanoparticle tracking analysis (NTA) and dynamic light scattering (DLS). Analyst 138(4):1212–1218

Ji X, Zhang W, Li X, Yu H, Dong H (2017) A novel hybrid method combining ASP with PECVD for in-situ low temperature synthesis of vertically aligned carbon nanotube films. Diam Relat Mater 77:16–24

Jin Y, Kannan S, Wu M, Zhao JX (2007) Toxicity of luminescent silica nanoparticles to living cells. Chem Res Toxicol 20(8):1126–1133

Kaur M, Ali A (2011) Lithium ion impregnated calcium oxide as nano catalyst for the biodiesel production from karanja and jatropha oils. Renew Energy 36(11):2866–2871

Khan I, Saeed K, Khan I (2017) Nanoparticles: properties, applications and toxicities. Arab J Chem. https://doi.org/10.1016/j.arabjc.2017.05.011

Kim HH, Kim HJ (2006) Preparation of carbon nanotubes by DC arc discharge process under reduced pressure in an air atmosphere. Mater Sci Eng B 133(1–3):241–244

Konwarh R, Karak N, Rai SK, Mukherjee AK (2009) Polymer-assisted iron oxide magnetic nanoparticle immobilized keratinase. Nanotechnology 20(22):225107

Kucinskis G, Bajars G, Kleperis J (2013) Graphene in lithium ion battery cathode materials: a review. J Power Sources 240:66–79. https://doi.org/10.1016/j.jpowsour.2013.03.160

Kwon SS, Jeon SH, Shon JK, Kim DH, Chang IS, Kim JM (2007) Preparation and stabilization of chitosan-lipase composite within mesoporous silica material. Solid State Phenom, Trans Tech Publ:1717–1720

Lee Y-C, Lee K, Oh Y-K (2015) Recent nanoparticle engineering advances in microalgal cultivation and harvesting processes of biodiesel production: a review. Bioresour Technol 184:63–72. https://doi.org/10.1016/j.biortech.2014.10.145

Li Y, Ruan Z, Zheng M, Deng Q, Zhang S, Zheng C, Tang H, Huang F, Shi J (2018) *Candida rugosa* lipase covalently immobilized on facilely-synthesized carbon nitride nanosheets as a novel biocatalyst. RSC Adv 8(26):14229–14236

Lin V, Mahoney P, Gibson K (2009) Nanofarming technology extracts biofuel oil without harming algae. News released from Office of Public Affairs

Liu X-M, Huang Z, Oh S, Zhang B, Ma P-C, Yuen MMF, Kim J-K (2012) Carbon nanotube (CNT)-based composites as electrode material for rechargeable Li-ion batteries: a review. Compos Sci Technol 72(2):121–144. https://doi.org/10.1016/j.compscitech.2011.11.019

Logeswari P, Silambarasan S, Abraham J (2015) Synthesis of silver nanoparticles using plants extract and analysis of their antimicrobial property. J Saudi Chem Soc 19(3):311–317. https://doi.org/10.1016/j.jscs.2012.04.007

López-Serrano A, Olivas RM, Landaluze JS, Cámara C (2014) Nanoparticles: a global vision. Characterization, separation, and quantification methods. Potential environmental and health impact. Anal Methods 6(1):38–56

Lu AH, Salabas EL, Schüth F (2007) Magnetic nanoparticles: synthesis, protection, functionalization, and application. Angew Chem Int Ed 46(8):1222–1244

Lu J, Yang J-x, Wang J, Lim A, Wang S, Loh KP (2009) One-pot synthesis of fluorescent carbon nanoribbons, nanoparticles, and graphene by the exfoliation of graphite in ionic liquids. ACS Nano 3(8):2367–2375. https://doi.org/10.1021/nn900546b

Luz Martínez S, Romero R, López JC, Romero A, Sánchez Mendieta VC, Natividad R (2010) Preparation and characterization of CaO nanoparticles/NaX zeolite catalysts for the transesterification of sunflower oil. Ind Eng Chem Res 50(5):2665–2670

Ma S, Livingstone R, Zhao B, Lombardi JR (2011) Enhanced Raman spectroscopy of nanostructured semiconductor phonon modes. J Phys Chem Lett 2(6):671–674

Madhuvilakku R, Piraman S (2013) Biodiesel synthesis by TiO_2–ZnO mixed oxide nanocatalyst catalyzed palm oil transesterification process. Bioresour Technol 150:55–59. https://doi.org/10.1016/j.biortech.2013.09.087

Magrez A, Kasas S, Salicio V, Pasquier N, Seo JW, Celio M, Catsicas S, Schwaller B, Forró L (2006) Cellular toxicity of carbon-based nanomaterials. Nano Lett 6(6):1121–1125

Marchand P, Hassan IA, Parkin IP, Carmalt CJ (2013) Aerosol-assisted delivery of precursors for chemical vapour deposition: expanding the scope of CVD for materials fabrication. Dalton Trans 42(26):9406–9422

Margulis-Goshen K, Magdassi S (2012) Organic nanoparticles from microemulsions: formation and applications. Curr Opin Colloid Interface Sci 17(5):290–296. https://doi.org/10.1016/j.cocis.2012.06.005

Maria KH, Mieno T (2015) Synthesis of single-walled carbon nanotubes by low-frequency bipolar pulsed arc discharge method. Vacuum 113:11–18. https://doi.org/10.1016/j.vacuum.2014.11.025

Maruyama S, Kojima R, Miyauchi Y, Chiashi S, Kohno M (2002) Low-temperature synthesis of high-purity single-walled carbon nanotubes from alcohol. Chem Phys Lett 360(3–4):229–234

Maruyama T, Kondo H, Ghosh R, Kozawa A, Naritsuka S, Iizumi Y, Okazaki T, Iijima S (2016) Single-walled carbon nanotube synthesis using Pt catalysts under low ethanol pres-

sure via cold-wall chemical vapor deposition in high vacuum. Carbon 96:6–13. https://doi.org/10.1016/j.carbon.2015.09.010
Matinise N, Fuku XG, Kaviyarasu K, Mayedwa N, Maaza M (2017) ZnO nanoparticles via Moringa oleifera green synthesis: physical properties & mechanism of formation. Appl Surf Sci 406:339–347. https://doi.org/10.1016/j.apsusc.2017.01.219
Meland H, Johannessen T, Arstad B, Venvik HJ, Rønning M, Holmen A (2006) Preparation of low temperature water-gas shift catalysts by flame spray pyrolysis. In: Gaigneaux EM, Devillers M, De Vos DE et al (eds) Studies in surface science and catalysis, vol 162. Elsevier, pp 985–992. https://doi.org/10.1016/S0167-2991(06)81006-2
Meysami SS, Dillon F, Koós AA, Aslam Z, Grobert N (2013a) Aerosol-assisted chemical vapour deposition synthesis of multi-wall carbon nanotubes: I. Mapping the reactor. Carbon 58:151–158
Meysami SS, Koós AA, Dillon F, Grobert N (2013b) Aerosol-assisted chemical vapour deposition synthesis of multi-wall carbon nanotubes: II. An analytical study. Carbon 58:159–169
Meysami SS, Koos AA, Dillon F, Dutta M, Grobert N (2015) Aerosol-assisted chemical vapour deposition synthesis of multi-wall carbon nanotubes: III. Towards upscaling. Carbon 88:148–156
Miura S, Yoshihara Y, Asaka M, Hasegawa K, Sugime H, Ota A, Oshima H, Noda S (2018) Millimeter-tall carbon nanotube arrays grown on aluminum substrates. Carbon 130:834–842. https://doi.org/10.1016/j.carbon.2018.01.075
Mokhena TC, Luyt AS (2017a) Electrospun alginate nanofibres impregnated with silver nanoparticles: preparation, morphology and antibacterial properties. Carbohydr Polym 165:304–312
Mokhena TC, Luyt AS (2017b) Development of multifunctional nano/ultrafiltration membrane based on a chitosan thin film on alginate electrospun nanofibres. J Clean Prod 156:470–479
Montero JM, Wilson K, Lee AF (2010) Cs promoted triglyceride transesterification over MgO nanocatalysts. Top Catal 53(11–12):737–745
Moreno-Vega A-I, Gomez-Quintero T, Nunez-Anita R-E, Acosta-Torres L-S, Castaño V (2012) Polymeric and ceramic nanoparticles in biomedical applications. J Nanotechnol 2012:1
Mottana A (2014) 1913–2013 – the centennial of X-ray absorption spectroscopy (XAS): evidences about a question still open. J Electron Spectrosc Relat Phenom 196:14–19. https://doi.org/10.1016/j.elspec.2013.12.004
Mtibe A, Mokhothu TH, John MJ, Mokhena TC, Mochane MJ (2018) Chapter 8. Fabrication and characterization of various engineered nanomaterials. In: Mustansar Hussain C (ed) Handbook of nanomaterials for industrial applications. Elsevier, Amsterdam, pp 151–171. https://doi.org/10.1016/B978-0-12-813351-4.00009-2
Mubarak N, Abdullah E, Jayakumar N, Sahu J (2014) An overview on methods for the production of carbon nanotubes. J Ind Eng Chem 20(4):1186–1197
Muehlethaler C, Leona M, Lombardi JR (2015) Review of surface enhanced Raman scattering applications in forensic science. Anal Chem 88(1):152–169
Nemade K (2014) Waghuley S (2014) synthesis of MgO nanoparticles by solvent mixed spray pyrolysis technique for optical investigation. Int J Met 2014:1
Netto CG, Andrade LH, Toma HE (2009) Enantioselective transesterification catalysis by *Candida antarctica* lipase immobilized on superparamagnetic nanoparticles. Tetrahedron Asymmetry 20(19):2299–2304
Noureddini H, Gao X, Philkana R (2005) Immobilized Pseudomonas cepacia lipase for biodiesel fuel production from soybean oil. Bioresour Technol 96(7):769–777
Novoselov KS, Geim AK, Morozov SV, Jiang D, Zhang Y, Dubonos SV, Grigorieva IV, Firsov AA (2004) Electric field effect in atomically thin carbon films. Science 306(5696):666–669
Okitsu K (2013) UV-vis spectroscopy for characterization of metal nanoparticles formed from reduction of metal ions during ultrasonic irradiation. In: UV-VIS and photoluminescence spectroscopy for nanomaterials characterization. Springer, Berlin, pp 151–177
Omraei M, Sheibani S, Sadrameli SM, Towfighi J (2013) Preparation of biodiesel using KOH-MWCNT catalysts: an optimization study. Ind Eng Chem Res 52(5):1829–1835. https://doi.org/10.1021/ie301418y

Patil PS (1999) Versatility of chemical spray pyrolysis technique. Mater Chem Phys 59(3):185–198. https://doi.org/10.1016/S0254-0584(99)00049-8

Pena MDPS, Gottipati A, Tahiliani S, Neu-Baker NM, Frame MD, Friedman AJ, Brenner SA (2016) Hyperspectral imaging of nanoparticles in biological samples: simultaneous visualization and elemental identification. Microsc Res Tech 79(5):349–358

Pizarro AVL, Park EY (2003) Lipase-catalyzed production of biodiesel fuel from vegetable oils contained in waste activated bleaching earth. Process Biochem 38(7):1077–1082

Ramos AP (2017) 4 - dynamic light scattering applied to nanoparticle characterization. In: Da Róz AL, Ferreira M, de Lima Leite F, Oliveira ON (eds) Nanocharacterization techniques. William Andrew Publishing, Oxford, pp 99–110. https://doi.org/10.1016/B978-0-323-49778-7.00004-7

Rolland JP, Maynor BW, Euliss LE, Exner AE, Denison GM, DeSimone JM (2005) Direct fabrication and harvesting of monodisperse, shape-specific nanobiomaterials. J Am Chem Soc 127(28):10096–10100. https://doi.org/10.1021/ja051977c

Romero G, Moya SE (2012) Chapter 4. Synthesis of organic nanoparticles. In: de la Fuente JM, Grazu V (eds) Frontiers of nanoscience, vol 4. Elsevier, Amsterdam, pp 115–141. https://doi.org/10.1016/B978-0-12-415769-9.00004-2

Scheffer A, Engelhard C, Sperling M, Buscher W (2008) ICP-MS as a new tool for the determination of gold nanoparticles in bioanalytical applications. Anal Bioanal Chem 390(1):249–252

Seo DJ, Cho MY, Park SB (2006) Preparation of titania nanoparticles of anatase phase by using flame spray pyrolysis. In: Rhee H-K, Nam I-S, Park JM (eds) Studies in surface science and catalysis, vol 159. Elsevier, Amsterdam, pp 761–764. https://doi.org/10.1016/S0167-2991(06)81708-8

Shah S, Solanki K, Gupta MN (2007) Enhancement of lipase activity in non-aqueous media upon immobilization on multi-walled carbon nanotubes. Chem Cent J 1(1):30

Shao Y, Wu C, Wu T, Yuan C, Chen S, Ding T, Ye X, Hu Y (2018) Green synthesis of sodium alginate-silver nanoparticles and their antibacterial activity. Int J Biol Macromol 111:1281–1292. https://doi.org/10.1016/j.ijbiomac.2018.01.012

Sigfridsson K, Björkman J-A, Skantze P, Zachrisson H (2011) Usefulness of a nanoparticle formulation to investigate some hemodynamic parameters of a poorly soluble compound. J Pharm Sci 100(6):2194–2202. https://doi.org/10.1002/jps.22440

Solanki K, Gupta M (2011) Simultaneous purification and immobilization of *Candida rugosa* lipase on superparamagnetic Fe_3O_4 nanoparticles for catalyzing transesterification reactions. New J Chem 35(11):2551–2556

Sree R, Babu NS, Prasad PS, Lingaiah N (2009) Transesterification of edible and non-edible oils over basic solid Mg/Zr catalysts. Fuel Process Technol 90(1):152–157

Su Y, Yang Z, Wei H, Kong ES-W, Zhang Y (2011) Synthesis of single-walled carbon nanotubes with selective diameter distributions using DC arc discharge under CO mixed atmosphere. Appl Surf Sci 257(7):3123–3127. https://doi.org/10.1016/j.apsusc.2010.10.127

Su Y, Wei H, Li T, Geng H, Zhang Y (2014) Low-cost synthesis of single-walled carbon nanotubes by low-pressure air arc discharge. Mater Res Bull 50:23–25. https://doi.org/10.1016/j.materresbull.2013.10.013

Tamilselvi P, Yelilarasi A, Hema M, Anbarasan R (2013) Synthesis of hierarchical structured MgO by sol-gel method. Nano Bulletin 2(1):130106

Tsang SC, Yu CH, Gao X, Tam K (2006) Silica-encapsulated nanomagnetic particle as a new recoverable biocatalyst carrier. J Phys Chem B 110(34):16914–16922

Ullah F, Nosheen A, Hussain I, Banon A (2009) Base catalyzed transesterification of wild apricot kernel oil for biodiesel production. Afr J Biotechnol 8(14)

van der Pol E, Coumans FAW, Sturk A, Nieuwland R, van Leeuwen TG (2014) Refractive index determination of nanoparticles in suspension using nanoparticle tracking analysis. Nano Lett 14(11):6195–6201. https://doi.org/10.1021/nl503371p

Vander Wal R, Berger G, Ticich T (2003) Carbon nanotube synthesis in a flame using laser ablation for in situ catalyst generation. Appl Phys A 77(7):885–889

Venkat Reddy CR, Oshel R, Verkade JG (2006) Room-temperature conversion of soybean oil and poultry fat to biodiesel catalyzed by nanocrystalline calcium oxides. Energy Fuel 20(3):1310–1314

Verziu M, Cojocaru B, Hu J, Richards R, Ciuculescu C, Filip P, Parvulescu VI (2008) Sunflower and rapeseed oil transesterification to biodiesel over different nanocrystalline MgO catalysts. Green Chem 10(4):373–381

Vicente G, Bautista LF, Rodríguez R, Gutiérrez FJ, Sádaba I, Ruiz-Vázquez RM, Torres-Martínez S, Garre V (2009) Biodiesel production from biomass of an oleaginous fungus. Biochem Eng J 48(1):22–27

Wang J-T (2011) Chapter 7. CVD and its related theories in inorganic synthesis and materials preparations. In: Xu R, Pang W, Huo Q (eds) Modern inorganic synthetic chemistry. Elsevier, Amsterdam, pp 151–171. https://doi.org/10.1016/B978-0-444-53599-3.10007-1

Wang L, Yang J (2007) Transesterification of soybean oil with nano-MgO or not in supercritical and subcritical methanol. Fuel 86(3):328–333

Wang Y, Hu S-y, Guan Y-p, Wen L-b, Han H-y (2009) Preparation of mesoporous nanosized KF/CaO–MgO catalyst and its application for biodiesel production by transesterification. Catal Lett 131(3–4):574–578

Wen L, Wang Y, Lu D, Hu S, Han H (2010) Preparation of KF/CaO nanocatalyst and its application in biodiesel production from Chinese tallow seed oil. Fuel 89(9):2267–2271

Williams DN, Ehrman SH, Holoman TRP (2006) Evaluation of the microbial growth response to inorganic nanoparticles. J Nanobiotechnol 4(1):3

Xiang B, Wang P, Zhang X, Dayeh SA, Aplin DPR, Soci C, Yu D, Wang D (2007) Rational synthesis of p-type zinc oxide nanowire arrays using simple chemical vapor deposition. Nano Lett 7(2):323–328. https://doi.org/10.1021/nl062410c

Xie W, Ma N (2009) Immobilized lipase on Fe_3O_4 nanoparticles as biocatalyst for biodiesel production. Energy Fuel 23(3):1347–1353

Xu H, Miao X, Wu Q (2006) High quality biodiesel production from a microalga Chlorella protothecoides by heterotrophic growth in fermenters. J Biotechnol 126(4):499–507

Yacob AR, Mustajab M, Samadi NS (2009) Calcination temperature of nano MgO effect on base transesterification of palm oil. World Acad Sci Eng Technol 56:408–412

Yang G, Tian-Wei T, Kai-Li N, Fang W (2006) Immobilization of lipase on macroporous resin and its application in synthesis of biodiesel in low aqueous media. Chin J Biotechnol 22(1):114–118

Yu J, Yu X (2008) Hydrothermal synthesis and photocatalytic activity of zinc oxide hollow spheres. Environ Sci Technol 42(13):4902–4907. https://doi.org/10.1021/es800036n

Yu J, Wang W, Cheng B, Su B-L (2009) Enhancement of photocatalytic activity of Mesporous TiO2 powders by hydrothermal surface fluorination treatment. J Phys Chem C 113(16):6743–6750. https://doi.org/10.1021/jp900136q

Zhang XL, Yan S, Tyagi RD, Surampalli RY (2013) Biodiesel production from heterotrophic microalgae through transesterification and nanotechnology application in the production. Renew Sust Energ Rev 26:216–223. https://doi.org/10.1016/j.rser.2013.05.061

Zhao X, Ohkohchi M, Inoue S, Suzuki T, Kadoya T, Ando Y (2006) Large-scale purification of single-wall carbon nanotubes prepared by electric arc discharge. Diam Relat Mater 15(4–8):1098–1102

Zhao J, Wei L, Yang Z, Zhang Y (2012) Continuous and low-cost synthesis of high-quality multi-walled carbon nanotubes by arc discharge in air. Physica E 44(7):1639–1643. https://doi.org/10.1016/j.physe.2012.04.010

Zhong YL, Swager TM (2012) Enhanced electrochemical expansion of graphite for in situ electrochemical functionalization. J Am Chem Soc 134(43):17896–17899

Zhou C, Krueger AB, Barnard JG, Qi W, Carpenter JF (2015) Characterization of nanoparticle tracking analysis for quantification and sizing of submicron particles of therapeutic proteins. J Pharm Sci 104(8):2441–2450. https://doi.org/10.1002/jps.24510

Nanotechnology: An Application in Biofuel Production

6

Veer Singh, Vipul Kumar Yadav, and Vishal Mishra

Abstract

Nanotechnology has been extensively studied for biofuel production. Fossil fuels are available in a limited amount on earth; therefore, it is needful to build up a new alternative to fulfill the requirement of energy sources in the near future. Biofuel such as biodiesel, biohydrogen, biomethanol, and bioethanol may replace the need for fossil fuel in the future. These fuels are produced from cost-effective and renewable sources such vegetable oils that come from different biological sources such as algal biomass, microbial biomass, and plant biomass. Nanomaterials have emerging applications in biofuel production due to their unique structural behavior such as small size (nanoscale size) of nanomaterials. These nanomaterials enhance biofuel production by improving the biosynthesis pathways. This manuscript mainly focuses on the various types of nanoparticles, synthesis methods of nanoparticles, and the emerging application of nanoparticles in biofuel production.

Keywords

Nanoparticles · Biofuel · Biodiesel · Biohydrogen · Bioethanol

6.1 Introduction

Fossil fuels such as coal, petroleum ions, and liquid petroleum gases are available in nature in a limited amount because these sources of energy are nonrenewable. These fuels have towering cost and cause several types of environmental pollutions such as air, soil, and water pollution. When fossil fuels burn, it produces energy and

V. Singh · V. K. Yadav · V. Mishra (✉)
School of Biochemical Engineering, IIT (BHU), Varanasi, India
e-mail: vishal.bce@itbhu.ac.in

M. Srivastava et al. (eds.), *Nanomaterials in Biofuels Research*,
Clean Energy Production Technologies, https://doi.org/10.1007/978-981-13-9333-4_6

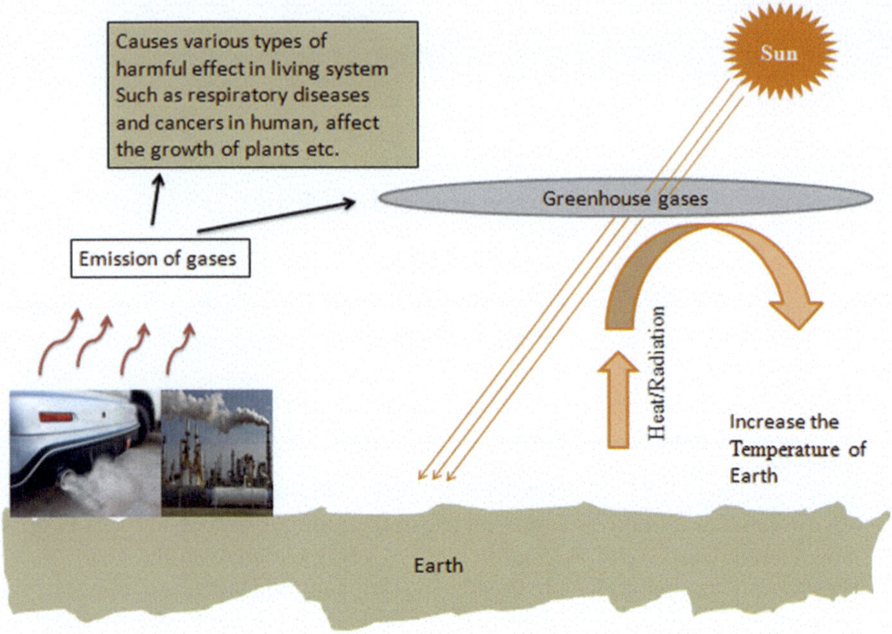

Fig. 6.1 Diagrammatic representation of the emission of gases during the burning of fossil fuel and the role of these gases in environmental pollution

various types of gases including carbon dioxide, sulfur dioxide, and other various types of gases. Carbon dioxide is the main greenhouse gas which maintains the earth temperature. Increase in the level of CO_2 causes increase in the temperature of the atmosphere which affects the distribution of living organisms (Fig. 6.1). Air pollution is main component of pollution which causes the depletion of the ozone layer. The ozone layer protects the earth from ultraviolet radiations. UV radiation causes various types of health effects in humans and other living organisms. Water pollution by these fuel waste materials is responsible for the addition of many types of toxic substances in the water sources which affect the distribution and life of living aquatic organisms. These wastes generated from these fuels also cause the loss of fertility in the soil. The availability of fossil fuel is limited, and after a certain time, these sources may be exhausted, making biofuel a potential candidate as an energy source. Therefore, biofuel production is very needful in the future as an alternative energy source (Palaniappan 2017).

Biofuel is referred to as energy sources generated from different types of biomass. These biomasses come from living organisms such as algae, bacteria, plants, and agricultural waste. Biofuel can terminate the demand for the fossil fuel in the future and minimize the environmental effects generated during the combustion of fossil fuel (Voloshin et al. 2015; Dragone et al. 2010; Demirbas 2009).

A recent approach for biofuel production is based on various types of microorganisms such as microalgae or cyanobacteria. Microalgae contain a huge amount of lipid which is used as raw material for biofuel production. The past few decades also

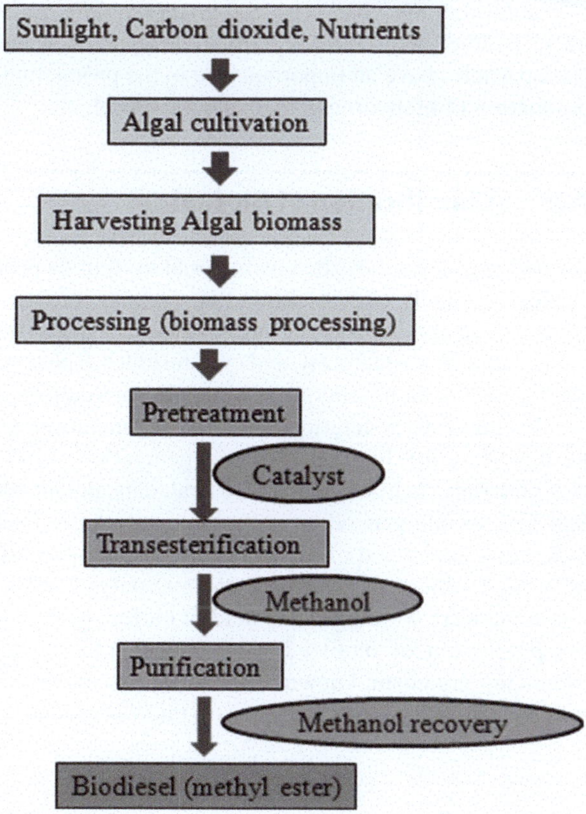

Fig. 6.2 Biodiesel production from algal cultivation

represent the production of biofuel from plant materials. Present research mainly focuses on the production of biofuel from microalgae. Plants and algae can synthesize sugar molecule by the process of photosynthesis. Photosynthesis is a process in which green plant and algae synthesize the carbon-based materials in the presence of environmental CO_2 and sunlight. Photosynthetic products require the growth of plant and production of biomass (Fig. 6.2). Plant and algal biomass are used as raw sources for the production of biofuel (Demirbas 2009; Heiman 2016; Babel and Kurniawan 2004; Barakat 2011).

Nanoscience and nanotechnology have also a great potential for remediation of various types of pollutants from contaminated water (Eccles 1999; Leung et al. 2000; Kurniawan et al. 2005; Rickerby and Morrison 2007; Brumfiel 2003). Nanostructures have more efficiency and fast adsorption rate compared to other conventional methods. Nanoparticles have a broad surface area due to the nano-range (small) particle size. Varieties of nanoparticles have been considered as an inexpensive and environmentally safe adsorbent material for adsorption of several types of contaminants such as toxic metallic pollutants and azo dyes. The few applications of nanotechnology in bioenergy production and wastewater treatment are given in Fig. 6.3 (Sadegh et al. 2016; Theron et al. 2008; Dil et al. 2017; Savage and Diallo 2005).

Nanotechnology has prospective applications in biofuel manufacturing industries. Biofuel is the best alternative option of energy source in the future. Nanoparticles have an important role in the production of biofuel such as enhanced production of biohydrogen methane, biodiesel, etc.

6.2 Classification of Biofuel

Biofuel is classified by the raw material used in its production and by its efficiency. Biofuel is mainly classified into two groups, namely, primary and secondary. The primary category of biofuel is classified as a natural biofuel source derived from plants, animal waste, forest, and crop residues. The second type of biofuel is directly synthesized from microorganism such as microalgae. The second type of biofuel is divided into four generations (the fourth generation is basically correlated with the third generation) (Fig. 6.3). The first generation of biofuel is produced from starch-rich materials such as sugarcane, wheat, oats, potato, sweet potato, corn, and animal fats. The second generation of biofuel is produced from various types of plant species, grass, and wood of plant. The third generation of biofuel is an advanced type of biofuel which is produced from microalgae and few microbial species. Many researchers are working in the field of biofuel production and are trying to improve the production of biofuel by the genetic modification in third-generation fuel-producing organism, known as the fourth-generation biofuel. The yield of biofuel production depends on the type of microorganism which has been selected for

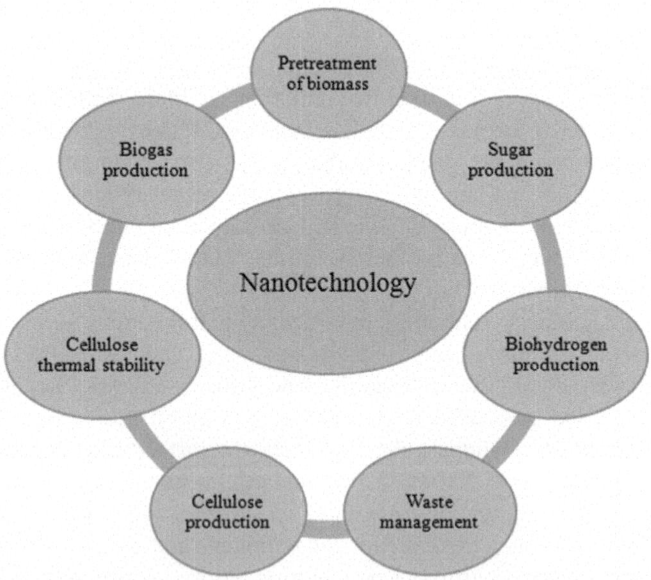

Fig. 6.3 Diagrammatic representation of application of nanotechnology in various fields

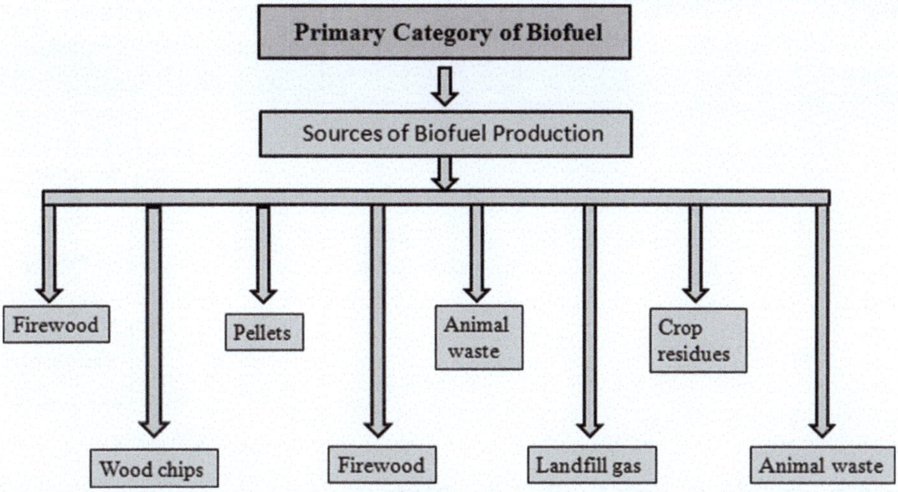

Fig. 6.4 (**a**) The primary category of biofuel and their sources of production. (**b**) The schematic diagram represents the second category of biofuel, generations of biofuel, and sources of biofuel production

biofuel production and optimum conditions required for the growth of the microorganism. Media compositions also affect the production of biofuel. Genetic engineering techniques are also used for the modification in genetic level which may be responsible for enhancing the production of biomass. These biomasses are used for the production of biofuel (Dragone et al. 2010; Abdelaziz et al. 2013). Various types of biofuels are shown in Fig. 6.4 (a and b).

6.3 Production of Biofuel

Biomass such as plant, algae, microbial biomass, and other waste materials is considered as suitable raw material for biofuel production. A variety of algal and microbial (fungal and bacterial) species are used for the production of bioethanol and biohydrogen. These products are partially mixed with fossil fuel (4–5%) in diesel and petrol engines. In the next few years, biofuel may be used as an additive in petroleum industry. Hence, biofuel can be considered as the best alternative of petroleum by-products (Ong and Bhatia 2010; Chanakya et al. 2013; Swain 2014). Pacific biodiesel is the first biodiesel plant commissioned in the United States in 1996, and this plant mainly focuses on biodiesel production from cooking oils. Biodiesel production was boosted after 2001 due to the price hike of petroleum oils. Biodiesel has many advantages like emission of less toxic waste, inexpensive, and independence from fossil fuels (Chanakya et al. 2013; Swain 2014). Various types of techniques and unit operations involved in biodiesel production (Fig. 6.5) are shown in Fig. 6.5.

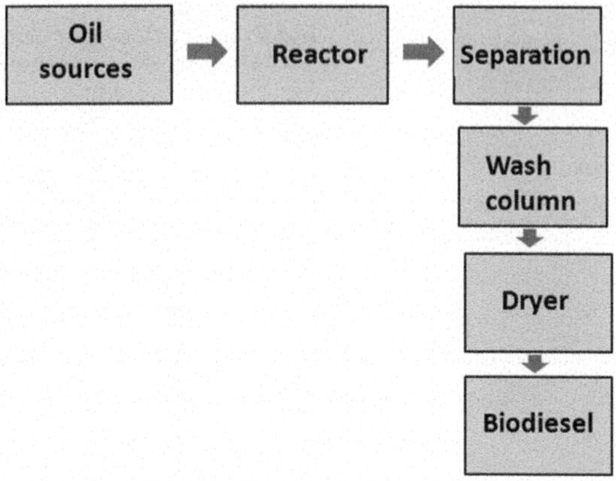

Fig. 6.5 Schematic diagram represents the stages of biodiesel production

6.3.1 Production Techniques for Biofuel

Biofuel production methods depend on the materials used for the production of biofuel, reactor types, reactor volume, and reactor parameters such as temperature, pH, mixing, etc. A variety of biological materials and production methods have been used for the production of biofuel processes. A few examples have been given in Table 6.1. Hydrotreatment is a technique in which vegetable oil and other lipid materials are used for biodiesel production purposes. It is a complex reaction process in which various types of reaction are involved (Swain 2014; Torres-Ortega et al. 2017).

6.3.2 Algal Biodiesel

Algae are the photosynthetic organisms which fix gaseous CO_2 in the form of algal biomass (Maeda et al. 1995). Algal biomass is considered as raw material for the bioenergy production. Microalga is the best alternative option for the production of diverse types of biofuel such as biohydrogen, bioethanol, biodiesel, etc. due to its high yield, fast growth rate, and high lipid and sugar contents and is easy to grow (cultivated on both arable and nonarable land) (Reddy et al. 2008). However, biodiesel production from microalgae is much expensive due to the tremendous requirement of energy source and maintenance of the growth condition for the cultivation of algal species. Algal biofuel production is a very complex process. A variety of mechanisms are involved in the production process (Fig. 6.6). Many researchers are working for the reduction of production cost and enhancement of the production of biofuel (Chanakya et al. 2013; Harayama 2012).

Table 6.1 Techniques of biofuel production and used raw materials for production processes

Raw material	Technique	Product	References
Vegetable oils	Transesterification	Biodiesel	Alptekin et al. (2014)
Animal fat waste	Transesterification, gas chromatography, bench-scale reactor	Ethylic biodiesel	CunhaJr et al. (2013a)
Lard oil	Transesterification, methanolysis	Biodiesel	Ezekannagha et al. (2017)
Microalgae	Full-scale photobioreactor	Biodiesel	Faried et al. (2017)
Bacteria (*Clostridium tyrobutyricum* JM1)	Immobilization and packed bed bioreactor	Biohydrogen	Singh et al. (2010)
Mixed culture (*Clostridium butyricum* and *Rhodopseudomonas faecalis* RLD-53)	Combined dark and photofermentation process	Biohydrogen	Singh et al. (2010)
Lignocellulosic materials and animal manure	Pretreatment techniques such as milling and extrusion and followed by microbial reaction in batch and continuous reactor	Biogas	Alptekin et al. (2014) and CunhaJr et al. (2013a)

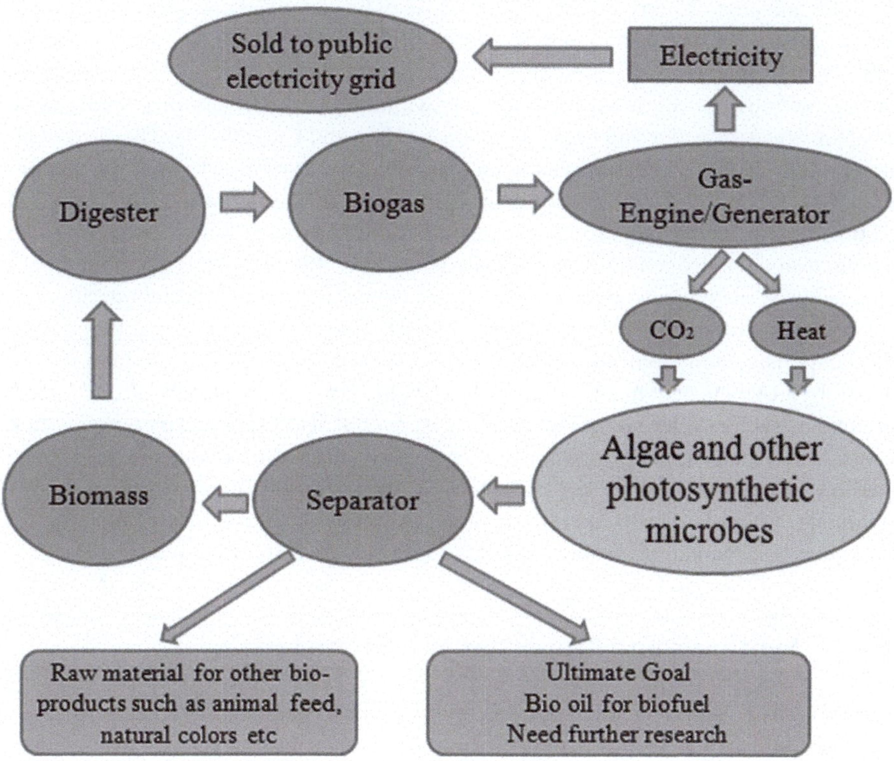

Fig. 6.6 The schematic diagram represents the processing of algae for the production of biofuel

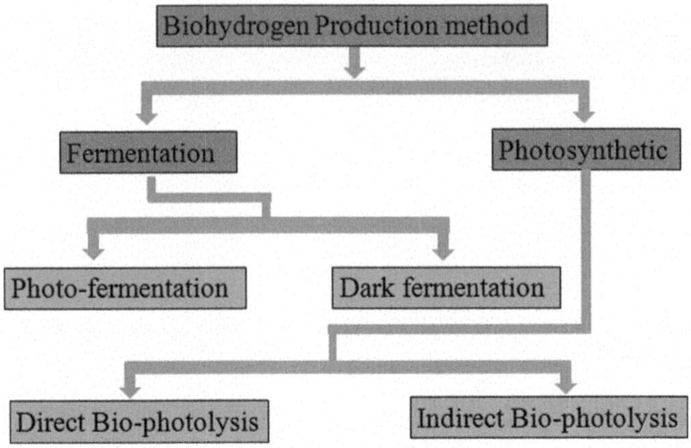

Fig. 6.7 The schematic diagram represents the various methods of biohydrogen production

6.3.3 Biohydrogen

Biohydrogen can be the best option of energy source in the future. It is an eco-friendly and cost-effective biofuel. Biohydrogen has many advantages compared to other fuels. It has high efficiency due to its high octane number. Nowadays, about 99% of hydrogen is generated from fossil fuels (Shaishav et al. 2013; Jo et al. 2006). Various conventional methods are available for hydrogen production such as electrolysis of water and coal gasification (Hsia and Chou 2014). These conventional methods have few disadvantages like requirement of high temperature (more than 840 °C) and environmentally unsafe processes (Shaishav et al. 2013; Hsia and Chou 2014). Electrolysis of water is an eco-friendly process but electricity for the production of hydrogen adds extra cost. Therefore, this process is possible where electricity is inexpensive because electricity covers the major cost (about 80%) of hydrogen production (Karthic and Shiny 2012). This manuscript is mainly focused on inexpensive and environmentally safe methods of biohydrogen production.

Biohydrogen is generated during the photolytic reaction in photosynthetic microorganisms such as microalgae and plants (Fig. 6.7). It also generated by fermentation process in anaerobic microorganisms. Various types of microorganisms have been studied for biohydrogen production such as *Chlamydomonas moewusii*, *Scenedesmus obliquus*, *Rhodobacter sphaeroides* (photosynthetic bacteria), and *Enterobacter aerogenes* (fermentative bacteria) (Fabiano and Perego 2002; Kars et al. 2006).

6.4 Synthesis and Properties of Nanomaterials

It is most important to know about the synthesis mechanism and basic properties of nanomaterials before discussing their use in the production of biofuel and waste management processes. A range of nanomaterials are available such as nanoparticles, nanotubes, nanosheets, etc. Two major methods have been used for the formation of different types of nanomaterials. The first is the top-down approach. In this method, bulk materials such as iron, gold, etc. are broken down into nanoscale-size particles. The second method is the down-up approach. In this approach, small-size particles come together and form nanomaterials (Biswas et al. 2012; Serrano et al. 2009).

Other methods are also used for the synthesis of nanomaterials. These methods include coprecipitation, laser ablation, self-assembly, and phase separation methods. These methods produce various types of nano-size particles. Nanomaterials produced as dry powder materials are used for suitable application (Yu and Xie 2012). A nanoporous gold particle from Au and Ag alloy foil dealloys in nitric acid at 25 °C (Qiu et al. 2008). This is considered as dealloying and thermal alloying method, respectively. In the coprecipitation method, nanomaterials are synthesized by precipitation in the presence of gravitational forces and magnetic field (Kalantari et al. 2012). Carbon nanotubes (CNTs) are synthesized by various methods. The arc discharge method is one of the best methods for the synthesis of good-quality nanotubes which require electricity and carbon sources (De Volder et al. 2013). The chemical vapor deposition method requires heat and carbon source for the synthesis of nanotubes also (Saifuddin et al. 2013). The electro-spinning technique is used for the production of nanofibers. This method requires a polymer solution, high-voltage electric supply, spinneret, and collecting plates (Bhardwaj and Kundu 2010; Sill and Recum 2008).

Nanosheets mainly are produced by thermal exfoliation technique. Nanographene sheets produced from graphite powder react with hard acids such as HNO_3, H_2SO_4, and HCl and heating at high temperature (1050 °C). Argon gas is also required in this process (Kishore et al. 2012).

After the synthesis of nanomaterials, another important process is surface functionalization. Surface functionalization is a process which enhances the property of nanomaterials. This process also provides many other properties to nanomaterials such as stability and biocompatibility. It also enhances the binding capacity of nanomaterials with various types of enzymes (Pavlidis et al. 2010).The functionalization method requires addition of functional groups with nanomaterials. The resources used for functionalization are biopolymers such as gelatin, carbohydrates, polypeptides, lipids, etc. A variety of synthetic polymeric materials are also used as functionalization materials (Wang et al. 2012). Adding the functional groups with nanomaterials responsible for the enhancement in the properties of nanomaterials provides different surface charges, provides a link between functional groups of more than one nanomaterial, and minimizes the pore entrance size which is also applicable for enzyme immobilization (Lee et al. 2009).

Nanomaterials have many different properties compared to other bulk materials. They have different electrical, heating conductivity, light reflection, and catalytic properties. These advanced properties of nanomaterials are responsible for the various applications of nanomaterials in drug delivery system, environmental engineering, and bioenergy production (Savolainen et al. 2016).

6.5 Application of Nanotechnology in Biofuel Production

Nanoparticles enhance the production of biofuel by increasing the catalytic reaction in the production process. Various types of nanomaterials have been reported for biofuel production (Ramsurn and Gupta 2013; Gordon and Seckbach 2012). This section of the chapter is mainly focused on the application of several types of nanoparticles and their role in the production of biodiesel, biohydrogen, bioethanol, biogas, and bioethanol.

6.5.1 Biohydrogen Production

Biohydrogen is produced from various types of microorganisms such as anaerobic bacteria, microalgae, etc. These microorganisms generate molecular hydrogen through various metabolic routes (Das et al. 2008). The biohydrogen production is extremely dependent on several ranges of parameters such as pH, temperature, retention time, and media compositions (Lukajtis et al. 2018; Nagaragan et al. 2017). It is also reported that nanoparticles enhance the production of biohydrogen in the microbial system by increasing the electron transfer rate in the microbial cell (Serrano et al. 2009; Ali et al. 2017; Zhang and Shen 2007).

6.5.1.1 Dark Fermentation for Production of Biohydrogen

Dark fermentative biohydrogen production is an inexpensive and eco-friendly process in which production of biohydrogen depends on various renewable feedstock and microbial species (Nagaragan et al. 2017). Different types of nanomaterials have application in the field of biohydrogen production. Gold (Au) nanoparticles (5 nm) enhance the substrate utilization efficiency by 56% and increase the yield of biohydrogen production up to 46% (Zhang and Shen 2007). Gold nanoparticles enhance the biohydrogen production due to their small size and large surface area which are responsible for the binding of microbial cells to active sites. Gold nanoparticles also enhance the activity of enzymes of biohydrogen synthesis machinery which is responsible for the production of biohydrogen.

Nanoparticles are beneficial to the microorganism when it uses the optimum concentration of nanoparticles. High concentration of silver nanoparticles inhibits the growth of microorganisms and decreases the production of biohydrogen. Silver nanoparticles are reported as an inducer for microbial biohydrogen production. It increases severalfolds the hydrogen production yield. Addition of silver nanoparticles enhances the substrate utilization up to 62% which is responsible for enhancing the biohydrogen production up to 2.48 mol H_2/mole substrate. These nanoparticles reduce the lag phase of bacterial and algal growth and activate acetic reaction. The acetic reaction is the main pathway for biohydrogen production (Zhao et al. 2013; Yang and wang 2018).

Silica nanoparticles have emerging application in the dark fermentation process for the production of biohydrogen. Silica nanoparticles combine with iron oxide and produce nanocomposite ($Fe_3O_4@SiO_2$) (detailed synthesis of composites is explained in Fig. 6.8). These newly synthesized nanocomposites/nanoconjugates

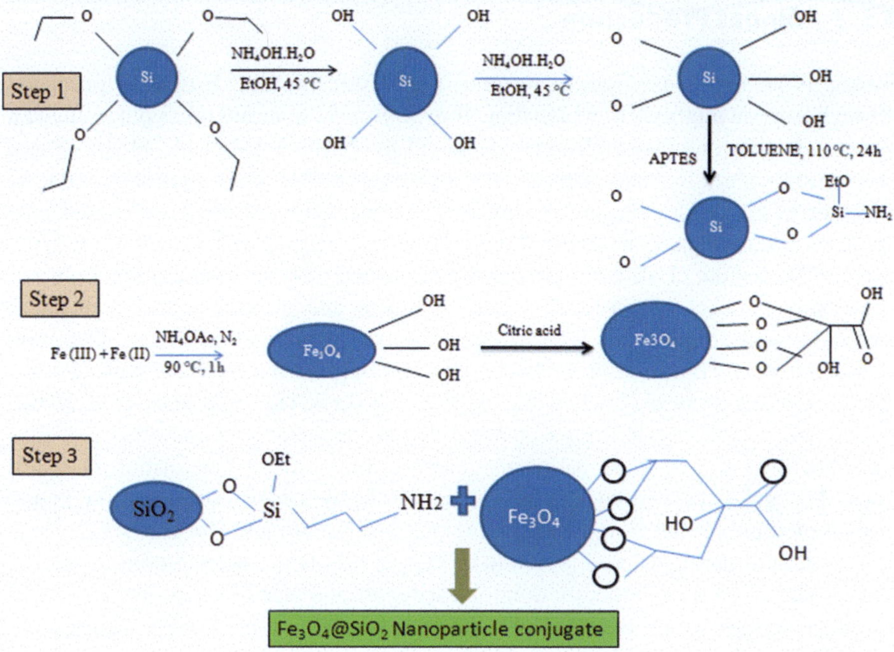

Fig. 6.8 The schematic diagram represents the synthesis of Fe3O4@SiO2 nanoparticle conjugate

have emerging application in biohydrogen synthesis due to more catalytic activity and stability. These nanomaterials are stable at high temperature and show a low level of toxic effects (Venkata et al. 2008; Abbas et al. 2014; Kunzmann et al. 2011).

6.5.1.2 Biohydrogen Production by the Photofermentation Process

Nanoparticles enhance the production of biohydrogen in photosynthetic microorganisms. The photosynthetic microorganisms used for the production of biohydrogen are microalgae, photosynthetic bacteria, etc. Addition of nanoparticles in growth medium enhances the growth, physiological process, photosynthetic efficiency, protein synthesis, lipid synthesis, and nitrogen metabolism in microbial species (Eroglu et al. 2013). Nanoparticles also enhance the activity of various enzymes which are involved in various applications such as glutamate dehydrogenase, glutamine synthetase production, etc. These microbial enzymes are also required for the metabolic activity of microorganisms (Yang et al. 2006; Mishra et al. 2014).

A variety of nanoparticles have been applied for biohydrogen production. The optimum concentration of Ag and Au nanoparticles enhances the photosynthetic activity of *Chlorella vulgaris* (Eroglu et al. 2013). Zerovalent nanoparticles also increase the activity of biohydrogen production by transfer of electrons from Fe° (Eq. 6.1) (Liu et al. 2012).

$$Fe + 2H^{+} \rightarrow Fe^{2+} + H_2 \tag{6.1}$$

6.5.2 Biogas Production

Biogas is an inexpensive and environmentally safe biofuel. Biogas is produced through various methods in which digestion of organic material takes place due to a variety of microorganisms (Romero-Guiza et al. 2016; Aryal et al. 2018). Biogas production occurs in many steps. The four important steps are hydrolysis (breakdown process), acidogenesis (alcohol production), acetogenesis (acetic acid production), and methanogenesis (production of methane gas) (Mao et al. 2015; Buitron et al. 2014; Sekoai et al. 2016). The hydrolysis step of biogas production contains conversion of large biomolecules (carbohydrates and proteins) into small sugar molecules and amino acids. The acidogenesis step produces hydrogen, CO_2, and alcohols from the products of the first hydrolysis step. Various types of pathways are involved in this process (Lochynsha and Frankowski 2018; Angelidaki et al. 2018). Acetogenesis pathways of biogas synthesis produce acetic acid and other constituents. The final step of these pathways generates methane gas by various methanogens. The growth of methanogens requires low hydrogen concentration in the growth medium (Khan et al. 2017; Leonzio 2016; Kadam and Panwar 2017).

Addition of nanoparticles has led to emerging results in the anaerobic condition. Nanoparticles act as an electron donor or acceptor and also enhance the activity of various enzymes which is responsible for biogas production (Romero-Guiza et al. 2016; Liu et al. 2013; Yang et al. 2013; Abdelsalam et al. 2016). Nanomaterials enhance the substrate utilization rate at various steps. Su et al. (Su et al. 2013) reported that adding the zerovalent iron nanoparticles ensures to improve the production of biogas from waste materials (Su et al. 2013; Karri et al. 2005).

6.5.3 Biodiesel Production

Biodiesel has much promising application in the future due to the emission of fewer pollutants, is eco-friendly, and is produced from edible as well as nonedible oils (De Araujo et al. 2013; Mohammadshirazi et al. 2014). Nonedible oils are considered as microbial oil and plants oil converted into biodiesel through various mechanisms. This process is considered as very beneficial because these nonedible oils are waste and are present in huge amount in nature (Rathore and Madras 2007; Bankovic Ilic et al. 2012).

Nanomaterials have promising results in biodiesel production. Nanoparticles enhance the catalytic reaction during transesterification, thereby improving the production of biodiesel (Chen et al. 2018; Lee et al. 2015). Chen et al. (2018) have reported that the biodiesel production yield was enhanced in the presence of Fe_3O_4/ZnMg(Al)O nanoparticles (Chen et al. 2018). Ca^{2+}doping in nanomaterials such magnetic nanoferrites has an emerging effect in biodiesel production. This process enhances the biodiesel production yield up to 85% from soybean cooking oils (Dantas et al. 2017).

Various researchers have reported that biodiesel production is also enhanced by using carbon-based nanoparticles. Carbon-based nanoparticles have been used as catalyst in various reactions such as transesterification of vegetable oils (Konwar et al. 2014; Zhang et al. 2014). Ion nanoparticles were incorporated on carbon nanotubes which are responsible for enzyme immobilization. Immobilization improves the efficiency of biofuel production and reduces the production cost. Nanostructure provides emerging immobilization support due to nanoscale size and large surface area. Microbial enzymes such as lipase from *Pseudomonas cepacia* are immobilized on the surface of nanoparticles and enhance the production of biofuel due to enhanced transesterification reaction (Goh et al. 2012).

Functionalization of the nanoparticle process also increases the production of biodiesel. Ion-silica nanoconjugates such as Fe_3O_4/SiO_2 have emerging application in biodiesel production. Nanoconjugates can increase the production of biodiesel up to 97.1%. In this process, various types of cooking and algal oils have been used. Algal oils have a high yield of production in the presence of these ion-silica nanocomposites (Chiang et al. 2015).

6.5.4 Bioethanol Production

It is one of the most important alternatives to fossil fuel. It is considered as an inexpensive and eco-friendly biofuel. Bioethanol has various advantages such as high evaporation enthalpy and high octane number (Saini et al. 2015; Sarkar et al. 2012; Limayen and Ricke 2012). Currently, bioethanol is produced from edible and nonedible vegetable oils, waste materials, and algal and bacterial biomass. Microalgae produced a large amount of bioethanol. Genetic engineering in microorganisms is also applied to improve production of bioethanol (Kuhad and Singh 1993; Balan 2014).

Various types of nanomaterials have emerging application in the production of bioethanol. Cherian et al. (2015) investigated that MnO_2 nanoparticles increase the bioethanol production by using sugarcane leaves. It catalyzes the mechanism of bioethanol production during various steps. In this process, sugarcane leaves are converted into bioethanol. MnO_2 nanoparticles provide the broad surface area which is responsible for the binding of enzymes on their active sites and causes the improvement in ethanol production (Cherian et al. 2015).

Bioethanol production using immobilized microbial cells on the surface nanoparticles has been reported in many literatures. Ivanova et al. (Ivanova et al. 2011) have investigated that yeast cells produced additional ethanol when immobilized on the surface of magnetic nanoparticles (Ivanova et al. 2011; Lee et al. 2011).

Lee et al. (2011) also reported that immobilized *S. cerevisiae* cells are responsible for more ethanol production. *S. cerevisiae* cells were immobilized on calcium alginate. These immobilized cells enhanced the bioethanol production yield up to 100% but production yield of suspended cells was 88%. In immobilization process, the cells can be reused up to several times and there are less chances of contamination. Therefore, immobilization is an emerging method used in microbial bioethanol production. Its application with nanotechnology can give better results at commercial-level bioethanol production (Galazzo and Bailey 1990).

6.6 Conclusion

This concludes that nanomaterials may play an important part in the enhancement of bioenergy production. Nanomaterials have promising application in biofuel production due to their small size, large surface area-to-volume ratio, and good catalytic properties which are responsible for enhancing the production of various types of biofuel such as biohydrogen, biodiesel, and bioethanol. This manuscript has described the synthesis of various types of nanomaterials and their application to enhance the production of biofuel. Recent trends on the use of nanotechnology in biofuel production were discussed in this chapter. From an up-to-date study in nanotechnology, the authors conclude that these approaches can be used in biofuel industries to enhance biofuel production and minimize the cost of biofuels due to unique structural properties of nanomaterials.

Acknowledgment The authors of this manuscript would like to give thanks to the IIT (BHU), Varanasi, for providing financial and technical support.

References

Abbas M, Rao B, Nazrul Islam M, Naga S, Takahashi M, Kim C (2014) Highly stable silica encapsulating magnetic nanoparticles (Fe_3O_4/SiO_2) synthesized using single surfactanting-polyol process. Ceram Int 40:1379–1385

Abdelaziz AEM, Leite GB, Hallenbeck PG (2013) Addressing the challenges for sustainable production of algal biofuels:II. Harvesting and conversion to biofuels. Environ Technol 34:1807–1836

Abdelsalam E, Samer M, Attia Y, Abdel-Hadi M, Hassan H, Badr Y (2016) Comparison of nanoparticles effects on biogas and methane production from anaerobic digestion of cattle dung slurry. Renew Energy 87:592–598

Ali A, Mahar R, Soomro R, Sherazi S (2017) Fe_3O_4 nanoparticles facilitated anaerobic digestion of organic fraction of municipal solid waste for enhancement of methane production. Energy Sources, Part A 37:1815–1815

Alptekin E, Canakci M, Sanli H (2014) Biodiesel production from vegetable oil and waste animal fats in a pilot plant. Waste Manag 34:2146–2154

Angelidaki I, Treu L, Tsapekos P, Luo G, Campanaro S, Wenzel H (2018) Biogas upgrading and utilization: current status and perspectives. Biotechnol Adv 36:452–466

Aryal N, Kvist T, Ammam F, Pnat D, Ottosen LDM (2018) An overview of microbial biogas enrichment. Bioresour Technol 264:359–369

Babel S, Kurniawan TA (2004) Cr(VI) removal from synthetic wastewater using coconut shell charcoal and commercial activated carbon modified with oxidizing agents and/or chitosan. Chemosphere 54:951–967

Balan V (2014) Current challenges in commercially producing biofuels from lignocellulosic biomass. ISRN Biotechnol 2014:1–31

Bankovic Ilic IB, Stamenkovic OS, Veljkovic VB (2012) Biodiesel production from non-edible plant oils. Renew Sust Energ Rev 16:3621–3647

Barakat MA (2011) New trends in removing heavy metals from industrial waste water. Arab J Chem 4:361–377

Bhardwaj N, Kundu SC (2010) Electrospinning:a fascinating fiber fabrication technique. Biotechnol Adv 28:325–347

Biswas A, Bayer IS, Biris AS, Wang T, Dervishi E, Faupel F (2012) Advances in top-down and bottom-up surface nanofabrication: techniques, applications & future prospects. Adv Colloid Interf Sci 170:2–27

Brumfiel G (2003) Nanotechnology: a little knowledge. Nature 424:246–248

Buitron G, Kumar G, Martinez-Arce A, Moreno G (2014) Hydrogen and methane production via a two-stage processes (H_2-SBR+CH_4-UASB) using tequila vinasses. Int J Hydrog Energy 39:19249–19255

Chanakya HN, Mahapatra DM, Sarada R, Abitha R (2013) Algal biofuel production and mitigation potential in India. Mitig Adapt Strateg Glob Chang 18:113–136

Chen Y, Liu T, He H, Liang H (2018) Fe_3O_4/ZnMg(Al)O magnetic nanoparticles for efficient biodiesel production. Appl Organomet Chem 32:1–10

Cherian E, Dharmendirakumar M, Baskar G (2015) Immobilization of cellulose onto MnO_2 nanoparticles for bioethanol production by enhanced hydrolysis of agricultural waste. Chin J Catal 36:1223–1229

Chiang Y, Dutta S, Chen C, Huaang Y, Lin KWJ (2015) Functionalized Fe3O4@ silica core-shell nanoparticles as microalgae harvester and catalyst for biodiesel production. ChemSusChem 8:789–794

CunhaJr A, Feddern V, Pra MCD, Higarashi MM, de Abreu PG, Coldebella A (2013a) Synthesis and characterization of ethylic biodiesel from animal fat wastes. Fuel 105:228–234

CunhaJr A, Feddern V, Pra MCD, Higarashi MM, de Abreu PG, Coldebella A (2013b) Synthesis and characterization of ethylic biodiesel from animal fat wastes. Fuel 105:228–234

Dantas J, Leal E, Mapossa A, Cornejo D, Costa A (2017) Magnetic nanocatalysts of $Ni_{0.5}Zn_{0.5}Fe_2O_4$ doped with Cu and performance evaluation in transesterification reaction for biodiesel production. Fuel 191:463–471

Das D, Khanna N, Veziroglu NT (2008) Recent development in biological hydrogen production processes. Chem Ind Chem Eng Q 14:57–67

De Araujo CDM, De Andade CC, De Souza E, Silva E, Dupas FA (2013) Biodiesel production from used cooking oil: a review. Renew Sust Energ Rev 27:445–452

De Volder MFL, Tawfick SH, Baughman RH, Hart AJ (2013) Carbon nanotubes: present and future commercial applications. Science 339:535–539

Demirbas A (2009) Potential, economic and environmental impacts of biofuels: a review. Appl Energy 86:S108–S117

Dil EA, Ghaedi M, Asfaram A (2017) The performance of nanorods material as adsorbent for removal of azo dyes and heavy metal ions: application of ultrasound wave, optimization and modeling. Ultrason Sonochem 34:792–802

Dragone G, Femande B, Vicente AA, Teixeira JA (2010) Third generation biofuel from microalgae. In: Mendez-Vilas A (ed) Current research, technology and education topics in applied microbiology and microbial biotechnology. Formatex; 2010, pp 1355–1366

Eccles H (1999) Treatment of metal-contaminated wastes: why select a biological process? Trends Biotechnol 17:462–465

Eroglu E, Eggers PK, Winslade M, Smith SM, Raston CL (2013) Enhanced accumulation of microalgal pigments using metal nanoparticles solutions as light filtering devices. Green Chem 15:3155–3159

Ezekannagha CB, Ude CN, Onukwuli OD (2017) Optimization of the methanolysis of lard oil in the production of biodiesel with response surface methodology. Egypt J Pet 26:1001–1011

Fabiano B, Perego P (2002) Thermodynamic study and optimization of hydrogen production by *Enterobacter aerogenes*. Int J Hydrogen Energy 27:149–156

Faried M, SamerM AE, Yousef RS, Attia YA, Ali AS (2017) Biodiesel production from microalgae: processes, technologies and recent advancements. Renew Sust Energ Rev 79:893–913
Galazzo J, Bailey J (1990) Growing *Saccharomyces cerevisiae* in calcium-alginate beads induces cell alterations which accelerates glucose conversion to ethanol. Biotechnol Bioeng 36:417–426
Goh WJ, Makam VS, Hu J, Kang L, Zheng M, Yoong SL, Udalagama CNB, Pastorin G (2012) Iron oxide filled magnetic carbon nanotube-enzyme conjugates for recycling of amyloglucosidase: toward useful applications in biofuel production process. Langmuir 28:16864–16873
Gordon R, Seckbach J (2012) The science of algal fuels. Springer, Dordrecht
Harayama S (2012) Biofuel production in algae: towards its commercialization. Research report-education course, Faculty of Science and Engineering, Chao University, Japan
Heiman K (2016) Novel approaches to microalgal and cyanobacterial cultivation for bioenergy and biofuel production. Curr Opin Biotechnol 38:183–189
Hsia SY, Chou YT (2014) Optimization of biohydrogen production with biomechatronics. J Nanomater 2014:1–11
Ivanova V, Petrova P, Hristov J (2011) Application in the ethanol fermentation of immobilized yeast cells in matrix of alginate/magnetic nanoparticles, on chitosan-magnetic microparticles and cellulose-coated magnetic nanoparticles. Int Rev Chem Eng 3:289–299
Jo JH, Lee DS, Park JM (2006) Modeling and optimization of photosynthetic hydrogen gas production by green alga *Chlamydomonas reinhardtii* in sulfur-deprived circumstance. Biotechnol Prog 22:431–437
Kadam R, Panwar NL (2017) Recent advancement in biogas enrichment and its applications. Renew Sust Energ Rev 73:892–903
Kalantari M, Kazemeini M, Tabandeh F, Arpanaei A (2012) Lipase immobilisation on magnetic silica nanocomposite particles: effects of the silica structure on properties of the immobilised enzyme. J Mater Chem 22:8385
Karri S, Sierra-Alvarez R, Field J (2005) Zero valent iron as an electron-donor for methanogenesis and sulfate reduction in anaerobic sludge. Biotechnol Bioeng 92:810–819
Kars G, Gunduz U, Yücel M, Turker L, Eroglu I (2006) Hydrogen production and transcriptional analysis of nifD, nifK and hupS genes in *Rhodobacter sphaeroides* O.U.001 grown in media with different concentrations of molybdenum and iron. Int J Hydrogen Energy 31:1536–1544
Karthic P, Shiny J (2012) Comparisons and limitations of biohydrogen production processes. Res J Biotechnol 7:59–71
Khan IU, Othman MHD, Hashim H, Matsuura T, Rezaei-Dasht Arzhandi M, Azelee IW (2017) Biogas as a renewable energy fuel–a review of biogas upgrading. Energy Convers Manag 150:277–294
Kishore D, Talat M, Srivastava ON, Kayastha AM (2012) Immobilization of-galactosidase onto functionalized graphene Nanosheets using response surface methodology and its analytical applications. PLoS One 7:e40708
Konwar LJ, Boro J, Deka D (2014) Review on latest developments in biodiesel production using carbon-based catalysts. Renew Sust Energ Rev 29:546–564
Kuhad R, Singh A (1993) Lignocellulose biotechnology: current and future prospects. Crit Rev Biotechnol 13:151–172
Kunzmann A, Andersson B, Vogt C, Feliu N, Ye F, gabrielsson S (2011) Efficient internalization of silica-coated iron oxide nanoparticles of different sizes by primary human macrophages and dendritic cells. Toxicol Appl Pharmacol 253:81–93
Kurniawan TA, Chan GYS, Lo WH, Babel S (2005) Comparisons of low-cost adsorbents for treating wastewaters laden with heavy metals. Sci Total Environ 366:409–426
Lee CH, Lin TS, Mou CY (2009) Mesoporous materials for encapsulating enzymes. Nano Today 4:165–179
Lee K, Choi I, Kim Y, Yang D, Bae H (2011) Enhanced production of bioethanol and ultrastructural characteristics of reused *Saccharomyces cerevisiae* immobilized calcium alginate beads. Bioresour Technol 102:8191–8198
Lee Y, Lee K, Oh Y (2015) Recent nanoparticles engineering advances in microalgal cultivation and harvesting processes of biodiesel production: a review. Bioresour Technol 184:63–72

Leonzio G (2016) Upgrading of biogas to bio-methane with chemical absorption process:simulation and environmental impact. J Clean Prod 131:364–375

Leung WC, Wong MF, Chua H, Lo W, Leung CK (2000) Removal and recovery of heavy metals by bacteria isolated from activated sludge treating industrial effluents and municipal wastewater. Water Sci Technol 41:233–240

Limayen A, Ricke S (2012) Lignocellulosic biomass for bioethanol production: current perspectives, potential issues and future prospects. Prog Energy Combust Sci 38:449–467

Liu Y, Zhang Y, Quan X, Li Y, Zhao Z, Meng X (2012) Optimization of anaerobic acidogenesis by adding Fe° powder to enhance anaerobic wastewater treatment. Chem Eng J 192:179–185

Liu Z, Zhou Y, Maszenam A, Ng W, Liu Y (2013) pH-dependent transformation of Ag nanoparticles in anaerobic processes. Environ Sci Technol 47:12630–12631

Lochynsha M, Frankowski J (2018) The biogas production potential from silkworm waste. Waste Manag 79:564–570

Lukajtis R, Holowacz I, Kucharska K, Glinka M, Rybarczyk A (2018) Hydrogen production from biomass using dark fermentation. Renew Sust Energ Rev 91:665–694

Maeda K, Owada M, Kimura N, Omata K, Karube I (1995) CO_2 fixation from the flue gas on coal-fired thermal power plant by microalgae. Energy Convers Manag 36:717–720

Mao C, Feng Y, Wang X, Ren G (2015) Review on research achievement of biogas from anaerobic digestion. Renew Sust Energ Rev 45:540–555

Mishra A, Kumari M, Pandey S, Chaudhary V, Gupta KC, Nautiyal CS (2014) Biocatalytic and antimicrobial activities of gold nanoparticles synthesized by *Trichoderma* sp. Bioresour Technol 166:235–242

Mohammadshirazi A, Akram A, Rafiee S, Kalhor EB (2014) Energy and cost analyses of biodiesel production from waste cooking oil. Renew Sust Energ Rev 33:44–49

Nagaragan D, Lee D, Kondo A, Chang J (2017) Recent insights into biohydrogen production by microalgae – from biophotolysis to dark fermentation. Bioresour Technol 227:373–387

Ong YK, Bhatia S (2010) The current status and perspectives of biofuel production via catalytic cracking of edible and non-edible oils. Energy 35:111–119

Palaniappan K (2017) An overview of applications of nanotechnology in biofuel production. World Appl Sci J 35:1305–1311

Pavlidis IV, Tsoufis T, Enotiadis A, Gournis D, Stamatis H (2010) Functionalized multi-wall carbon nanotubes for lipase immobilization. Adv Eng Mater 12:B179–B183

Qiu H, Xu C, Huang X, Ding Y, Qu Y, Gao P (2008) Adsorption of laccase on the surface of nanoporous gold and the direct electron transfer between them. J Phys Chem C 112:14781–14785

Ramsurn H, Gupta R (2013) Nanotechnology in solar and biofuels. ACS Sustain Chem Eng 1:779–797

Rathore V, Madras G (2007) Synthesis of biodiesel from edible and non-edible oils in supercritical alcohols and enzymatic synthesis in supercritical carbon dioxide. Fuel 86:2650–2659

Reddy BVS, Ramesh S, Kumar AA, Wani SP (2008) Biofuel crops research for energy security and rural development in developing countries. Bioenergy Res 1:248–258

Rickerby D, Morrison M (2007) Nanotechnology and the environment: a European perspective. Sci Technol Adv Mater 8:19–24

Romero-Guiza M, Vila J, Mata-Alvarez J, Chimenos J, Astals S (2016) The role of additives on anaerobic digestion: a review. Renew Sust Energ Rev 58:1486–1499

Sadegh H, Ghoshekandi RS, Masjedi A, Mahmoodi Z, Kazemi M (2016) A review on carbon nanotubes adsorbents for the removal of pollutants from aqueous solutions. Int J Nano Dimens 7:109

Saifuddin N, Raziah AJ, Junizah AR (2013) Carbon nanotubes: a review on structure and their interaction with proteins. J Chem 2013:1–18

Saini J, Saini R, Tewar L (2015) Lignocellulosic agriculture waste as biomass feedstocks for second-generation bioethanol production: concepts and recent development. 3 Biotech 5:337–353

Sarkar N, Ghosh S, Bannerjee S, Aikat K (2012) Bioethanol production from agriculture waste: an overview. Renew Energy 37:19–27

Savage N, Diallo MS (2005) Nanomaterials and water purification: opportunities and challenges. J Nanopart Res 7:331–342
Savolainen K, Pylkkänen L, Norppa H, Falck G, Lindberg H, Tuomi T, Vippola M, Alenius H, Hämeri K, Koivisto J, Brouwer D, Mark D, Bard D, Berges JE, Posniak M, Farmer P, Sekoai PT, Yoro K, Daramola MO (2016) Batch fermentative biohydrogen production process using immobilized anaerobic sludge from organic solid waste. Environment 38:1–10
Sekoai PT, Yoro K, Daramola MO (2016) Batch fermentative biohydrogen production process using immobilized anaerobic sludge from organic solid waste. Environment 38:1–10
Serrano E, Rus G, García-Martínez J (2009) Nanotechnology for sustainable energy. Renew Sust Energ Rev 13:2373–2384
Shaishav S, Singh RN, Satyendra T (2013) Biohydrogen from algae: fuel of the future. Int Res J Environ Sci 2:44–47
Sill TJ, Recum HAV (2008) Electrospinning:applications in drug delivery and tissue engineering. Biomaterials 29:1989–2006
Singh R, Krombach F, Bihari P, Kasper G, Seipenbusch M (2010) Nanotechnologies, engineered nanomaterials and occupational health and safety – a review. Saf Sci 48:957–963
Su L, Shi X, Gua G, Zhao A, Zhao Y (2013) Stabilization of sewage sludge in the presence of nanoscale zero-valent iron (nZVI): abatement of odor and improvement of biogas production. J Mater Cycles Waste Manage 15:461–468
Swain KC (2014) Biofuel production in India: potential, prospectus and technology. J Fundam Renew Energy Appl 4:129–132
Theron J, Walker J, Cloete T (2008) Nanotechnology and water treatment: applications and emerging opportunities. Crit Rev Microbiol 34:43–69
Torres-Ortega CE, Fengqi You JG, Rong BG (2017) Optimal synthesis of integrated process for co-production of biodiesel and hydrotreated vegetable oil (HVO) diesel from hybrid oil feedstocks. Comput Aided Chem Eng 40:673–678
Venkata MS, Mohanakrishna G, Reddy S, David RB, Rama RKS, Sarma PN (2008) Self-immobilization of acidogenic mixed consortia on mesoporous material (SBA-15) and activated carbon to enhance fermentative hydrogen production. Int J Hydrog Energy 33:6133–6142
Voloshin RA, Kreslavski VD, Zharmukhamedov SK, Bedbenow VS, Ramakrishna S, Allakhverdiev SI (2015) Photoelectrochemical cells based on photosynthetic systems: a review. Biofuel Res J 6:227–235
Wang L, Li J, Jiang Q, Zhao L (2012) Water-soluble Fe_3O_4 nanoparticles with high solubility for removal of heavy-metal ions from waste water. Dalton Trans 41:4544–4551
Yang G, wang J (2018) Improving mechanisms of biohydrogen production from grass using zero valent ion nanoparticles. Bioresour Technol 266:413–420
Yang F, Hong F, You W, Liu C, Gao F, Wu C (2006) Influence of nano-anatase TiO_2 on the nitrogen metabolism of growing spinach. Biol Trace Elem Res 110:179–190
Yang Y, Zhang C, Hu Z (2013) Impact of metallic and metal oxide nanoparticles on wastewater treatment and anaerobic digestion. Environ Sci Process Impacts 15:39–48
Yu W, Xie H (2012) A review on nanofluids: preparation, stability mechanisms and applications. J Nanomater 2012:1–17
Zhang Y, Shen J (2007) Enhancement effect of gold nanoparticles on biohydrogen production from artificial wastewater. Int J Hydrog Energy 32:17–23
Zhang D, Wei D, Ding W, Zhang X (2014) Carbon-based nanostructured catalyst for biodiesel production by catalytic distillation. Catal Commun 43:121–125
Zhao W, Zhang Y, Du B, Wei Q, Zhao Y (2013) Enhancement effect of silver nanoparticles on fermentative biohydrogen production using mixed bacteria. Bioresour Technol 142:240–245

Nanomaterial Synthesis and Mechanism for Enzyme Immobilization

7

Veena Paul, Prasad Rasane, Kajal Dhawan, and Abhishek Dutt Tripathi

Abstract

Enzymes play a significant role in different industries related to food, pharmaceutical, agriculture, and other sectors. Although enzymes find tremendous application in all sectors, their limited reusability and higher cost of production possess hurdle in their wide commercial acceptance. By using different immobilization strategies, this problem can be overcome; however, conventional strategies used for enzyme immobilization, viz., entrapment, encapsulation, and adsorption, are not efficient enough to meet the industry demands. Nowadays, the application of nanotechnology is quite a useful tool in different applications specifically in drug delivery and tissue engineering. The nanoparticle can be effectively and efficiently utilized in the enzyme immobilization to enhance its reusability while retaining its activity prevailing in the free form. This chapter focuses on the application of the novel nanotechnological tools currently used in the enzyme immobilization at the commercial level.

Keyword

Nanomaterial · Nanomaterial synthesis · Enzyme immobilization · Biofuels

V. Paul · A. D. Tripathi (✉)
Centre of Food Science and Technology, Institute of Agricultural Sciences,
Banaras Hindu University, Varanasi, Uttar Pradesh, India

P. Rasane · K. Dhawan
Department of Food Technology, Lovely Professional University, Phagwara, Punjab, India

M. Srivastava et al. (eds.), *Nanomaterials in Biofuels Research*,
Clean Energy Production Technologies, https://doi.org/10.1007/978-981-13-9333-4_7

7.1 Introduction

Nanomaterials have become the most active field of research in recent years due to its novel characteristics as compared to traditional materials, thus providing opportunities for new technological innovations and biomedical applications. These nanostructured materials have wide applications in numerous fields like electronics, optics, electrochemistry, biomedical science, and magnetism. These miniature particles are known to possess enhanced magnetic, optical, electronic, and mechanical properties as compared to larger counterparts of the same chemical composition. Also, the need to synthesize fine materials to enhance the speed of information transmission, decrease costs, and increase the storage of nonmaterial has stimulated their development in this field (Fernández-Pacheco et al. 2006). Table 7.1 enlists typical nanomaterials (Rao and Cheetham 2001). There has been explosive growth in nanotechnology field owing to the accessibility of new techniques of

Table 7.1 Methods of synthesis of nanomaterials

Synthesis method	Main principle	References
Sol-gel process	Nanoparticle synthesis by hydrolysis and condensation of inorganic salts or alkoxides in the presence of catalyst (mineral base or acid), followed by aging, drying, and calcinations	Rahman and Padavettan (2012)
Hydrothermal synthesis	Nanoparticle synthesis in an autoclave under controlled temperature and pressure conditions. Solvent used is aqueous	Qiu et al. (2011a)
Facile microwave-assisted hydrothermal synthesis	Nanoparticle synthesis in microwave synthesizer under controlled temperature and pressure conditions. Solvent used is aqueous	Qiu et al. (2011b)
Solvothermal synthesis	Nanoparticle synthesis in microwave synthesizer under controlled temperature and pressure conditions. Solvent used in non-aqueous	Xia et al. (2003)
Gas-phase combustion synthesis	Synthesis of nanoparticles in flame	Aruna and Mukasyan (2008)
Sonochemical synthesis	High-intensity ultrasound waves are used for nanomaterial synthesis	
Microwave synthesis	Nanoparticle synthesis in microwave reactors with optimum irradiation strength, reaction temperature, and inner pressure of the vessel	Bilecka and Niederberger (2010)
Arc-discharge synthesis	Single step for the formation of well-coated and homogenous particles in a chamber where a product is deposited on the cathode due to temperature gradients	Arora and Sharma (2014)
Solution combustion synthesis	The initial reaction mixture is an aqueous solution	Mukasyan and Dinka (2007)
Condensed-phase combustion synthesis	The initial reactants are in the solid state	
Chemical vapor condensation (CVC)	Nanoparticles are synthesized by decomposition of metal-organic precursors in high-temperature flame	Rahman and Padavettan (2012)

nanomaterial synthesis in addition to characterization of these nanomaterials. The synthesis involves organic, biological, and inorganic systems with control of size, structure, and shape.

Recently, synthesis of nanomaterials has been given much attention, particularly in terms of its shape and size as these two factors primarily influence the performance of nanomaterials. Size of nanomaterials mainly affects its distribution in the biological system, while its shape influences its interaction with the target molecule. Besides shape and size, surface area, morphology, and surface functionality of nanomaterials must be taken into consideration as these factors help to inhibit microbial growth. In order to synthesize nanomaterials with desired characteristics, it is essential to select proper synthesis routes. For example, gas-phase routes like laser ablation, arc discharge, chemical vapor deposition, and plasma torch have been used in bulk synthesis of specific nanoparticle morphology like nanorods and nanotubes (Jiang et al., 2014). Also, control over process parameters like applied voltage, temperature of the reactor, gas composition, nature of starting material, and input power helps in determining the shape and size of nanomaterials.

Several approaches are used for the synthesis of these nanomaterials. Some of these are template-based synthesis, arc discharge synthesis, microwave synthesis, microfluidic synthesis, solution combustion synthesis, sol-gel synthesis, hydrothermal synthesis, solvothermal synthesis, sonochemical synthesis, mechanical milling, microemulsion synthesis, chemical vapor condensation, one-pot synthesis, etc.

The synthesized nanomaterial is used for various applications. Enzyme immobilization is an essential application of the nanomaterial. Various methods like adsorption, cross-linking, entrapment, and membrane confinement are utilized to develop immobilized enzymes. Immobilized enzymes have broad industrial applications in the bioenergy industry and are therefore of substantial commercial value.

This chapter aims to discuss various methods of synthesis of nanomaterials and their characterization techniques in details. It also aims to explore the various enzyme immobilization techniques used to immobilize the enzymes using synthesized nanomaterials. The chapter also highlights the uses of immobilized enzymes in the bioenergy industry.

7.2 Different Methods of Nanomaterial Synthesis

7.2.1 Sol-Gel Synthesis

The sol-gel process has become an essential area of recent research as it has found an interesting aspect in the synthesis of nanomaterials. However, this process is largely dependent on various experimental conditions that induce condensation reactions and hydrolysis rate. These conditions are time of reaction, catalyst's nature, concentration and type of starting materials, temperature and pH, concentration and nature of the solvent, water-alkoxide ratio, drying method, and aging (Jafarzadeh et al. 2009). The chief merit of this process is that it controls the size,

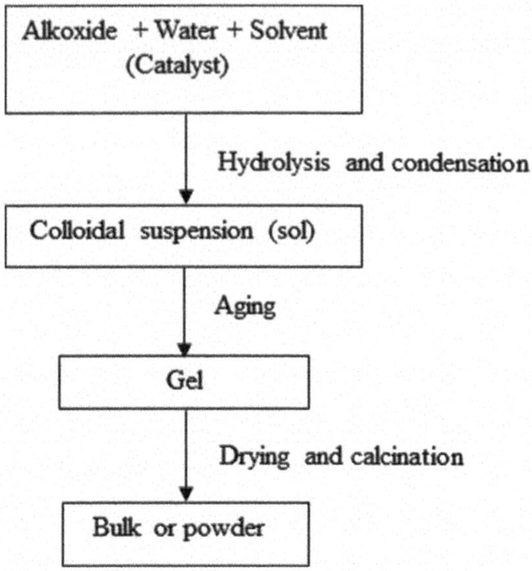

Fig. 7.1 Sol-gel process. (Chen and Mao 2007)

morphology, and distribution of particle by carefully monitoring and manipulating the parameters of the reaction process. It also has the ability to produce homogenous and pure products at mild conditions. The process involves hydrolysis and condensation of inorganic salts or alkoxides in the presence of a catalyst (mineral base or acid). This leads to the formation of the colloidal suspension. After complete polymerization, there is a loss of solvent, and liquid sol turns into a solid gel. Further drying forms the fine, porous, and low-density material (Fig. 7.1) (Chen and Mao 2007). Each parameter of reaction is optimized, and a combination of optimal conditions is set to produce small, homogenous, and monodispersed nanoparticles. Drying leads to solid material formation. The different processes used for drying are freeze-drying, supercritical drying, thermal drying, and spray drying. Drying leads to agglomeration that is governed by coalescence and collision of nanoparticles. In addition, aging occurs during drying leading to complex agglomeration due to polycondensation reactions. These reactions increase in the presence of a catalyst and water; hence, it is important to remove water during the drying process. In addition to condensation reactions, hydrodynamic effect, capillary drag, and Brownian motion during drying also lead to agglomeration in an aqueous system (Rahman and Padavettan 2012). Alcohol dehydration reduces agglomeration in the synthesis of silica nanoparticles (Rahman et al. 2008). Out of all the chemical synthesis processes, sol-gel offers better homogeneity, phase-pure powder at low temperature, controlled stoichiometry, high purity, and high flexibility (Gnanam and Rajendran 2011).

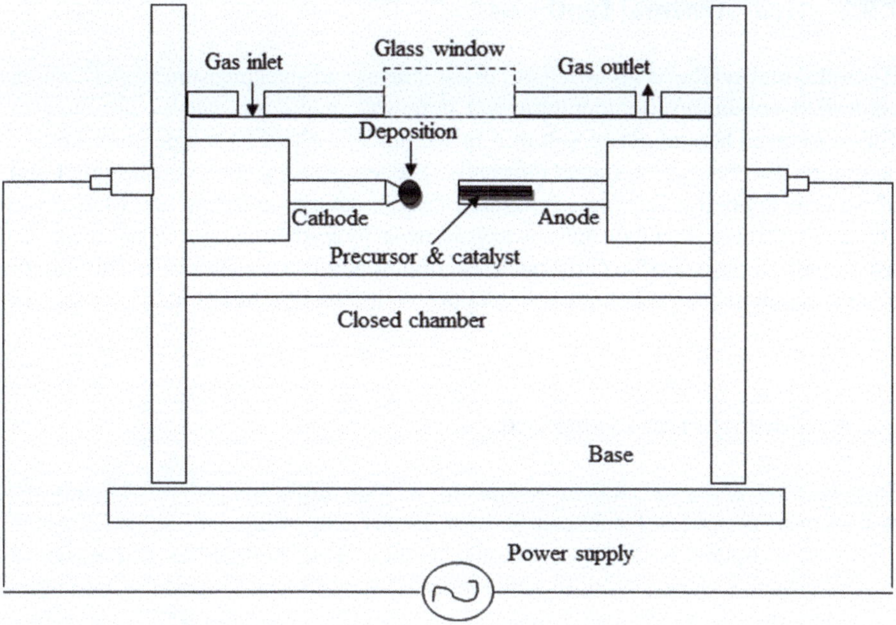

Fig. 7.2 Arc-discharge setup

7.2.2 Arc-Discharge Method

Arc-discharge method, developed by Kratschmer-Huffman, involves a single step in the formation of well-coated and homogenous particles. The apparatus comprises of an evacuated cylindrical chamber filled with helium gas and two electrodes: a stationary anode and a movable cathode (Fig. 7.2). The anode is filled with precursor along with a catalyst. The lifetime of catalyst plays a crucial role in arc-discharge synthesis as they are easily poisoned due to amorphous carbon accumulation (Su and Zhang 2015). The cathode is usually a pure graphite rod in a water cooling system. The pressure inside the chamber is adjusted to reduce fluctuations during the synthesis. This is because any fluctuation in pressure will affect the product to be synthesized. Initially, the anode and cathode are in contact, and on applying a constant current between the two, gradual separation occurs until an arc is formed, which leads to helium plasma ionization. A gap of 1–2 mm is maintained between the electrodes for a steady discharge. The precursor on the anode sublimes and condenses on the cooler parts of the system. The condensed particles are collected on the cathode due to temperature gradient. During the process, the temperature of the system reaches 3000 °C or 4000–6000 K due to the arc current. The arc-discharge method depends on various parameters such as catalysts, type of power supply, electrode geometry, environment, temperature, and pressure (Fernández-Pacheco et al. 2006; Arora and Sharma 2014).

7.2.3 Hydrothermal Synthesis

Hydrothermal synthesis is carried out in autoclaves and steel pressure vessels, under controlled conditions (temperature and pressure) in the absence or presence of Teflon liners. The amount of solution in the autoclave and its temperature largely determine the pressure produced internally. The temperature above the boiling point of water is applied achieving a pressure of vapor saturation (Chen and Mao 2007; Qiu et al. 2011a). In facile microwave-assisted hydrothermal synthesis, a microwave synthesizer is used to carry out a reaction rather than an autoclave. This equipment is maintained at a definite temperature with 5 to 30 min of holding time (Qiu et al. 2011b).

7.2.4 Solvothermal Synthesis

Solvothermal synthesis uses a non-aqueous solvent, unlike hydrothermal synthesis. A variety of organic solvents can be used so the temperature can be raised much higher as compared to the hydrothermal method. This method has better control over size, shape, and crystallinity of nanoparticles which can be used for its synthesis with narrow dispersity and distribution of size (Chen and Mao 2007). The solvothermal method involves the addition of a reagent (like amines) to a solvent in appropriate ratios. The catalyst regulates the growth of crystals in the solvent. The mixture is then kept to react in an autoclave at appropriate temperature and pressure conditions. The main advantage is the solubility of materials by the application of heat and pressure at a critical point. However, this method gives a poor yield and non-uniform-sized materials, which have less purity and poor morphological uniformity. In addition, the solvents used are not environmentally friendly, making the process complex. The use of autoclaves also hinders the study of mechanisms of growth (Xia et al. 2003).

7.2.5 Combustion Synthesis (CS)

Combustion synthesis (CS), also known as self-propagating high-temperature synthesis (SHS), is a cost-efficient and useful technique for nanoparticle production. The combustion synthesis is divided into further categories on the basis of physical nature of initial reaction medium. In conventional SHS, the initial reactants are in the solid state so it is also known as condensed phase combustion. The heterogeneity scale for reactants ranges from 10 to 100 μm, and the reaction temperature is quite high (less than 2000 K). Due to this, SHS cannot be used for the synthesis of large surface area nanostructures (Aruna and Mukasyan 2008). The conventional SHS is characterized by ignition of the reaction mixture by external thermal source and propagation of high-temperature reaction wave through the mixture. It leads to solid material formation without using any additional energy (Mukasyan et al. 2007). To overcome the drawback of SHS, it is combined or modified using a number of other processes such as intensive milling, mechanical activation, and chemical dispersion and using additives, carbon, and reaction fuel. Chemical dispersion is the method of dissolving the SHS powder in dilute acid to remove the impurities and defects in crystals. When an additive is used, it is termed as alkali metal molten salt-assisted combustion. The metal particles are protected from agglomeration and growth as their nucleation occurs in molten salt (NaCl). The salt melts (1083 K) due to the heat generated by the combustion reaction between reducing metal and transition metal oxide of the alkali metal salt. In carbon combustion synthesis, carbon is employed as a fuel in the reaction unlike metals in conventional SHS. This makes the reaction gaseous, and due to the release of carbon dioxide at a high rate, highly porous powders of particle size ranging from 50 to 800 nm can be synthesize easily. In gas-phase combustion synthesis, nanomaterials are synthesized in the flame (Aruna and Mukasyan 2008).

Volume combustion synthesis (VCS) is another mode of CS, which involves the uniform heating of the entire sample until the self-initiation of reaction through the entire volume. However, this method cannot be used for weakly exothermic reactions requiring preheating before ignition since the mode is less controllable. Therefore, conventional CS and reactive solution are combined to form a process called solution combustion synthesis (SCS). SCS is an attractive, effective, novel, rapid, and versatile process for nanomaterial synthesis. The main features of SCS, which make it an effective method, are that firstly, the initial reaction medium is an aqueous solution, thereby allowing reactant mixing at the molecular level and permitting uniform and accurate formulation of composition as desired on the nanoscale. Secondly, the reaction temperature (T_c) is high ensuring high purity and crystallinity of the product. Finally, it has a short duration process and allows the rapid synthesis of nanosize materials as it does not require high-temperature product calcinations. It favors the synthesis of nanosize powders having a high surface area due to the formation of various gases, which also inhibits the growth of particle size (Mukasyan et al. 2007; Mukasyan and Dinka 2007).

7.2.6 Microwave Synthesis

The microwave synthesis is a fast technique for nanomaterial synthesis and has great potential in case of nanomaterials whose growth is highly affected by slight changes in the reaction conditions. This technique allows the controlled and efficient heating by manipulating the microwave irradiation conditions. The setup mainly involves the microwave reactors available commercially for the chemical synthesis and use of microwave irradiations (frequency −0.3 to 300 GHz; wavelength −1 nm to 1 m). The matter is heated using microwave dielectric heating that involves the absorption of microwave energy by a specific material and its conversion to heat energy. This forms the principle of microwave synthesis. The microwave photon energy is too low, i.e., 1.0×10^{-5} eV (1 J mol^{-1}) at 2.45 GHz frequency to break the chemical bonds. Thus, the process depends upon the effective heating of material and not on chemical reaction induction. The microwave heating mechanism involves dipolar polarization and ionic conduction. When the samples are irradiated with microwaves, ions or dipoles try to reorient in the electric field continuously as EM radiation produces an oscillating field. This causes molecular friction and dielectric loss, depending on orientation and disorientation time scales, leading to the production of heat. If the dipole fails to realign, there is no heating. Dielectric loss is due to the lagging of polar molecular rotations behind the electric field in liquid at 2.45 GHz frequency. The ionic conduction process involves the production of heat due to the collision of dissolved charged particles, oscillating back and forth under microwave irradiation effect, with the surrounding molecules. This process produces more heat as compared to dipolar polarization. The advantage of microwave synthesis is that it provides uniform internal heating as compared to other synthesis processes. Other advantages include control of reaction parameters for better product quality, selective heating, high product yield, high rates of heating to speed up the reaction rate, high-throughput synthesis, indirect reactant, and heating source contact and atomization. The limitations include the use of costly microwave reactors and scale up the problem as microwave irradiation has a short depth of penetration in a liquid medium (Bilecka and Niederberger 2010).

7.2.7 Experimental Tools and Characterization of Nanomaterials

Characterization of individual nanostructure requires extreme accuracy, sensitivity, and high level of resolution at the atomic level. The various structural characterization techniques used widely are electron microscopy (EM), scanning electron microscopy (SEM), transmission microscopy (TEM), small-angle X-ray scattering (SAXS), X-ray diffraction (XRD), and scanning probe microscopy (SPM). The chemical characterization techniques include ionic spectrometry, optical spectroscopy, and electron spectroscopy.

7.2.8 Structural Characterization

The structural characterization of nanostructures involves the techniques that analyze the surface properties of these nanomaterials. The following techniques are used in the structural characterization of nanomaterials.

7.2.9 X-Ray Diffraction (XRD)

X-ray diffraction is a strain-sensitive and nondestructive technique that has been used widely for determining the lattice parameters, geometry, composition, identification of unknown materials, thickness, crystallinity, and also the defects and stresses of nanoparticles and nanowires (Cao 2004). In this technique, a monochromatic X-ray beam of wavelength range of 0.7–2 Å generated by bombardment of electrons on a metal in a tube is incident on the surface of the crystal. The electron

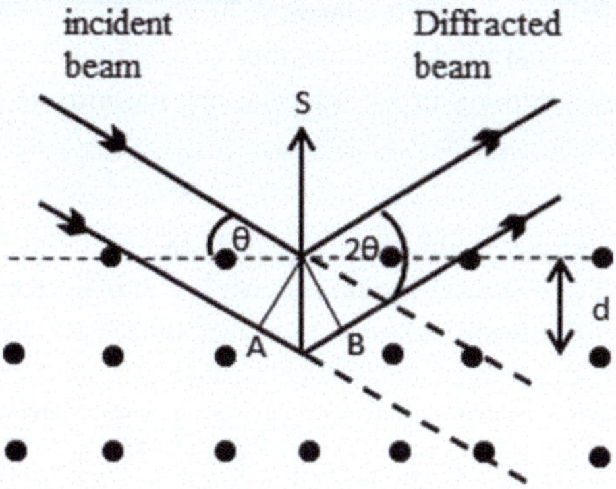

Fig. 7.3 X-ray diffraction. *S* scattering. (Adapted from Moram and Vickers 2009)

cloud of the atoms of crystal scatters these rays, and constructive interference occurs which follows Bragg's law as path difference (nλ) equals 2d sin θ (Fig. 7.3).

According to Bragg's law, "the spacing between the planes of atoms from which diffraction is occurring (d) to the angle (θ) at which the incident monochromatic beam must probe the plane to give constructive interference." The 2θ angle, corresponding to the intensity of X-rays, can be measured experimentally, and the crystal acts as a 3-D diffraction surface. d is interplanar spacing in the crystalline phase and λ is the wavelength of X-rays. The real space-associated crystal planes produce reciprocal space-associated diffraction spots, forming a diffraction pattern. The position and shape of spots are related inversely to the size of crystallites and spacing of crystal planes (Moram and Vickers 2009). The main advantage of XRD is that it does not involve the preparation of large samples, as it is a nondestructive technique. The disadvantage is the low intensity of X-rays and small diffraction, which makes it less sensitive. However, XRD measures diffraction peak positions accurately making it suitable for both homogeneous and inhomogeneous strain. Homogenous strains shift the diffraction peak positions leading to change in d-spacing which corresponds to lattice constant change under strain. In case of inhomogeneous strain, broadening of a diffraction peak occurs which increase with sin θ within a single crystallite. The broadening of peak also occurs due to finite crystal size, but it is independent of sin θ (Cao 2004).

7.2.10 Small-Angle X-Ray Scattering (SAXS)

SAXS is an incredible asset for nanomaterial characterization at large distances or small angles and has become popular worldwide. In this technique, the X-rays are scattered which produces strong diffraction peaks from constructive interference. However, the scattering in SAXS is due to the inhomogeneity in size ranging from several nanometers to ten nanometers (Jeng et al. 2010). It is an effective tool for determination of size and shape of the object, surface structure, and size distribution. It also determines the relative position of a particle from which pair potential and equation of states can be inferred. It is a nondestructive technique and provides structural data for large volumes of samples. According to SAXS theory, the intensity scattered by a number of noninteracting nanoparticles having a uniform electron density in a homogeneous medium (Li et al. 2016) is given by

$$I(q) = I_0 N (P - P_0)^2 F^2(q)$$

where I_0 is the intensity of incident X-ray, $I(q)$ is SAXS intensity, N is the number of noninteracting nanoparticles, P is uniform electron density, P_0 is electron density of the homogeneous medium, and $F(q)$ is form factor which is expressed as

$$F(q) = 4\Pi R^3 \frac{\{\sin(qR) - qR\cos(qR)\}}{(qR^3)}$$

where R = radius of sphere, q = [$4\Pi \sin(\theta/2)$]/λ, λ = wavelength of X-ray, and θ = angle between primary and scattering beam of X-ray (Cao 2004).

7.2.11 Electron Microscopy (EM)

Electron microscopy helps to characterize and visualize the nanomaterials due to its high resolving power as it uses a low wavelength electron beam, which is below the nm range. There are two main techniques of EM, namely, scanning EM and transmission EM.

7.2.12 Scanning Electron Microscopy (SEM)

This technique employs low energy electron beam ranging from 1 to 30 keV, which scans the sample surface and detects the scattered electrons using detectors which select unique ranges of energy from the scattered signal (Dudkiewicz et al. 2011). In this technique, the electron source is focused into a beam and is incident on the sample mounted on aluminum or carbon stubs. The electrons strike and penetrate the sample resulting in various reactions leading to photons and electron emission from the sample. The cathode ray tube (CRT) collects the electrons and photons emitted and produces the images. SEM produces three types of images, namely, secondary electron images, backscattered electron images, and elemental X-ray maps. The interaction between an atom and a high-energy primary electron results either in elastic scattering with nucleus or inelastic scattering with electrons. During inelastic scattering, energy is transferred from the primary electron to another electron, and if the transferred energy is large, then another electron present in the sample emits it. If the energy of emitted electron is less than 50 eV, it is known as a secondary electron. The electrons, which are scattered elastically and have the same energy as the primary electron, are called backscattered electrons, and it increases with atomic number (Cao 2004). The secondary electrons (SEs) having lower frequency provide data on topological surface, and high-energy backscattered electrons (BSEs) provide information mapping contrast which depends upon the atomic number of elements in a sample. The image produced has a visible contrast of light and heavy elements. The BSE detectors give contrast, when the atomic number difference is equal to or less than 0.1, and thus have superiority of simple data interpretation over other techniques. The best resolution was reported for platinum in which 6–9 nm was achieved at 25 keV (Brodusch et al. 2013). BSE imaging is the main feature of SEM (Dudkiewicz et al. 2011). The resolving power of SEM

$$R = \frac{\lambda}{2NA}$$

where λ = electron wavelength and NA = numerical aperture of objective and condenser lens.

7.2.13 Transmission Electron Microscopy (TEM)

This technique uses electrons with energy ranging from 100 KeV to 1 MeV, which is projected onto the sample with the help of a condenser lens. The thickness of the sample should be less than 200 nm for proper penetration. As the electron beam penetrates, the electrons experience scattering in the sample, which is generally of two types, viz., elastic and inelastic scattering. Depending on the type of scattering, various types of data are acquired. In the case of elastic scattering, there is no loss of energy, and hence diffraction patterns are obtained, whereas inelastic interactions lead to spatial variation in transmitted electron intensity. Inelastic interactions occur due to heterogeneity, viz., dislocations, defects, density variations, boundaries, etc., and it causes complex absorption. TEM has high magnification due to the application of small electron wavelength λ (Cao 2004). This resolution largely depends upon sample thickness and voltage at which the electron beam is accelerated. The higher the voltage at which the electron beam is accelerated, the better the resolution of the image (Dudkiewicz et al. 2011). The electron wavelength is given by de Broglie relationship:

$$\lambda = \frac{h}{\sqrt{2mqv}}$$

where m = mass and q = charge of electron, h = Planck's constant, and v = voltage at which the electron beam is accelerated.

The advantage of TEM is its high resolution ranging from 50 to 10^6 and greater electron penetration due to high-energy electron which interacts weakly with sample matter as compared to low-energy electrons. It makes TEM suitable for operation at high voltage with thicker samples. Another advantage offered by TEM is that it provides both diffraction information and image from the same sample by altering the intensity of the intermediate ocular lens. The only shortcoming is the limited depth resolution of TEM as the scattering information from a 3D sample is projected on a 2-D detector (Cao 2004).

7.2.14 Scanning Probe Microscopy (SPM)

Scanning probe microscopy is a recently developed tool for the characterization of nanomaterials. This technique has revolutionized in the last few years due to its wide applications at nanoscale characterization. This technique provides three-dimensional real space images, which give it an edge over other characterization techniques. It uses two types of probing forces depending on which two members of SPM, viz., scanning tunneling microscopy (STM) and atomic force or friction force microscopy (AFM/FFM). Both techniques provide three-dimensional topographic images in real spaces with high atomic resolution. It helps to characterize all types of solid surfaces whether electrically conductive or not or hard or soft (Alex and Gruverman 2004).

Scanning tunneling microscopy, developed by Binning and Rohrer in 1981 (Bhushan 2008), is based on the phenomenon of electron tunneling. The sample electrons cannot move through the insulator due to the energy barrier. On applying the voltage across the surfaces, energy barrier is changed, and the electrons move across the surfaces by tunneling under the influence of a driving force. The movement of electrons produces a small amount of current given by

$$I = e^{-2kz}$$

where z = distance between two surfaces or insulator thickness and k is given by

$$k = \sqrt{2m}\left(V - E\right)/h$$

where m = mass of electron, E = energy of electron, V = insulator potential, and h = Planck's constant.

The conductive tip is placed on the top of the sample at some height, and as the tip moves over the sample, its position is used to obtain a topographic map (Fig. 7.4). The tunneling current is kept constant by adjusting the distance between the conductive tip and sample. The limitation of STM is that it can only be used for a surface, which is electrically conductive as it depends on tunneling current. STM can be operated in two different ways (modes). Constant voltage mode involves keeping the position of the tip constant, and constant current mode involves keeping the distance between the tip and sample constant. Generally, the height is maintained between 0.2 and 0.6 nm, producing a current of 0.1–10 nA. The image has a resolution of 0.01 and 0.002 nm in XY and Z direction, respectively, and thus produces a 3-D image (Cao 2004; Xu et al. 2009).

Atomic force or friction force microscopy, developed in 1985, was introduced as a modification for STM as it can be used for dielectric materials (Bhushan 2008). Both long-range and short-range forces operate in AFM involving van der Waals forces between the tip and sample. It has three components, namely, induced dipole, permanent dipoles, and electronic polarization. The principle is based on the deflection of cantilever by a photodiode which records the laser beam position reflected off the top of cantilever (Fig. 7.5) (Bhushan 2001; Cao 2004).

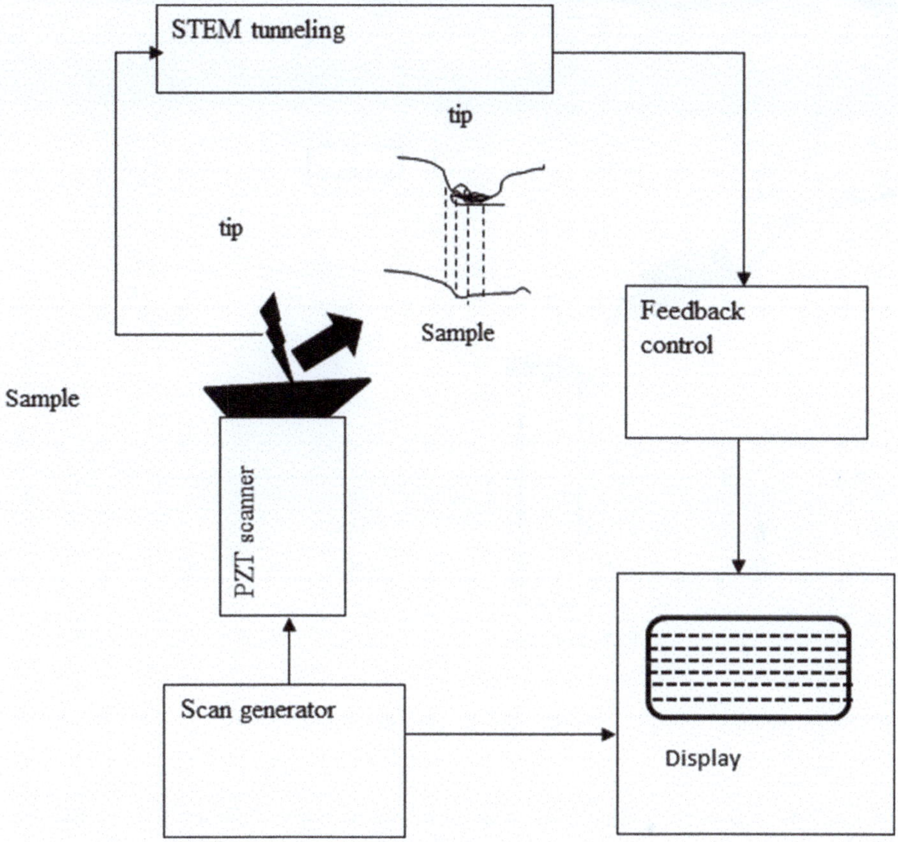

Fig. 7.4 Structure of scanning tunneling microscope

Near-field scanning optical microscopy (NSOM) is another member of SPM that involves the irradiation of sample through a tapered sub-wavelength aperture in probe. This leads to the formation of waves corresponding to the wavelengths of the diameter of aperture only. These waves are unable to spread in free space and are called evanescent waves. However, these transfer the radiative energy on the other side of the screen by winding around the aperture. Three zones of energy density are formed depending on the distance from the screen. At a distance of 2–5 nm, the intensity is constant and is known as evanescent wave regime. At a distance of 5–500 nm, the intensity decreases proportionally to $s^{-3.7}$ (distance from aperture). This is known as near-field regime and has a very small energy density ranging from 10^{-4} to 10^{-6}. In this regime, the evanescent waves evanescent. The energy density decreases by s^{-2} at a distance larger than wavelength and is known as far-field regime. Hence, it is important to keep the sample-tip distance in the nm range (near-field regime) so the intensity is sufficient for detection by the system. During the process, two types of images are produced: topographic image by scanning force

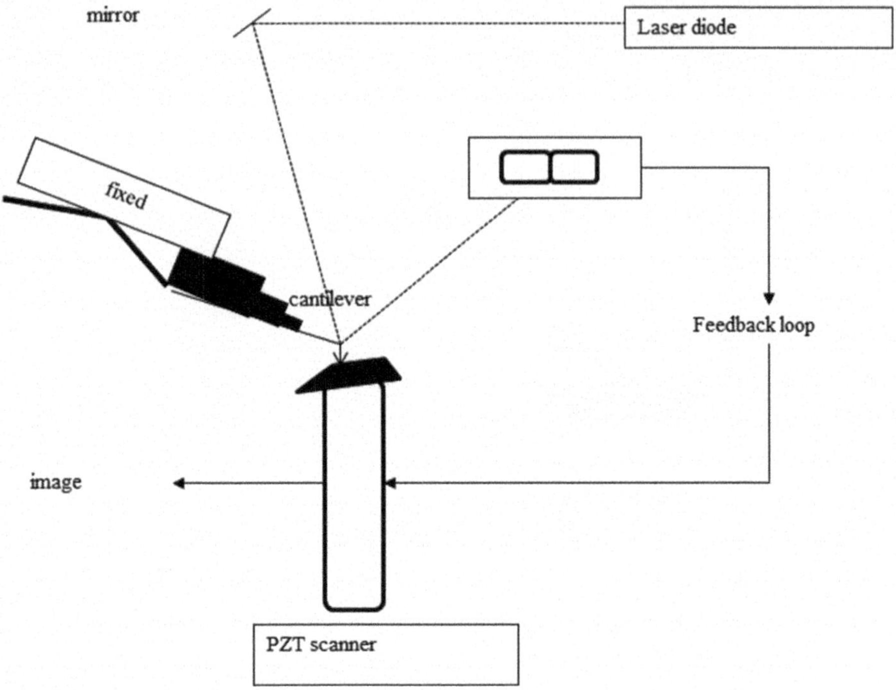

Fig. 7.5 Principle of atomic force microscope. (Adapted from Cao 2004)

microscopy and optical image by near field microscopy. The setup is similar to AFM and allows access to sub-wavelength scales ranging from 50 to 100 nm by breaking the diffraction limit (Cao 2004; Rao and Cheetham 2001).

Magnetic force microscopy (MFM) is another member of SPM and was discovered in 1987 by Martin and Wickramasinghe. It is based on the measurement of force between the magnetized tip and magnetic or diamagnetic sample. In MFM, a large volume of the sample interacts with a larger volume of sample, so the resolution is poor (20 nm) as compared to STM or AFM. It helps in the magnetic characterization of nanomaterials involving the orientation and magnitude of every spin in the sample and its 3-D location (De Lozanne 2006).

7.2.15 Chemical Characterization

The chemical characterization of nanostructures involves the techniques that are used to determine the atoms and compounds present inside and at the surface of the sample and their spatial distribution. The most popular methods are described below.

7.2.16 Optical Spectroscopy

The techniques of optical spectroscopy are divided into absorption spectroscopy, emission spectroscopy, and vibrational spectroscopy. In absorption spectroscopy, the excitation of electrons occurs from ground state to excited state (absorption) and in emission spectroscopy; the relaxation of electrons occurs from the excited state to ground states (emission). These techniques help in determining electronic structures of molecules, ions, and atoms of crystals of sample. In vibrational spectroscopy, the photons interact with the sample particles, and energy is transferred between them through vibrational excitation and de-excitation. This technique determines the types of chemical bonds in the sample.

7.2.16.1 Photoluminescence and UV/Vis Spectroscopy

Photoluminescence is a phenomenon of light emission by a material through the relaxation of electrons as they shift back from excited states to ground state by the emission of some energy in the form of light. When light strikes the sample, electrons are excited which produces luminescence that can be collected and passed by a lens through the optical spectrometer to a photon detector. The major benefit of this technique is high sensitivity which allows the analysis of very small quantities, ranging from nanograms to parts per trillion (Kumar 2013).

7.2.16.2 Raman Spectroscopy

The technique is an effective tool for determining the scattering of light by photons and to analyze the electronic and structural properties of the sample. It is a type of vibrational technique. It involves a combination of high-frequency radiations with chemical bond vibrations. This technique is more sensitive for determining the strength, length, and distribution of chemical bonds excluding the chemical composition. When the photon is incident on the sample, it interacts with the chemical bonds in the material leading to the excitation of chemical bond to higher energy level. Two types of scattering occur: Rayleigh scattering and Stokes scattering. The former is due to the reradiation at a frequency same as that of incident light used for excitation. The latter is due to the excitation of vibrational modes due to the energy transfer. Further reradiation will have lower frequency as compared to incident light. A third type of scattering, known as anti-Stokes scattering, occurs when the existing exciting vibrations combine and add their energies to incident beam (Kumar 2012).

7.2.17 Electron Spectroscopy

This technique involves the energy levels of emitted photons or electrons from the atoms. When the photon strikes, there is release of energy either in the form of X-rays or fluorescent X-rays or Auger electrons. Every element possesses a peculiar and characteristic electronic structure with predefined energy levels. Also, the Auger spectral lines and X-rays are specific for each element in the periodic table. Thus, the determination of composition and chemical character of a sample can be done by measurement of X-rays and Auger electron energies emitted by them. On this basis, types of electron spectroscopy are discussed below.

7.2.17.1 Energy Dispersive X-Ray Spectroscopy (EDS)

The fundamental principle of this technique is the X-ray emission when an outer shell electron dissipates its energy to complete the inner shell electron vacancy. The excess energy is released in the form of an X-ray. When an electron or photon is incident on an unexcited atom, electron vacancy is created by the ejection of an electron from the inner shell of the atom (Cullity and Stock 2014).

7.2.17.2 Auger Electron Spectroscopy (AES)

Pierre Auger discovered the Auger electron spectroscopy and it is a nonradiative process. The Auger process occurs when the Auger electron, a third electron from a further electron shell, is ejected to fill the electron vacancy in the inner shell. These Auger electrons have very small mean free path in solids. This Auger effect largely influences the fluorescence yields and X-ray line width broadening by shortening the vacancy state lifetime (Carlson 2013).

7.2.17.3 X-Ray Photoelectron Spectroscopy (XPS)

It involves the use of low-energy X-rays for the electron ejection from an atom through photoelectric effect:

$$E_E = hv - E_B$$

where E_E = energy of ejected electron, hv = energy of incident electron, and E_B = bound electron state.

The ejected electron energy by photoelectric effect is known as photoelectron energy, and as the binding energy is element specific, the chemical composition can be determined by measuring the photoelectron energies.

7.2.18 Ionic Spectrometry

7.2.18.1 Rutherford Backscattering Spectrometry (RBS)

This technique is discovered by Rutherford for the characterization of thin films, and the basic principle involves the use of beams of high energy, generally in MeV, of ions with low mass. In this technique, the collimated alpha particles are incident on a thin film, and almost all these incident particles reappear on the far side of thin film with small reduction in energy and slight alteration in direction. This implies that the incident beam is transmitted with minor losses. Very few alpha particles interact with the nucleus of the atoms of the thin film, and as a result, they are lost with large changes in energy and direction. In the case of thick samples, the particles are scattered backward at an angle of 90°, and they are detected. The backward scattering is due to the repulsion between ion and nucleus called coulomb repulsion, and this is called Rutherford backscattering (RBS). The collision is elastic and unresponsive to chemical bonding of atoms of sample. By measuring the backscattered ion energy and number, the atomic data and its concentration can be determined as the backscattered energy depends on energy and mass of incident ion and mass of target atom. It also depends upon the scattering angle (Chu 2012).

7.2.18.2 Secondary Ion Mass Spectrometry (SIMS)

This technique is used for analyzing exceedingly lesser concentration of elements or atoms present in a solid which is a major limitation of other analytical techniques. It can detect all the elements of periodic table, unlike RBS which can only detect those elements whose spectra do not overlap. In this technique, ions are incident on the sample surface, and it scatters most of the neutral atoms and a few positive as well as negative ions from the outermost shell. These ions have mass which is analyzed to determine the type and concentration of the atomic species in the sample. There are generally two types of SIMS, viz., static and dynamic. Static SIMS involves data collection before the incidence of surface bombardment and thus determines the surface structure. Dynamic SIMS depends on scattering rates of ions and thus helps in-depth analysis (Cao 2004).

7.3 Enzyme Immobilization

Enzymes play a vital role in biocatalysis owing to their high substrate specificity and simple production and posing lesser threats to the environment. For significant development, the utilization of enzymes becomes finite, as they are not as cost-effective because of their high cost and low recyclable factor. Throughout the biochemical reaction, in the presence of an organic solvent at a higher temperature and acidic or alkaline pHs, these enzymes are unable to reuse as these cannot conserved from the reaction medium and affects the molecular structure of enzymes (Naqash et al., 2019). To overcome this, immobilization of enzymes with practical potency and enhanced reliability is more straightforward despite their high cost. Whole cells and enzymes are used as immobilized biocatalysts frequently (Kawaguti et al., 2006). Immobilized enzyme refers as a catalyst that constricts to a certain extent; however, their catalytic activities are preserved (Brena and Francisc 2006). Immobilized enzymes are essential, as they can be reused and able to sequester an enzyme, enhance stability and favorable variations in temperature and pH, and maintain activity for an extended duration (Soozanipour and Taheri-Kafrani 2018). Immobilized enzyme can be attained by binding the enzyme enclosed by a carrier matrix. The enzyme molecule and the product formed ought to move naturally in and out of the part to that enzymes were controlled. Biocompatible carrier matrices that are insoluble in water used ought to have a vast surface area with functional groups (Naqash et al., 2019). Carrier matrices used for enzyme immobilization should have the following properties (Singh 2009) as shown in Fig. 7.6.

Selection of carrier matrices for the enzyme immobilization should have properties as mentioned earlier. These carrier matrices can be categorized as organic (natural and synthetic) and inorganic. Natural organic carrier matrices include agar,

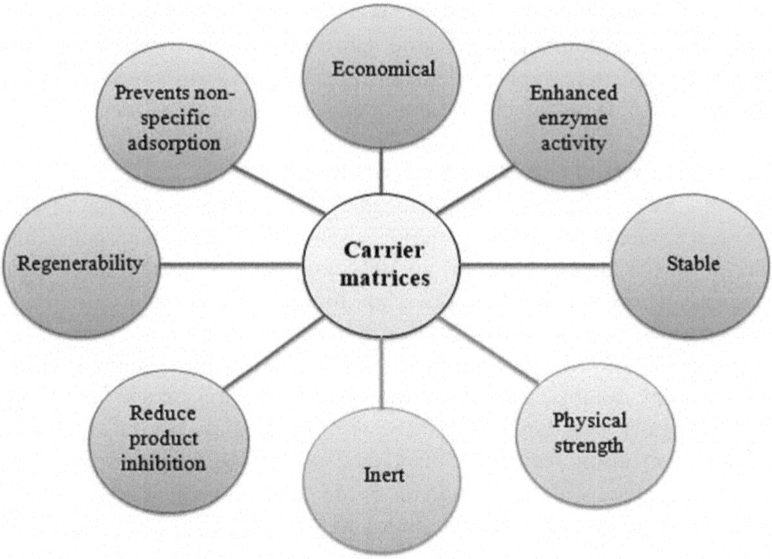

Fig. 7.6 Properties of carrier matrices

cellulose, agarose, dextran, chitin, albumin, alginate, and collagen, while synthetic organic carrier matrices include polynorbornenes, polystyrene, polymethacrylate, polyacrylate, polyacrylamide, polyamide, polyisobutylene, and polyglycerol. Inorganic carrier matrices include bentonite, silica, glass, metals, and hydroxyapatite. Popat et al. (2011) reported that eco-friendly carrier matrices are biocompatible, biodegradable, efficient, and economical and prevent ethical issues. Eco-friendly carrier matrices such as coconut fibers show increased cation exchange property and better water-holding capacity; kaolin shows increased retainability of the enzyme upon acetylation; cellulose has the irreversible binding ability, and mesoporous materials have functionalized thiol group.

Immobilization of enzymes onto carrier matrices results in higher productivity due to their long-term reusability, with further definite control of the degree of reaction, the potential of mechanization, and steady operation and abolition of enzyme inactivation (Pundir 2015). An immobilized enzyme has potential merit: (i) easy to handle, (ii) low contamination, (iii) productive recovery of enzyme, (iv) reusable, (v) minimum chances of allergenicity, (vi) increased enzyme stability, and (vi) higher productivity and low cost. Immobilized enzymes also have some demerits that restrict its application such as high cost of carrier matrices, reduced enzyme activity, and steady mass transfer by the carrier matrices. At present, immobilized enzymes like lipase, glucose isomerase, and lactase contribute to extensive production (Gutarra et al., 2016). Expansion in enzyme immobilization thus leads to the use of nanomaterials. These nanomaterials enhance the immobilization because of their increased surface area in comparison to the conventional carrier matrices. Enzyme immobilization may be implemented either reversibly or irreversibly.

Reversible immobilization allows adsorption directly to the surface by encouraging the interaction through hydrophobic interactions, hydrogen bonding, electrostatic affinity interactions, and van der Waals forces, thus enabling the recovery of carrier matrices. In irreversible immobilization, enzymes are covalently attached to the surface or entrapped within a carrier matrix to halt its leaching; however, the carrier matrices cannot be reused. Specific properties of carrier matrices are essential to attain an optimal enzyme immobilization. Up-to-date advancement in nanotechnology has enhanced multiple alternatives for immobilized enzyme. Nanomaterials embellish better substitutes because of their small surface area, increased diffusivity, and low mass transfer rate. The amalgamation of nanomaterials with enzymes forms nanobiocatalyst, which contributes to green technology by replacing the use of chemical catalysts (Naqash et al., 2019). Enzyme immobilization on nanomaterials is an encouraging abstraction for its application in the area of bioenergy.

Moreover, a basic perceptive of biomolecular actions among the enzyme and the nanomaterial surroundings is necessary for its accomplished application. These enzymes show a necessary attribute, by facilitating specific three-dimensional configurations intended for their catalytic action. Enzyme active site is the primary aspect of stabilizing the conversion among the substrate and its product. Varying the whole conformation of the molecule affects biological activity. Usually, the freeze-dried enzyme is used which shows low activity and stability.

Conversely, immobilization of the enzyme on a carrier matrix may provide benefits like elevated catalytic activity, recovery of enzyme, and continuous operation. A study by Chibata (1978) on proteins resulted in 60% better stability on immobilization with 24% unaltered and 16% low stability. Nowadays, the use of nanomaterials is intended for elevating enzyme retention with increased biocatalytic activity and efficiency. For enzyme immobilization, nanomaterials such as nanofibers, mesoporous media, nanorods, carbon nanotubes, and nanocomposites are used resulting in increased stability and quantity of immobilized enzymes owing to their enhanced surface area and biocompatibility. Production of active biological systems by the interaction of enzymes and nanomaterials involves the expansion and characterization of nanomaterial and assessment of biological activity (Andreescu et al. 2008).

7.4 Techniques for Enzyme Immobilization

For immobilization of enzyme, various methods are reported. These techniques can be categorized into either reversible or irreversible or based on their chemical reaction. Each immobilization methods have definite merits and demerits. Characteristics of enzyme and carrier matrices are essential for the assortment of a suitable technique for enzyme immobilization (Adhikari 2019). There are four techniques of enzyme immobilization:

(i) Carrier-binding method or adsorption which uses synthetic polymers, polysaccharide derivatives, and glass as a carrier matrix

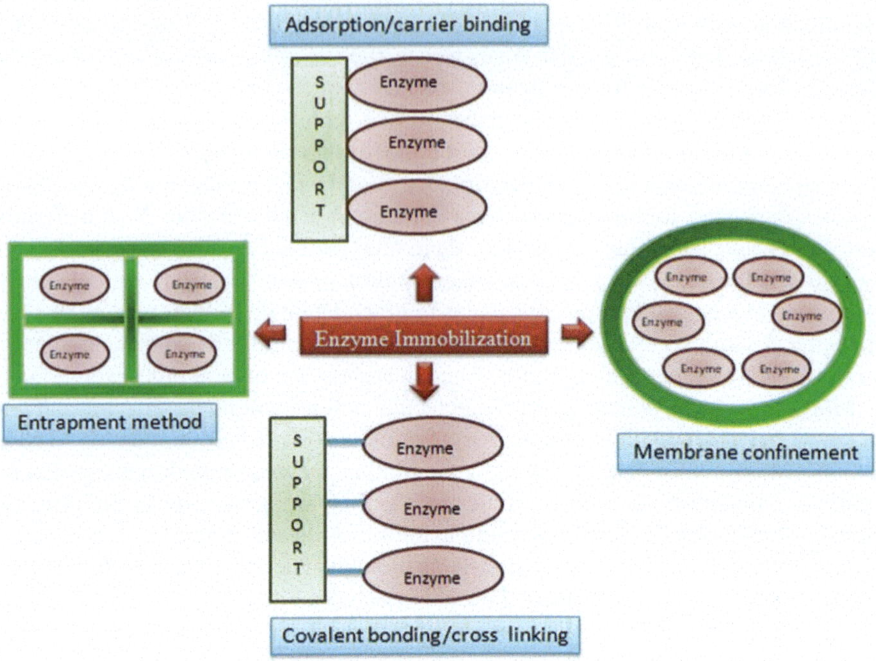

Fig. 7.7 Various methods of enzyme immobilization

(ii) Covalent bonding or cross-linking method which uses glutaraldehyde, hexamethylene diisocyanate, and bis(diazo)benzidine
(iii) Entrapment method which uses polymers like collagen and κ-carrageenan for entrapping enzyme
(iv) Membrane confinement method which uses microcapsules as carrier matrix

Figure 7.7 represents the different mode of enzyme immobilization

1. *Adsorption/carrier-binding method*

Adsorption or carrier-binding method of enzyme immobilization is a simple method where an adsorbent is being mixed with the enzyme to be immobilized under the appropriate environment and is kept for incubation to wash off loosely bound proteins. A carrier matrix bounds the enzymes through different types of bonding such as hydrogen, ionic, covalent, and van der Waals forces. Selection of the carrier matrix depends on its affinity with the enzyme molecule, as not all enzymes can bind every carrier matrices. It also relies on the particle size of the carrier matrix, active group present on the surface area, and pore size. Carrier matrices are assembled as void porous fiber membrane for adsorption of the enzyme to be immobilized. This method for the production of immobilized enzyme becomes a reversible process as soon as the hydrogen ion concentration, temperature, ionic potency, and polarity of the solvent are reformed (Sneha et al., 2019). Thus, this helps in the noninvasive and straightforward recovery of the enzyme (Qi et al., 2011). This method helps to prevent the physical and chemical modification of enzyme by joining the enzymes inward or outward to the surface of the carrier matrix. Silicone is used to prevent the leakage of the enzyme adsorbed to the customized adsorbent. For instance, *Candida antarctica* lipase B (novozyme 435) prevents leakage of the enzyme by coating the enzyme surface by a silicon polymer (Sneha et al., 2019). Immobilization of lipase by using octyl-agarose and octadecyl-sepabeads has been studied resulting in 97% immobilization. As compared to free lipase, octadecyl-sepabeads attained tenfold increased enzyme stability, whereas the enzyme stability of octyl-agarose is decreased (Cunha et al., 2007). Immobilization of *Candida rugosa* lipase with biodegradable poly(3-hydroxybutyrate-co-hydroxyvalerate) revealed that after 4 h at 50 °C, the residual activity is 94% with reusability till 12 cycles (Cabrera-Padilla et al., 2012). Mateo et al. (2000) observed that coating of polyethylenimine covalently on various adsorbent carrier matrices like silica, agarose, and polymeric resins resulted in effective and strong D-amino acid oxidase β-galactosidase and lipase immobilization. The adsorption between the carrier matrices and enzyme molecule is not stable, whereas enzyme immobilized by covalent bonding is a limitation of this method of immobilization. Sabbani et al. (2006) observed that during the process of biocatalysis, the enantiomeric ratio and reaction rates are higher with the reduction in particle size of carrier matrix Accurel. Mishra et al. (2011) studied the adsorption of urease with 1,4-butenediol diglycidyl resulting in high pH stability with 50% retention activity.

2. *Covalent-bonding/cross-linking method*

Enzyme immobilization by covalent bonding results in strong bonding of active groups on the enzyme with the carrier matrices. Chemical properties of carrier matrices and the active groups present on the enzyme molecules are an important factor for enzyme immobilization by covalent bonding (Agyei et al., 2015). Functional groups on the surface of the enzyme molecules are amino group, carboxylic group, hydroxyl group, and sulfhydryl group (Adhikari 2019). Immobilization of enzyme by cross-linking binds the enzyme to the carrier matrices by forming a three-dimensional structure and prevents desorption of the enzyme molecule. This method is an irreversible method that is advantageous, as the hydrogen ion concentration and substrate potency do not reverse. However, the limitation is that this method of immobilization is relatively expensive, and at the time of immobilization, it may block the active site of the enzyme molecule of which the enzyme rigidifies. With an increase in hydrogen ion concentration, the enzyme activity decreases. To overcome this, researches continue to produce a stable and robust immobilized enzyme for the food sector. For example, immobilization of plant protease enzyme (cardiosin A) on agarose-glutaraldehyde has been used as rennet enzyme for cheesemaking and processing of whey in the dairy industry (Agyei et al., 2015). Enzyme immobilization by chitosan and mesoporous silica as carrier matrices resulted in higher thermal stability and half-life of immobilized enzyme. When modified silica gel is used as a carrier matrix, the covalently immobilized enzymes act as highly stable and hyperactive biocatalysts. The process of silanization modifies silica gels where it eliminates the aldehyde groups. Carrier matrices used for immobilization may lead to a decrease in enzyme activity which might be overcome by immobilization by enzyme cross-linking. This immobilization process does not require carrier matrices, which are advantageous over carrier-bound immobilized enzyme system (Agyei et al., 2015). For example, the formation of a bond with an amino group immobilizes (Poulsen 1984). Cross-linking method of immobilization can be classified as (i) cross-linked dissolved enzymes (CLEs), where the enzymes to be immobilized cross-linked together; (ii) cross-linked crystalline enzymes (CLECs), where the crystals of enzymes cross-linked together; and (iii) cross-linking aggregated enzymes (CLEAs), where aggregates of enzymes are cross-linked together. CLECs and CLEAs are used commonly for the immobilization of enzyme because of their easy handling and reliable nature. CLEA method immobilizes enzymes such as chymotrypsin, trypsin, lipase, subtilisin, naringinase, and cellulases, which results in high enzyme stability and improved enzyme activity. This method is inexpensive as the carrier matrices are not used (Agyei et al., 2015). Enzymes cross-linked to electrospun nanofiber increases the porosity and surface area by increasing its residual activity. Cross-linking agents help to sustain the function and structural characteristics of enzymes. Glutaraldehyde is used as a bifunctional cross-linker as it is soluble in aqueous solvents (Datta et al., 2013).

Case Study

Cross-linked aggregates of enzymes can form in two steps: (i) protein precipitation and (ii) coagulation. First, the enzyme (5–25 mg of protein/ml) was mixed with ammonium sulfate to attain 60–90% of salt saturation. Second, the mixture was homogenized by using a magnetic stirrer for 2 h at 4 °C. Third, glutaraldehyde was added and mixed using a magnetic stirrer for 24 h at 4 °C. Fourth, centrifugation of the mixture was done at 14,000 rpm for 10 min to separate the enzyme aggregates. And lastly, the aggregates were washed using a buffer and stored at 4 °C (Gutarra et al., 2016).

3. *Entrapment method*

Enzymes are immobilized by entrapping it in a polymeric network. This network allows the substrate to pass through and preserves the enzymes. The carrier matrices create a barrier for mass transfer. Entrapment refers to entrapping the enzymes within a fiber or gel matrix. There are different methods of entrapment like temperature-induced gelation (like agarose and gelatin), ionotropic gelation of macromolecules with alginate (multivalent cation), and organic polymerization (like polysaccharide), for instance, entrapping amyloglucosidase or α-amylase with a polyacrylamide gel. Immobilization by entrapping reduces the blockage of enzyme active site (Adhikari 2019). Entrapment of enzyme also controls the thickness of the carrier matrix (Andreescu et al. 2008). The demerit of this process of immobilization is that the enzymatic reactions are retarded because of diffusional control of the product and the substrate and a loss of enzyme from the matrix pore. Enzymes are immobilized by confining it within the insoluble matrix. Thus, it prevents the leaching of the enzymes, but if the carrier matrix is not uniform or the pore size is increased, then leaching may occur. To overcome this, enzymes are premixed with the carrier matrix and form a gel with desired particle size. This method helps to maintain the structural integrity of the enzyme. Studies on the entrapment method of immobilization show that silica can be used as an entrapment material. Forsyth and Patwardhan (2013) reported that entrapment of *Candida antarctica* lipase in bio-inspired silica increases the efficiency of immobilization, enhances the enzyme activity, and betters thermal stability and pH (Agyei et al., 2015). Enzymes are immobilized by using activators like quinoline derivatives, organic salts, tetrathiafulvalene, and ferrocene that help transport the electrons prevailing at the enzyme active site to the surface of electrode (Andreescu et al. 2008). Commonly used carrier matrix for entrapment is alginate as it forms beads which can entrap various macromolecules (e.g., calcium-alginate beads). Shen et al. (2011) reported that use of alginate-gelatin-calcium hybrid carriers for the entrapment provides efficient encapsulation by hampering the leaching of enzyme and increasing the mechanical stability. Electrospun nanofibers when used as carrier matrix for entrapment show a significant impact on immobilization of enzyme (Sneha et al., 2019). Entrapment of *Candida rugosa* lipase in chitosan prevents the leakage and friability of enzyme and augments the entrapment efficiency (Datta et al., 2013). Ispas et al. (2009) studied the entrapment by mesoporous silica which provides uniform pore distribution, higher surface area, and enhanced adsorption capacity. Magnetite and lipase nanoparticles entrapped at the same time in biomimetic silica increase the enzyme activity. Entrapment of *Candida rugosa* lipase done in sol-gel matrices enhances their selective binding (Datta et al., 2013). κ-carrageenan used for entrapment of lipases showed high thermal stability.

4. *Membrane confinement*

Immobilization of enzymes by membrane confinement refers to encapsulation of enzyme in a carrier matrix. This method of immobilization is easy and low cost. Encapsulation can be done by two methods, viz., encapsulation in the matrix and gel technology. Natural or synthetic gels are used as a carrier matrix. Natural gels can be obtained from alginate, agar, or κ-carrageenan and synthetic gels from acrylamide. Gels obtained from acrylamide are used as carrier matrices for entrapment due to their high molecular weight which is cross-linked in the polymer network. These gels absorb the solvent by the process of osmosis or by capillary forces. The internal structure of the gel relies upon the cross-linker. Choice of cross-linkers should be accurate as they modify the pore size. When an optimized gel is used for entrapment, it separates the carrier matrices from diffusional resistances. Agar consists of agarose and agaropectins found in red algae cell wall. It can be degraded enzymatically in two types of agarases by hydrolysis, specifically β- and α-agarases which hydrolyze β-1, 4-linkages and α-1,-3 linkages in agarose, respectively. The α-agarase segregates the α-1-(1,3) bonding of agarose to yield oligosaccharides such as 3,6-anhydro-1-galactopyranose at the reducing end of the sugar (Datta et al., 2013). In contradiction, β-agarase segregates the β-d-(1,4) bonding of agarose to yield neoagro-oligosaccharides with d-galactopyranoside residues at the reducing end of the sugar. Neoagro-oligosaccharides obtained from β-agarase are of great interest due to their various biological functions, like inhibiting the growth of bacteria, retarding starch degradation, and improving the quality of food as low-calorie additives. Thus, agar-derived oligosaccharides have a large area of application in industries such as biodiesel (Kim et al., 2018).

7.5 Application of Nanomaterial-Immobilized Enzyme in Biofuel Production

Enzymes play an essential role in different industrial processes focusing on the improvement of enzyme's activity along with its efficiency and stability. Different applications of the nano-immobilized enzyme on bioenergy-based industry have been studied. Table 7.2 represents the immobilized enzyme application in the bioenergy-based industry.

Table 7.2 Immobilized enzyme application in bioenergy industry

S no	Immobilized enzyme	Nanomaterial	Carrier matrix	Application
1.	Cellulase	Activated magnetic nanoparticle	Activated magnetic nanoparticle	Enhancing enzymatic saccharification of pretreated hemp biomass
2.	Cellulase	Silica nanoparticle	Microcrystalline cellulose	Production of ethanol and glucose
3.	β-Glucosidase (*Aspergillus niger*)	Iron oxide nanoparticle	Cellobiose	Production of glucose
4.	β-Glucosidase (*Trichoderma reesei*)	Magnetic nanoparticle	Wheat straw, *Eucalyptus globules* pulp	Production of glucose
5.	Lipase (*Burkholderia* sp.)	Ferric silica nanocomposite	Olive oil and methanol	Biodiesel
6.	Lipase (*Rhizopus miehei*)	Silica nanoparticle	Triolein and methanol	Biodiesel
7.	Lipase (*Pseudomonas cepacia*)	Nanoporous gold	Soybean and methanol	Biodiesel
8.	Lipase (*Candida rugosa*)	Carbon nanotubes	Multi-walled carbon nanotube (MWCNT)	Enhanced enzyme activity
9.	Lipase (*Candida rugosa*)	Nanofibers	Polyvinyl alcohol	Equivalent esterification activity similar to Novozym 435

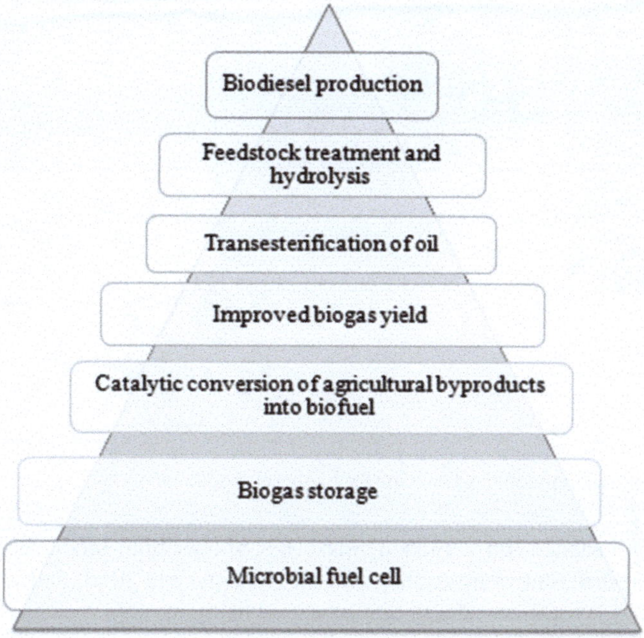

Fig. 7.8 Schematic view of the application of nano-immobilized enzymes in the bioenergy industry

In recent times, using nanomaterials as a substrate for enzyme immobilization shows an increasing trend for production of biodiesel. This nano-immobilized enzyme contributes to enhancing the stability and activity of the enzyme and reduces the expenses of the enzyme used. Figure 7.8 represents the schematic view of the probable application of nano-immobilized enzyme in the bioenergy industry.

There are reports on application of immobilized enzymes in biofuel production such as covalent lipase immobilization on magnetic nanoparticles for biodiesel production (Wang et al., 2009). The conversion rate of 100% was seen in the first three cycles during the conversion of soybean oil to biodiesel using immobilized lipase (*Pseudomonas* cepacia). This high conversion indicates that immobilized lipase (*Pseudomonas* cepacia) preserves stability and reuse of enzyme during the transesterification of soybean oil. Raita et al. (2015) stated that biodiesel produced from palm oil by using immobilized lipase (obtained from *Thermomyces lanuginosus*) on magnetic nanoparticles showed 97.2% yields with 80% enzyme activity retention (Fig. 7.9).

Karimi (2016) reported the conversion of waste cooking oil into biodiesel using immobilized *Burkholderia cepacia* lipase using superparamagnetic iron oxide nanoparticles with a yield of 91% in 35 hours. Fan et al. (2016) reported that immobilization of *Rhizomucor miehei*-derived lipase onto polyamidoamine grafted with magnetic MWCNT resulted in the conversion of waste cooking oil to biodiesel with a yield of 94%. Nano-immobilized enzyme resulted in increased production of

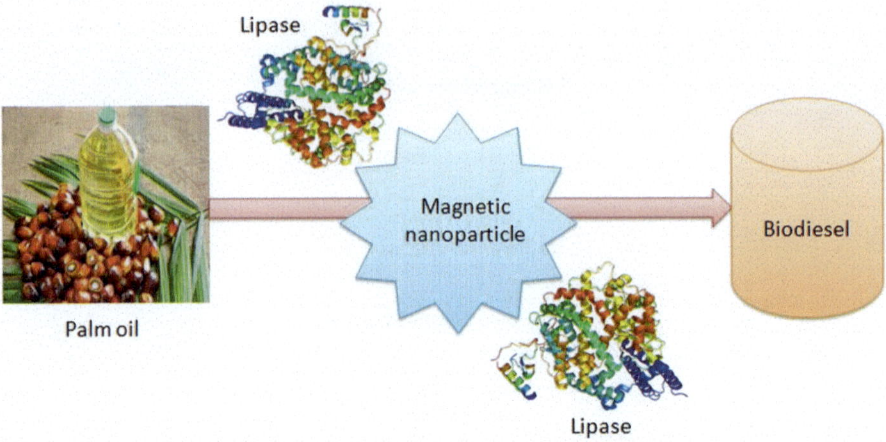

Fig. 7.9 Use of nano-immobilized enzyme for biodiesel production

biodiesel; for instance, immobilized lipase has shown high biodiesel production obtained from different sources. Soybean oil (Xie and Ma 2010) and waste grease (Ngo et al., 2013) for the production of biodiesel using immobilized lipase obtained from *Thermomyces lanuginosus* covalently bound to magnetic nanoparticles were previously reported. Nano-immobilized enzymes are efficient for the transesterification of the oil. For example, immobilized lipase (*Burkholderia* sp.) adsorbed on ferric-silica nanocomposites is used for the transesterification of olive oil (Tran et al., 2012). Cellulase immobilized by physical adsorption on silica nanoparticles resulted in enhanced ethanol production (Lupoi & Smith 2011). Nano-immobilized enzymes are used to perk up the yield of biogas. Studies have shown the increased and improved yield of biogas by using nanomaterial (Palaniappan 2017). Nano-immobilized enzyme application for the conversion of agricultural by-products to biofuels results in enhanced biodegradability and eco-friendly biofuels. Nanosized molybdenum balls are used for the conversion of palm by-products to biofuels. Various applications of nano-immobilized enzymes in the bioenergy industry are, for the storage of biogas, to recover the performance of microbial fuel cell using nanotechnology (Zhang et al., 2010).

7.6 Conclusions

In the modern era, nanotechnology has gained profound attention owing to its vast application in all the fields including pharmaceutical, agriculture, food, and biotechnology. Nanomaterials have been widely used in biomedical and bioprocess applications such as synthesis of biosensors, aerodynamic devices, and also enzyme immobilization. There is extensive research going on the synthesis of various nanomaterials for enzyme immobilization. Although significant efforts have been made in the nanomaterial synthesis and its characterization, it is still in the nascent phase. There is a need to look for an innovative approach for nanomaterial synthesis which should be efficient, less complicated, and economical. The nanomaterial-based enzyme immobilization has shown significant improvement in the enzyme activity and stability. This technology has been quite successful in the different enzyme-based applications as desired in detergent, pharmaceutical, drug, and agricultural-based products. The nanomaterial-based enzyme application has also shown tremendous application in bioenergy industry where it has been successfully used for biodiesel production along with feedstock treatment, biogas storage, and catalytic conversion of lipids into biodiesel. There is further scope of improvement in the approach adopted for nanomaterial synthesis so that it can be widely used at commercial level and should be cost-effective too.

References

Adhikari S (2019) Application of immobilized enzymes in the food industry. In: Enzymes in food biotechnology. Academic, New York, pp 711–721

Agyei D, Shanbhag BK, He L (2015) Enzyme engineering (immobilization) for food applications. In: Improving and tailoring enzymes for food quality and functionality, pp 213–235

Alex, Gruverman (2004) Nanostructures & nanomaterials: synthesis, properties & applications. Imperial College Press, London

Andreescu S, Njagi J, Ispas C (2008) Nanostructured materials for enzyme immobilization and biosensors. In: The new frontiers of organic and composite nanotechnology, pp 355–394

Arora N, Sharma NN (2014) Arc discharge synthesis of carbon nanotubes: comprehensive review. Diam Relat Mater 50:135–150

Aruna ST, Mukasyan AS (2008) Combustion synthesis and nanomaterials. Curr Opinion Solid State Mater Sci 12(3–4):44–50

Bhushan B (2001) Nano- to microscale wear and mechanical characterization using scanning probe microscopy. Wear 251(1–12):1105–1123

Bhushan B (ed) (2008) Nanotribology and nanomechanics: an introduction. Springer Science & Business Media

Bilecka I, Niederberger M (2010) Microwave chemistry for inorganic nanomaterials synthesis. Nanoscale 2(8):1358–1374

Brena B, Francisc BV (2006) Immobilization of enzymes: a literature survey. In: Immobilization of enzymes and cells. Humana Press, Totowa

Brodusch N, Demers H, Trudeau M, Gauvin R (2013) Acquisition parameters optimization of a transmission electron forward scatter diffraction system in a cold-field emission scanning electron microscope for nanomaterials characterization. Scanning 35(6):375–373

Cabrera-Padilla RY, Lisboa MC, Fricks AT, Franceschi E, Lima AS, Silva DP, Soares CM (2012) Immobilization of Candida rugosa lipase on poly (3-hydroxybutyrate-co-hydroxyvalerate): a new eco-friendly support. J Ind Microbiol Biotechnol 39(2):289–298

Cao G (2004) Nanostructures & nanomaterials: synthesis, properties & applications. Imperial college press, London

Carlson T (2013) Photoelectron and auger spectroscopy. Springer Science & Business Media

Chen X, Mao SS (2007) Titanium dioxide nanomaterials: synthesis, properties, modifications, and applications. Chem Rev 107(7):2891–2959

Chibata I (1978) Immobilized enzymes, research and development. Kodansha, Tokyo

Chu WK (2012) Backscattering spectrometry. Elsevier

Cullity BD, Stock SR (2014) Elements of X-ray diffraction. Pearson Education

Cunha AG, Fernández-Lorente G, Bevilaqua JV, Destain J, Paiva LM, Freire DM et al (2007) Immobilization of Yarrowia lipolytica lipase – a comparison of stability of physical adsorption and covalent attachment techniques. In: Biotechnology for fuels and chemicals. Humana Press, Totowa, pp 169–176

Datta S, Christena LR, Rajaram YRS (2013) Enzyme immobilization: an overview on techniques and support materials. 3 Biotech 3(1):1–9

De Lozanne A (2006) Application of magnetic force microscopy in nanomaterials characterization. Microsc Res Tech 69(7):550–562

Dudkiewicz A, Tiede K, Loeschner K, Jensen LHS, Jensen E, Wierzbicki R et al (2011) Characterization of nanomaterials in food by electron microscopy. TrAC Trends Analyt Chem 30(1):28–43

Fan Y, Wu G, Su F, Li K, Xu L, Han X, Yan Y (2016) Lipase oriented-immobilized on dendrimer-coated magnetic multi-walled carbon nanotubes toward catalyzing biodiesel production from waste vegetable oil. Fuel 178:172–178

Fernández-Pacheco R, Arruebo M, Marquina C, Ibarra R, Arbiol J, Santamaría J (2006) Highly magnetic silica-coated iron nanoparticles prepared by the arc-discharge method. Nanotechnology 17(5):1188

Forsyth C, Patwardhan SV (2013) Controlling performance of lipase immobilised on bioinspired silica. J Mater Chem B 1(8):1164–1174

Gnanam S, Rajendran V (2011) Synthesis of CeO_2 or α–Mn_2O_3 nanoparticles via sol–gel process and their optical properties. J Sol-Gel Sci Technol 58(1):62–69

Gutarra ML, Miranda LS, de Souza RO (2016) Enzyme immobilization for organic synthesis. In: Organic synthesis using biocatalysis. Academic Press, pp 99–126

Ispas C, Sokolov I, Andreescu S (2009) Enzyme-functionalized mesoporous silica for bioanalytical applications. Anal Bioanal Chem 393(2):543–554

Jafarzadeh M, Rahman IA, Sipaut CS (2009) Synthesis of silica nanoparticles by modified sol–gel process: the effect of mixing modes of the reactants and drying techniques. J Sol-Gel Sci Technol 50(3):328–336

Jeng US, Su CH, Su CJ, Liao KF, Chuang WT, Lai YH et al (2010) A small/wide-angle X-ray scattering instrument for structural characterization of air–liquid interfaces, thin films and bulk specimens. J Appl Crystallogr 43(1):110–121

Jiang H-B, Pan L-F, Liu P-F, Fang W-Q, Yang H-G (2014) Synthesis of well-defined functional crystals by high temperature gas-phase reactions. Chin Sci Bull 59:2135–2143

Karimi M (2016) Immobilization of lipase onto mesoporous magnetic nanoparticles for enzymatic synthesis of biodiesel. Biocatal Agric Biotechnol 8:182–188

Kawaguti HY, Manrich E, Sato HH (2006) Production of isomaltulose using *Erwinia* sp. D12 cells: culture medium optimization and cell immobilization in alginate. Biochem Eng J 29(3):270–277

Kim K, Lee OK, Lee E (2018) Nano-immobilized biocatalysts for biodisel production from renewable and sustainable resources. Catalysts 8(2):68

Kumar CS (ed) (2012) Raman spectroscopy for nanomaterials characterization. Springer Science & Business Media, Berlin

Kumar CS (ed) (2013) UV-VIS and photoluminescence spectroscopy for nanomaterials characterization. Springer, Berlin/Heidelberg, pp 231–285

Li T, Senesi AJ, Lee B (2016) Small angle X-ray scattering for nanoparticle research. Chem Rev 116(18):11128–11180

Lupoi JS, Smith EA (2011) Evaluation of nanoparticle-immobilized cellulose for improved yield in simultaneous saccharification and fermentation reactions. Biotechnol Bioeng 108:2835–2843

Mateo C, Abian O, Fernandez-Lafuente R, Guisan JM (2000) Reversible enzyme immobilization via a very strong and nondistorting ionic adsorption on support–polyethylenimine composites. Biotechnol Bioeng 68(1):98–105

Mishra N, Pithawala K, Bahadur A (2011) Byssus thread: a novel support material for urease immobilization. Appl Biochem Biotechnol 165(7–8):1568–1576

Moram MA, Vickers ME (2009) X-ray diffraction of III-nitrides. Rep Prog Phys 72(3):036502

Mukasyan AS, Dinka P (2007) Novel approaches to solution-combustion synthesis of nanomaterials. Int J Self-Propag High-Temp Synth 16(1):23–35

Mukasyan AS, Epstein P, Dinka P (2007) Solution combustion synthesis of nanomaterials. Proc Combust Inst 31(2):1789–1795

Naqash F, Masoodi FA, Rather SA (2019) Food enzymes and nanotechnology. In: Enzymes in food biotechnology. Academic, New York, pp 769–784

Ngo TP, Li A, Tiew KW, Li Z (2013) Efficient transformation of grease to biodiesel using highly active and easily recyclable magnetic nanobiocatalyst aggregates. Bioresour Technol 145:233–239

Palaniappan K (2017) An overview of applications of nanotechnology in biofuel production. World Appl Sci J 35(8):1305–1311

Popat A, Hartono SB, Stahr F, Liu J, Qiao SZ, Lu GQM (2011) Mesoporous silica nanoparticles for bioadsorption, enzyme immobilisation, and delivery carriers. Nanoscale 3(7):2801–2818

Poulsen PB (1984) Current applications of immobilized enzymes for manufacturing purposes. Biotechnol Genet Eng Rev 1(1):121–140

Pundir CS (2015) Enzyme nanoparticles: preparation, characterisation, properties and applications. William Andrew, Amsterdam

Qi B, Chen X, Su Y, Wan Y (2011) Enzyme adsorption and recycling during hydrolysis of wheat straw lignocellulose. Bioresour Technol 102(3):2881–2889

Qiu G, Huang H, Dharmarathna S, Benbow E, Stafford L, Suib SL (2011a) Hydrothermal synthesis of manganese oxide nanomaterials and their catalytic and electrochemical properties. Chem Mater 23(17):3892–3901

Qiu G, Dharmarathna S, Zhang Y, Opembe N, Huang H, Suib SL (2011b) Facile microwave-assisted hydrothermal synthesis of CuO nanomaterials and their catalytic and electrochemical properties. J Phys Chem C 116(1):468–477

Rahman IA, Padavettan V (2012) Synthesis of silica nanoparticles by sol-gel: size-dependent properties, surface modification, and applications in silica-polymer nanocomposites—a review. J Nanomater 2012:8

Rahman IA, Vejayakumaran P, Sipaut CS, Ismail J, Chee CK (2008) Effect of the drying techniques on the morphology of silica nanoparticles synthesized via sol-gel process. Ceram Int 34(8):2059–2066

Raita M, Arnthong J, Champreda V, Laosiripojana N (2015) Modification of magnetic nanoparticle lipase designs for biodiesel production from palm oil. Fuel Process Technol 134:189–197

Rao CNR, Cheetham AK (2001) Science and technology of nanomaterials: current status and future prospects. J Mater Chem 11(12):2887–2894

Sabbani S, Hedenström E, Nordin O (2006) The enantioselectivity of Candida rugosa lipase is influenced by the particle size of the immobilising support material Accurel. J Mol Catal B Enzym 42(1–2):1–9

Shen Q, Yang R, Hua X, Ye F, Zhang W, Zhao W (2011) Gelatin-templated biomimetic calcification for β-galactosidase immobilization. Process Biochem 46(8):1565–1571

Singh BD (2009) Biotechnology (expanding horizons). Kalyani, New Delhi

Sneha HP, Beulah KC, Murthy PS (2019) Enzyme immobilization methods and applications in the food industry. In Enzymes in food biotechnology. Academic, New York, pp 645–658

Soozanipour A, Taheri-Kafrani A (2018) Enzyme immobilization on functionalized graphene oxide nanosheets: efficient and robust biocatalysts. Methods Enzymol 609:371–403

Su Y, Zhang Y (2015) Carbon nanomaterials synthesized by arc discharge hot plasma. Carbon 83:90–99

Tran DT, Chen CL, Chang JS (2012) Immobilization of Burkholderia sp. lipase on a ferric silica nanocomposite for biodiesel production. J Biotechnol 158(3):112–119

Wang X, Dou P, Zhao P, Zhao C, Ding Y, Xu P (2009) Immobilization of lipases onto magnetic Fe_3O_4 nanoparticles for application in biodiesel production. ChemSusChem 2(10):947–950

Xia Y, Yang P, Sun Y, Wu Y, Mayers B, Gates B, Yin Y, Kim F, Yan H (2003) One-dimensional nanostructures: synthesis, characterization, and applications. Adv Mater 15(5):353–389

Xie W, Ma N (2010) Enzymatic transesterification of soybean oil by using immobilized lipase on magnetic nano-particles. Biomass Bioenergy 34(6):890–896

Xu K, Cao P, Heath JR (2009) Scanning tunneling microscopy characterization of the electrical properties of wrinkles in exfoliated graphene monolayers. Nano Lett 9(12):4446–4451

Zhang XL, Yan S, Tyagi RD, Surampalli RY, Zhang TC (2010) Application of nanotechnology and nanomaterials for bioenergy and biofuel production. In: Bioenergy and biofuel from biowastes and biomass. American Society of Civil Engineers, Reston, pp 478–496

Nanomaterial Synthesis and Mechanism for Enzyme Immobilization: Part II

8

Ankit Kumar Singh and Ida Tiwari

Abstract

Free enzymes do not possess properties of recovery and reusability, and also they are not stable at wide pH and temperature range. Therefore, new ways which can enhance enzyme stability and reusability should be developed, and hence, the immobilization technique is one such approach. These immobilization techniques offer such materials which have the ability to be active in the much wide range of pH and temperature, and also they are more stable than the free enzymes. Immobilization is carried out on the nanosized material either by adsorption, covalent coupling, entrapment, encapsulation or cross-linking. These nanomaterial-immobilized enzymes show several advances over the free enzymes because of large surface area-to-volume ratio, lower mass transfer resistance and high mobility. Several nanomaterials are used for immobilizing the enzymes; however, their recovery from the reaction mixture is very poor. Therefore, the magnetic nanomaterials are more attractively used in immobilization because the enzyme immobilized through magnetic nanomaterial has the tendency to be easily separated out from the reaction mixture. These nanomaterial-immobilized enzymes show wide range of applications in biotechnology, bioanalysis, biomedicine, pathology and biosensors.

Keywords

Nanomaterial · Enzyme · Immobilization · Biotechnology · Biomedicine · Nanocarrier

A. K. Singh · I. Tiwari (✉)
Department of Chemistry, Centre of Advanced Study, Institute of Science,
Banaras Hindu University, Varanasi, Uttar Pradesh, India
e-mail: idatiwari@bhu.ac.in

M. Srivastava et al. (eds.), *Nanomaterials in Biofuels Research*,
Clean Energy Production Technologies, https://doi.org/10.1007/978-981-13-9333-4_8

8.1 Introduction

Enzymes are protein molecules that enhance the rate of biochemical reactions, but they are not used up during the reactions. Enzymes are biological catalyst and show a lot of dominance over any chemical catalyst. Besides having high catalytic efficiency, enzymes are also chemoselective, regioselective and stereoselective, and hence, they are widely used in the industries (Gupta et al. 2011). Enzyme can catalyse several chemical and biochemical reactions. Mankind has used enzyme in the processing of food from several millennia. Various raw materials such as oils, fats, carbohydrates, proteins, vitamins, lignin and amino acids are used for the production of several bio-renewables. These compounds are also important in our daily life in the form of wood, paper, starch, rubber, fabrics and some other biorenewable materials. However, the enzymatic actions on these raw materials are needed for their transformation into biorenewables under some mild and sustainable conditions (Franssen et al. 2013).

The present-day development and success widely depend on the synthesis of chemicals and their production at industry level in the presence of enzymes. This success is due to the increased number of enzymes and their applications in a wide range of processes (Hanefeld et al. 2013). Enzymes show diverse applications in dairy products: in the preparation of wine, beer, fruit and vegetable juices and in the preparation of paper and pulp. Enzymatic pathways are cost-effective, sustainable and environment friendly. However, the free enzymes have very poor stability towards high temperature and pH, and hence, there are only restricted possibilities of their recovery as well as reusability. In order to remove these limitations and to turn the enzyme activity into industrial applications, numbers of approaches have been tried, and immobilization technique is one such tool for this purpose. These increase the stability of enzyme, its selectivity and also its reusability with the emerging tools. A number of immobilization techniques are available that can utilize newly developed materials that offer high biocatalytic activity and favourable environments that are necessary for the action of immobilized enzymes. In the presence of these favourable environments, activities and stability of enzymes are improved for any applications in comparison to free enzymes because of much wider range of pH and temperature available (Mateo et al. 2007; Goldberg and Kolibas 1990; Iyer and Ananthanarayan 2008; Brady and Jordaan 2009; Stepankova et al. 2013; Chibata 1978). A powerful approach for immobilization of enzymes is by the use of nanomaterials (Franssen et al. 2013). Materials that range from 1 to 100 nm in particle size and show different characteristics from their bulk precursors are called nanomaterials. According to ISO (2015), nanomaterials are defined as materials having external dimension in nanoscale, or materials having internal structure or their surface structure in the nanoscale. Nanomaterials have extremely small dimensions, and hence, they allow us to take advantage of unique physical, chemical, optical, electronic and mechanical properties that are present at nanoscale. For the immobilization of enzymes, nanomaterials are used because they can make up a novel and attractive matrices. Enzymes immobilized on these matrices are highly impactful in the today's research due to the high surface-to-volume ratio of nanomaterials and their Brownian motion. Carbon nanotubes, superparamagnetic nanoparticles and some mesoporous materials are used as some important category of matrices.

There are number of reports of micro- and nanomaterials that have various applications in biotechnology and biomedicine due to their incredible potential in delivery systems and targeting, and it also shows advantages in biosubstance binding (Inès and Dhouha 2015; Kuthati and Kankala 2015; Wang et al. 2015; Löhr et al. 1998). There are some examples of different processes occurring on industrial scale that are using immobilized enzymes including high-fructose corn syrup production (Gupta et al. 2011). A number of substrates such as fatty acids, carbohydrates, proteins, rubber and their building blocks are utilized in the creation of biorenewables in the presence of any immobilized enzymes. The renewable produced in this way shows a wide range of applications, that is, they are acting as building blocks for several industries such as in dairy, textile industries, pharmaceutical and polymer industry and the preparation of additives for food industries. They are also showing several bioanalytical and biomedical applications because they are utilizing immobilized antibodies or antigens. They may make use of some immobilized receptors or ligands and also a variety of immobilized cells. Immobilized antigens or antibodies are used in affinity chromatography, and immobilized receptors or ligands are used in biosensors. In the last few decades, biocatalysis has emerged as an important tool in order to meet the demand for green as well as sustainable synthesis of chemicals that mainly involves the production of pharmaceuticals, vitamins, flavours, fragrances and some other specific chemicals. Additionally, the system of the immobilized enzyme is useful either in aqueous or media with low water content in the field of biocatalysis or in resolution of racemates (Wang 2006; Chiang and Sung 2006; Mikhaylova et al. 2004; Kim et al. 2003; Gao et al. 2003; Dyal et al. 2003; Gardimalla et al. 2005; Zhang et al. 2008; Wang et al. 2009; Lee et al. 2008; Netto et al. 2009).

8.2 Synthesis of Nanomaterials

Nowadays, researchers and scientists are working together to develop some new materials having some superior properties, more functionality and lower cost than the on-hand materials. Several physical, chemical and physiological methods have been tried to boost up the performance and behaviour of nanomaterials. Several approaches have been tried for synthesizing nanomaterials in order to get control over distribution of nanoparticles and their particle size (Shibata et al. 1998).

8.2.1 Different Approaches for the Synthesis of Nanomaterials

There are two general approaches that are widely involved in the nanomaterial synthesis and the manufacture of different nanostructures. These two main approaches are represented in Fig. 8.1.

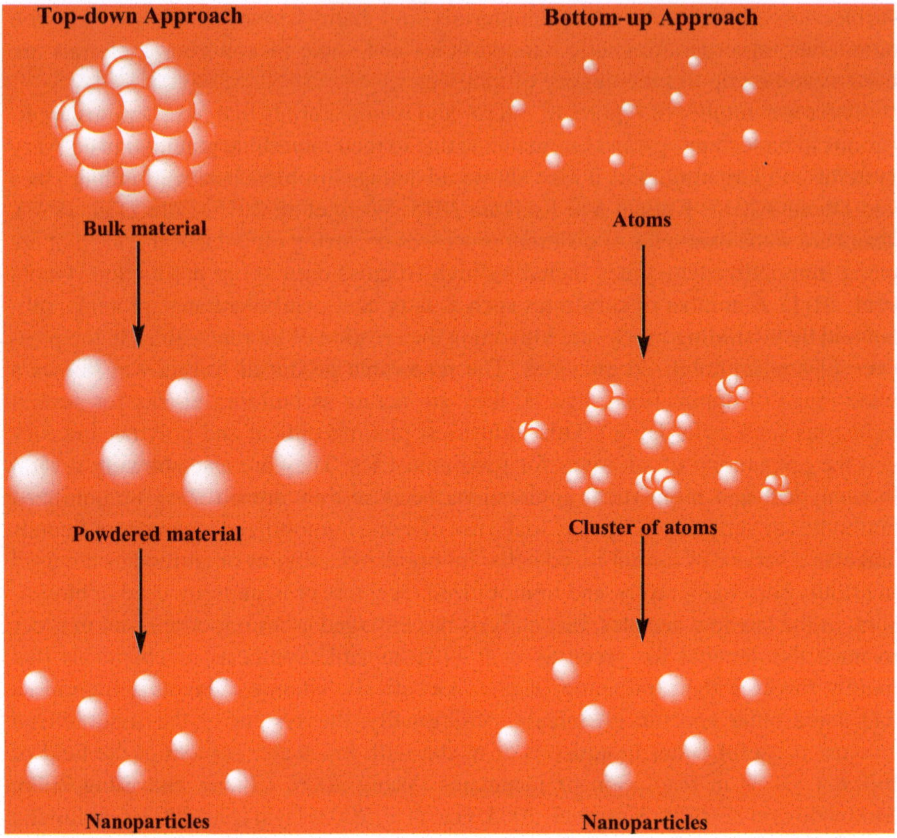

Fig. 8.1 Scheme showing top-down and bottom-up approaches that are used in the synthesis of nanoparticles

8.2.1.1 Top-Down Approach

This route generally entails breaking of larger particles into smaller ones, and this can be achieved by using several forces like crushing, milling or grinding. These approaches use larger or macroscopic structure as the initial material that is controlled during the processing of nanostructures from the outside. Although this approach is used in the synthesis of nanomaterials, this route is generally not appropriate for the preparation of materials having uniform shape. However, the major problem with this approach is the limitation in the surface configuration. Such limitations or imperfection would have a considerable impact on physical properties and surface chemistry of nanostructures and nanomaterials. These approaches are also slow and hence are not suitable for the production at large scale.

8.2.1.2 Bottom-Up Approach

Bottom-up approach involves the synthesis of material from the bottom: atom-by-atom, molecule-by-molecule or cluster-by-cluster. In these approaches, material components are firstly miniaturized up to their atomic level which later on construct the nanostructures by self-assembling. During the self-assembly process, numerous substantial forces are operating at nanoscale that is used for uniting the fundamental units into the larger established structure. This route is mostly used for the preparation of the nanoscale materials, and it shows the ability for the generation of a uniform size, shape and distribution of the nanomaterials. It has the ability to control the reaction to inhibit the growth of particles further. However, the bottom-up approach is not showing any novelty, but the fabrication and processing of nanostructures and nanomaterials can be easily achieved by these approaches. Researchers always feel challenging to control the particles size distribution, purity, morphology, quality and quantity to a better extent during the synthesis of nanomaterials through environment friendly and economical processes (Cao 2006).

8.2.2 Methods Involved in Nanomaterial Synthesis

The top-down and the bottom-up approaches that are used in the synthesis of nanomaterials offer several diversities in the synthesis methods which result in the diversities in the product. There are several methods, namely, physical, chemical, biological or hybrid, that are widely used in the synthesis of nanomaterials such as nanoparticles, nanotubes, colloids, thin films, quantum dots or nanorods. These different methods are shown in Fig. 8.2.

8.2.2.1 Physical Methods

Nanomaterials can be synthesized via physical methods through evaporation and condensation or by applying mechanical pressure and high-energy radiation or by using thermal or electrical energy for the abrasion of the material. These methods are generally based on the top-down approach. Physical methods that are generally used for the synthesis of the nanomaterials are laser pyrolysis, spray pyrolysis, high-energy ball milling, physical vapour deposition, melt mixing, laser ablation, sputter deposition, electric arc deposition, ion implantation, etc. These methods show several advantages as they can generate uniform monodisperse nanoparticles, and also they are free from the contamination through solvent. However, during the synthesis processes, profuse amount of waste is produced; hence, the physical methods may be treated as uneconomical (Dhand et al. 2015).

8.2.2.1.1 High-Energy Ball Milling (HEBM)

HEBM was first time used by John Benjamin in 1970 for the synthesis of alloys strengthened with the oxide dispersion. This method is capable for synthesizing nanoparticles having a lot of variation in their shape and sizes (Dhand et al. 2015; Xing et al. 2013). The HEBM route involves the transfer of kinetic energy from the moving balls to the milled material. Due to this transfer of kinetic energy, bond breaking will take place, and hence, the new surface will be generated by the breakdown of milled materials into the smaller particles (Fig. 8.3a). HEBM process is

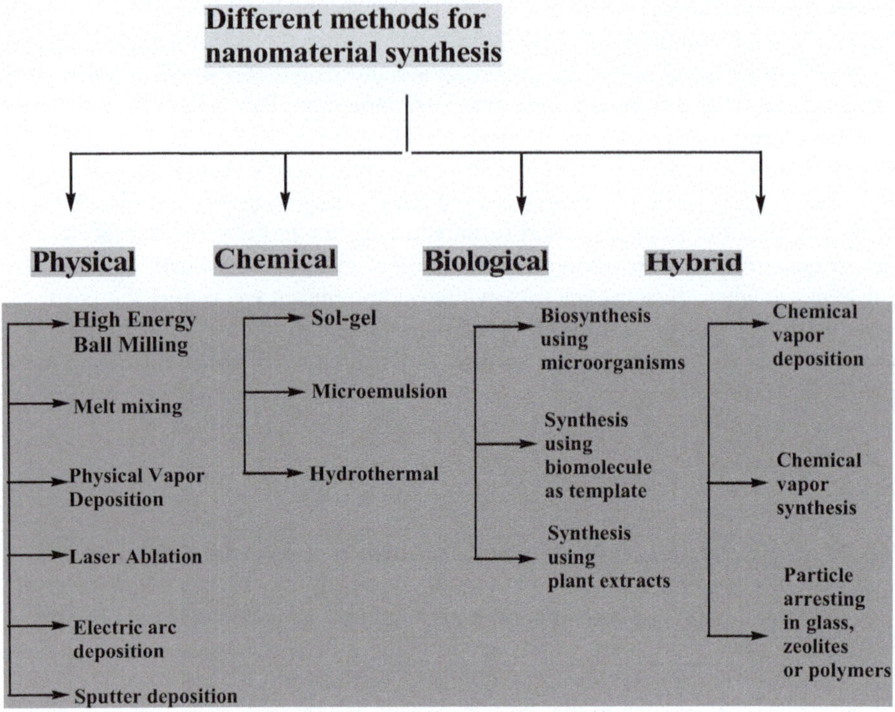

Fig. 8.2 Flowchart representing different methods for nanomaterial synthesis

also considered a mechanochemical process of synthesis because this method is based on the conditions of temperature and pressure. Nowadays, nanomaterial synthesis using HEBM method is carried out in either the presence or absence of surfactant (Fig. 8.3b) (Dhand et al. 2015; de Carvalho et al. 2013). However, the synthesis of nanoparticles through HEBM method in the presence of surfactant is widely used because in the presence of surfactant, we can get precise particle size and their morphology (Dhand et al. 2015; de Carvalho et al. 2013) (Fig. 8.3).

8.2.2.1.2 Melt Mixing

In this method, nanomaterial synthesis can be carried out by mixing the modified form of nanofillers with the polymer mechanically (Karak 2009). Melt mixing method is widely used because this method is highly compatible for different industrial processes and is also environment friendly (Lin et al. 2006). Melt mixing technique was used by Zuhal et al. for the synthesis of polypyrrole with polypropylene nanocomposite nanoparticles (Sevil and Zuhal 2010). Weiss et al. synthesized hybrid nanoparticles by mixing hydroxybenzoate (HBA) and hydroxynaphthanoate with the zinc salts of sulfonated ionomers of polystyrene. This involves the chemical bond formation between the polyester and the residual zinc acetate as it is carried out at high temperature (Lee et al. 2005).

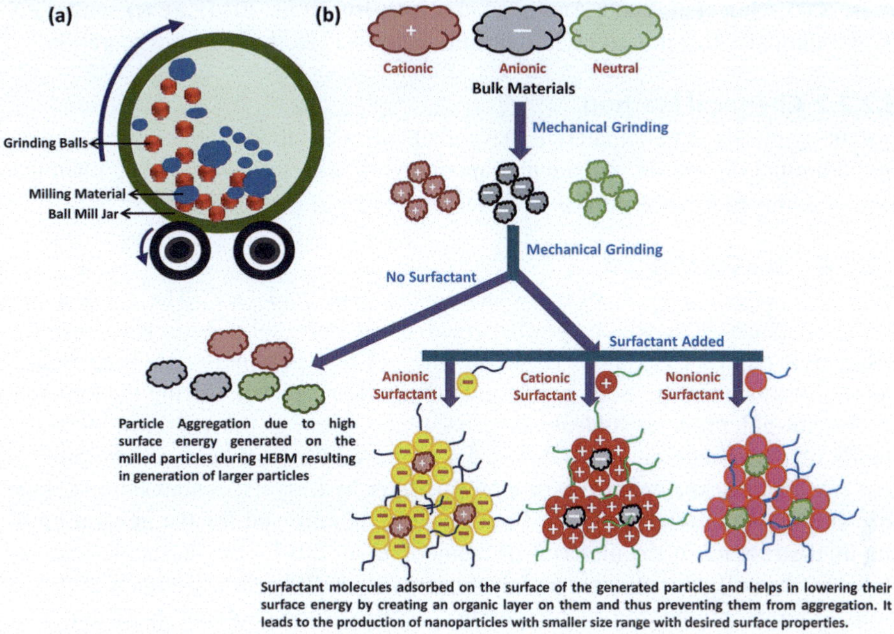

Fig. 8.3 (**a**) It represents the high-energy ball milling scheme of nanomaterial synthesis and (**b**) representation of nanoparticle synthesis through HEBM system in the presence and absence of surfactant. (Dhand et al. 2015)

8.2.2.1.3 Laser Ablation

In laser ablation method, vaporization of material is carried out by using laser light (Chen and Yeh 2002). After vaporizing the material, nanomaterial can be obtained in the colloidal solution form. It is a very easy and fast method for synthesis of the nanomaterials. It is also environment friendly and hence considered a green method. In this method, we are also able to introduce some organic moieties on the surface of the nanoparticles (Dhand et al. 2015; Singh and Gopal 2007).

8.2.2.1.4 Physical Vapour Deposition

Physical vapour deposition consists of group of processes that are used for the production of nanoparticles and for the deposition of thin layers of materials. These thin layers generally lie in the nanometre or micrometre range. In PVD technique, mainly three steps are involved: (1) the first step involves vaporization of the materials that are mainly from solid source; (2) in the second step, vaporized material is transported; and (3) the third step is the nucleation and growth for the generation of thin films and nanoparticles (Okuyama and Lenggoro 2003). Some of the frequently used PVD methods for synthesizing the nanomaterials (Hatakeyama et al. 2011; Veith et al. 2007; Bouchat et al. 2013; Asanithi et al. 2012; Ichida et al. 2014; Ghosh et al. 2007; Veith et al. 2005; Ramalingam et al. 2013; Zhang et al. 2004; Takahashi et al. 2004; Dmitrieva et al. 2006; Stellacci et al. 2002; Hsieh et al. 2010; Naessens et al. 2001; Ong et al. 2008; Andrea et al. 2009; Jing et al. 2014) are (i) sputtering, (ii) vacuum arc, (iii) pulsed laser deposition and (iv) electron beam evaporation.

8.2.2.2 Chemical Method

For the synthesis of nanomaterials, several chemical methods are also used such as hydrothermal synthesis, chemical vapour synthesis, sol-gel method, chemical vapour deposition, microemulsion technique and polyol synthesis.

8.2.2.2.1 Sol-Gel Method

The term sol gel consists of two components: 'sol' and 'gel'. The sol is a colloidal suspension in which the dispersed phases are solid particles while the dispersion medium is liquid. However, the gel is a polymer containing liquid. Hence, in this process, 'sols' are formed in the liquid phase. During the first step of the sol-gel process, hydrolysis takes place wherein the presence of water bonds of the precursor molecules is disintegrated (*Sol-gel science: the physics and chemistry of sol-gel processing* 1990). After hydrolysis, condensation occurs allowing nanomaterials to form. Finally, water is removed for the generation of the final structure of the material (Behnajady et al. 2011). Synthesis of nanoparticles by the sol-gel method is widely dependent on different experimental conditions; therefore, the reflux temperature, reflux time, solvent percentage or calcination temperature should be optimized.

8.2.2.2.2 Microemulsion Method

Microemulsions are homogenous, and transparent dispersion consists of mainly three components: polar phase, nonpolar phase and surfactant. Polar phase generally includes water, and the hydrocarbon liquid or oil is used as nonpolar phase. Surfactant molecules are used so that (i) separating layer may be created between the polar and the nonpolar phase and (ii) interfacial tension may be reduced between the microemulsion and the excess phase. This technique is mainly used for the preparation of inorganic nanomaterials that may be Au, Pt and Pd nanoparticles (metal nanoparticles); $BaCO_3$, $SrCO_3$ and $CaCO_3$ nanoparticles (metal salt nanoparticles); CdS, PbS, CuS, CdSe and Cu_2S nanoparticles (semiconducting metal sulphite nanoparticles); and ZrO_2, GeO_2, TiO_2, Fe_2O_3 and SiO_2 nanoparticles (metal oxide nanoparticles) (Solanki and Murthy 2011; Zhang et al. 2006). Au nanoparticles were synthesized from Au(III) ions by its reduction in the presence of alkaline solution of 2,7-dihydroxynaphthalene (DNP).

8.2.2.2.3 Hydrothermal Method

This method generally involves the fabrication of nanoparticles of some metal oxide, lithium iron phosphate and iron oxide. In this method, we have control over the different properties of the particles by varying the conditions of pressure and temperature (Hayashi and Hakuta 2010). This method is carried out in vessel called autoclave that is operated under high pressure, and the reaction is carried out in the aqueous solution. In hydrothermal method, by taking care of the reaction temperature, pressure, solution composition, solvent properties, etc., we have better control over the particle size as well as morphology (Abedini et al. 2013). Hydrothermal synthesis is generally carried out in two types of systems that are continuous hydrothermal and the batch hydrothermal processes. Du et al. synthesized Pt nanoparticles by hydrothermal method through one-pot synthesis (Du et al. 2014). Hydrothermal method is also widely used for the synthesis of iron oxide (Fe_3O_4, Fe_2O_3), copper oxide, silver oxide, zinc oxide and nickel oxide nanoparticles (Liu et al. 2015; Tadic et al. 2014; Sue et al. 2011).

8.2.2.3 Biological Method

The biological method or biosynthesis creates low toxicity and is cost-effective and also environment friendly. Hence, it is considered a green synthesis method. In these methods, several bio-organisms such as bacteria, viruses, fungi, yeast, plant extracts, etc. are involved in the production of metal and metal oxide nanoparticles. This biosynthesis method can be categorized into three major parts.

8.2.2.3.1 Biosynthesis Using Microorganisms

In this method, several microorganisms are used in the synthesis of nanoparticles. Target ions from their environment are captured by the microorganisms, and then, the metal ions are converted into their elemental form through several enzymes produced through cellular activities. Depending on the location of nanoparticle synthesis, it is further classified as extracellular and intracellular. In the intracellular method, metal ions are transported into the inside of the microbial cell for the formation of nanoparticles, whereas in extracellular synthesis method, metal ions are trapped on the surface of the cell followed by reduction in the presence of enzymes (Zhang et al. 2011). Bacteria can reduce the metal ions by utilizing several functional groups, proteins, enzymes and also some reducing sugars (Nanda and Saravanan 2009). Biosynthesis by utilizing fungi has several advantages over other microorganisms: (i) it can show high bioaccumulation, (ii) it is economically viable, and (iii) it also shows easy handling of biomass (Mukherjee et al. 2001).

8.2.2.3.2 Nanomaterial Synthesis Using Biomolecules as Templates

Several biomolecules such as nucleic acid, viruses, membranes, etc. are used as templates in the synthesis of nanomaterials. DNA is considered one of the most useful templates in the synthesis process as it shows very strong affinity with the transition metal ions. Au nanoparticles are formed by incorporating gold, Au (III) metal ions to DNA. In this process, reduction of Au (III) occurs, and hence, the Au atoms and metal cluster developed which will result in the formation of nanoparticles on the chain of the DNA (Zinchenko et al. 2014). Similarly, Ag nanoparticles were synthesized by Kundu et al. by using DNA as a template. These Ag nanoparticle clusters that have developed on the surface of the DNA also show good catalytic activities in the reduction of some aromatic nitro compounds (Kundu 2013; Kundu et al. 2008; Kundu and Nithiyanantham 2014).

Biological membranes are also used as templates in the synthesis of nanoparticles because they are having ultrafine pores in their structure. In the synthesis of Au nanoparticles, rubber membrane prepared from the *Hevea brasiliensis* trees is used as template where reduction of Au (III) takes place at 80 °C (Cabrera et al. 2013).

8.2.2.3.3 Nanomaterial Synthesis Using Plant Extracts

The generation of nanomaterials from the plant extracts is eco-friendly and a rapid process. Hence, this method is widely used for the synthesis of nanoparticles of noble metals, metal oxides and some bimetallic alloys (Iravani 2011). Gold nanotriangles were synthesized by using lemongrass leaf extract. For this preparation, Shankar et al. treated the lemongrass leaf extract with the aqueous solution of the AuCl ions (Shankar et al. 2014). In a similar way, the leaf extract of *Aloe vera*, *Tamarindus indica* and *Emblica officinalis* and leaf extract of some other plants are also used for the synthesis of the Au nanoparticles (Chandran et al. 2006; Ankamwar et al. 2005a; b). Some other nanoparticles such as Pd nanoparticles and Pt nanoparticles were also synthesized by using the extract taken out from different parts of diverse species of plants (Coccia et al. 2012; Sathishkumar et al. 2009). Shankar et al. also synthesized Ag nanoparticles from the leaf extract of *Azadirachta indica* and also from the fruit extract of *Emblica officinalis* (Ankamwar et al. 2005b; Shankar et al. 2004).

8.2.2.4 Hybrid Method

In this category of nanomaterial synthesis method, physical, chemical and biological are all involved. Hence, it is considered a hybrid method for the synthesis of the nanomaterial. It involves several methods such as chemical vapour deposition, chemical vapour synthesis or arresting of the particles in the glass or zeolites or polymers.

8.2.2.4.1 Chemical Vapour Deposition and Chemical Vapour Synthesis

In this method, solid films are deposited from the vapour phase through chemical reactions taking place at high temperature. This method was earlier used for the production of thin films. However, nanoparticles can also be produced by this method if the following conditions are optimized:

(i) In the hot wall reactors, the temperature should be kept high.
(ii) The partial pressure of monomers should be kept high so that high supersaturation can be achieved.
(iii) Residence time should be kept slow.

This modified process is then called chemical vapour synthesis (CVS) or chemical vapour reaction (CVR) or chemical vapour precipitation (CVP) or chemical vapour condensation (CVC). During the synthesis process, either solid, liquid or gaseous precursors are converted to their vapour form (Swihart 2003). Chemical vapour synthesis method is generally utilized for the generation of nanoparticles by using several material such as ZnO, iron oxide, silicon oxide and copper oxides, Cr-doped zinc oxide, Al-doped zinc oxide, etc. (Hartner et al. 2009; Jin et al. 2007; Suffner et al. 2010; Lahde et al. 2011; Lee et al. 2009). One disadvantage of this method is that high temperature is required for this method, and hence, there are difficulties in providing such high temperature.

8.2.3 Synthesis of Nanoparticles

Nanoparticles show wide range of applications as they are used in number of optical, electronic, and magnetic devices and also they are used as catalysts, adsorbents, sensors and ferrofluids. Most of these applications of nanoparticles depend on their size as well as their shape, and hence, the synthesis of nanoparticles should be carried out in such a way so that well-defined shape and size of the nanoparticles may be obtained. One of the very common techniques used is the formation of metallic colloid dispersions by the reduction of metal complexes. Sizes of nanoparticles depend on the type of reducing agent used during its synthesis. If the reducing agent is strong, then the rate of reaction will be fast. Hence, the nanoparticles formed are of smaller size, but the nanoparticles formed will be of larger sized when we use a weak reducing agent because it induces slower rate of reaction.

Aggregation of nanoparticles can be prevented by the use of some polymeric stabilizers. These polymeric stabilizers form a single layer on the nanoparticle surface, and in this way aggregation can be prevented. These polymeric stabilizers are also called as capping material. But when we use polymeric stabilizers, the growth of nanoparticles gets inhibited because the monolayer of polymer has the tendency to significantly affect the process of growth by blocking the growth sites. Polymeric stabilizers also have the tendency to affect the shape of nanoparticles when they are used in different amounts.

8.2.4 Synthesis of Nanowires, Nanorods and Nanotubes

Several monodimensional nanomaterials like nanowires, nanorods and nanotubes can be synthesized by different techniques:

(a) Spontaneous growth – It involves various methods such as evaporation-condensation, vapour-liquid-solid growth and stress-induced recrystallization
(b) Template-based synthesis – The methods used here are electroplating, electrophoretic deposition, colloid dispersion, melt or solution filling and chemical reaction.
(c) Electrospinning
(d) Lithography

In spontaneous growth due to the preferential direction for crystal growth, there is formation of nanowires and nanorods in their single-crystal form.

Long ribbon-like nanostructures were synthesized by Pan et al. (2001) by evaporating the metal oxide powders at high temperatures, and for this they used mainly zinc, tin, indium, cadmium and gallium in their oxide form having semiconducting behavior (Pan et al. 2001).

One method for nanowire synthesis involves filling a template having cylindrical holes of size ranging in nanoscale. It is one of the most straightforward and common methods. It involves synthesis of nanowires into the cylindrical pores present at thin membrane by reducing the metal salt. This method has one advantage of producing nanowires and nanorods of controlled size when there is availability of template with homogeneous pores, and for this requirement, there are a number of permeable templates available that can accomplish such necessities (Martin 1994; Thurn-Albrecht et al. 2000).

By using various techniques, several nanochannel silicon membranes are prepared having pores in perfectly structured form, and such membranes of silicon are widely used in the synthesis of modulated-diameter gold nanorods. By using electrodeposition techniques, this method has been adapted for the creation of nanorods having different metal segments. They are widely used for tagging molecules in analytical chemistry and in the field of biology (Ozin and Arsenault 2006).

Some nanorods have functional molecules introduced to its some selective locations, and these functional molecules which anchored nanorods have wide applications in the today's research. One of the applications of this method involves functionalization of gold nanorod end with thiolated DNA. The complimentary DNA is Rhodamine labelled and is coupled to the nanorods through its end. By using a similar approach, gold nanorods side are also functionalized with DNA, i.e. gold nanorods with DNA side functionalized are also created, and these are directed for self-assembling on complimentary DNA-functionalized gold surface sites with soft lithographic pattern.

8.3 Enzyme Immobilization

Our main aim for immobilization of enzyme is to improve the stability of enzyme and its handling as well as storage properties. Enzymes are utilized for their several applications where indigenous form of them is not possible. Biosensors based on enzyme reactions are used for their certain therapies, but the enzymes used for biotransformation in industries are based on the use of some stable biocatalysis. These necessities can be fulfilled by immobilizing the enzymes (Sheldon 2007; Tran and Balkus 2011; Garcia-Galan et al. 2011; Hanefeld et al. 2009). There are certain techniques available for effective immobilization of enzymes, and these can be achieved by (i) binding the enzyme molecules with solid carrier (support), (ii) embedding the enzyme in an open matrix (encapsulation) or (iii) cross-linking through physical or chemical interactions (Singh and Gopal 2007). These different techniques are represented in Fig. 8.4.

(a) Binding to a support (carrier) involves several interactions that may be physical (e.g. hydrophobic and van der Waals interactions), ionic or covalent (Sheldon and van Pelt 2013; Cao 2005). However, among these several interactions,

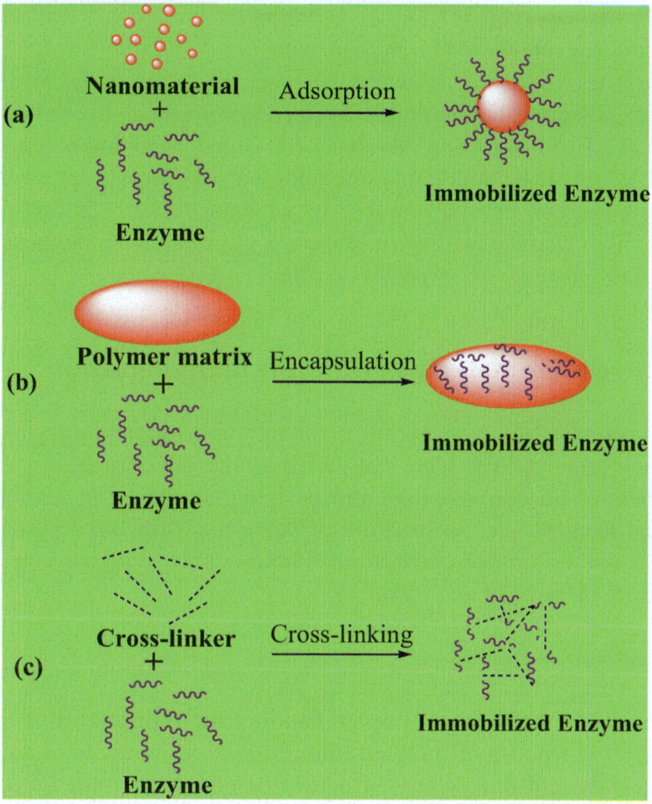

Fig. 8.4 Scheme showing different techniques for enzyme immobilization

physical binding has no sufficient strength to hold the enzyme with the carrier during the rigorous industrial environment such as high concentration of reactant and product and high ionic strength. Ionic interaction is somewhat stronger; however, the covalent bond between the enzyme and the carrier usually does not allow the leaching of enzyme through its surface. However, the covalent bonding between the enzyme and the support has one limitation, that is the enzymes become irreversibly deactivated making the enzymes and the support useless. Some examples of support in immobilizing the enzymes are synthetic resins, biopolymers such as polysaccharides or inorganic solids such as mesoporous silicas or zeolites.

(b) Encapsulation of enzyme can be performed in matrix of organic polymer such as polyacrylamide or in matrix of inorganic polymer such as silica sol gel or in a membrane device such as hollow fibre or a microcapsule. Entrapment can be done after synthesizing the polymer matrix when enzymes are being there. One example that shows this behaviour is that when immobilization of enzymes is performed in mesoporous silica which is already fabricated, the enzyme would be located inside the mesopores but this cannot be supposed as encapsulation;

however, when enzyme is being there during the process of synthesis of a silica sol gel, the enzyme would be encapsulated.

(c) The carrierless macroparticles can be prepared by cross-linking the enzyme aggregates or crystals having the bifunctional reagent. Cross-linked enzyme crystals (CLECs) (Roy and Abraham 2004) and cross-linked enzyme aggregates (CLEAs) (Sheldon 2011) are some carrier-free immobilized enzymes. These approaches provide number of advantages such as high stability of enzyme, high enzyme activity or low costs of production, and hence it has the potential to replace other expensive carrier.

8.3.1 Active Nanomaterials in Enzyme Immobilization

The major advantage of nanomaterials in the immobilization of enzyme is that surface area-to-volume ratio is high that increases the effectiveness of the enzyme. Large numbers of nanoparticles are already being employed in immobilizing the enzymes that leads to the construction of some nanostructured systems such as nanofibres, nanotubes, nanomaterials, nano hollow fibre materials and some nanoporous material (Misson et al. 2014).

(a) *Sol-gel networks*

Enzymes can be immobilized by embedding them through this sol-gel process. During the process of enzyme immobilization, enzyme is added first to the sol materials and due to this structure of enzyme would be frozen. This enzyme becomes entrapped into the sol-gel network and is called xerogels that are fine powdered substance and stable from the mechanical point of view (Avnir et al. 2006).

(b) *Protein matrices*

During the creation of protein-based matrices, enzymes are connected by covalent binding, and there is a generation of cross-linking between the enzymes by some bifunctional agents. These materials actually do not have any solid carrier, but they are called nanostructured immobilized enzyme. This method shows a lot of advantages as it is simple and cost-effective and can be applicable for a number of enzymes. The major benefit of this method is that immobilized enzyme can be created without any intermediate materials and also without the need for cleaning. However, this method shows some disadvantages also, that is, the enzymes are located on the outer surface, and hence, it may come in contact to the substrate without any problem, but the transport of material by the enzymes located inside the matrix is difficult. Another problem is that binding of cross-linked agents shows some negative effect on the structure of enzyme by disturbing the activity of enzyme, making it inactive.

(c) *Nanofibres*

Both inside and outside of the nanofibres created by electrostatic fibre production are used to immobilize the enzymes. After creating the nanofibres, the enzyme solution is allowed to mix with the polymer solution when the immobilization is carried out inside the nanofibres. The enzyme immobilization can show stability, security and some better quality for the enzyme functionality when immobilization is carried out inside the nanofibres in comparison to the immobilization of enzyme on the outer side of the nanofibres.

(d) *Nanotubes*

Nanotubes are used in electronic, optic fibres and in different sensors due to their mechanical, thermal and structural properties. These are among the largely used enzyme carriers that are especially used in the development of biofuel cells. Both the single-walled and multi-walled carbon nanotubes are broadly used for immobilizing the enzymes. Single-walled nanotubes are used because they show high surface area and hence act as successful carrier material for the enzymes, while the multi-walled nanotubes are used because of their ease in dispersion property. The interactions or the type of bonding involved in immobilization are covalent or secondary bonding. Adsorption occurs through H-bonding or hydrophobic or π-π interactions, and the advantage of adsorption is that original behaviour of both the enzymes and the support is conserved. But it may dissolve easily during the applications due to which catalyst may become free; however, the covalent bonding gives more robust and durable association between the enzyme and nanotube (Feng and Ji 2011).

(e) *Magnetic nanoparticles*

Successful immobilization of enzymes on magnetic nanoparticles is widely utilized for diagnostic purposes such as in measurement of blood sugar, biological transformations with stereoselectivity, etc. Nanosized magnetic particles show number of beneficial behaviour. Therefore, the cross-linked iron oxide particles, monocrystalline iron oxide nanoparticles and ultrasmall superparamagnetic iron oxide are used as an imaging agent in magnetic resonance imaging. Nowadays, magnetic nanoparticles are used for immobilizing the enzymes because they have the tendency to be easily separated out from the reaction mixture, they have the tendency to enhance the stability of bioelement and they can also increase the stability of enzymes. These magnetic nanoparticle-immobilized enzymes illustrate several applications in biotechnology or in analytical devices such as biosensors or in nanomedicine where nanoparticles are used for ease in diagnosis and disease treatments (Govan and Gunko 2014). Due to the advantage of magnetic nanoparticles in the ease of their separation, Liang et al. tried to immobilize the *Candida rugosa*

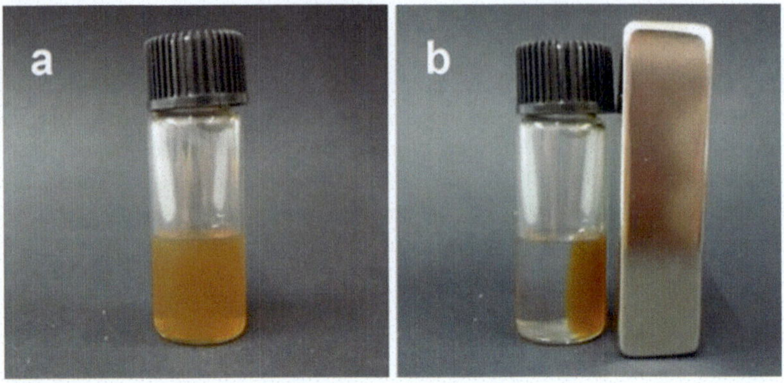

Fig. 8.5 The behaviour of magnetic nanoparticle-immobilized CRL enzyme, i.e. CA-Fe_3O_4@Zn/AMP CPs. (**a**) Before separation and (**b**) after separation. (Li et al. 2017)

lipase (CRL) enzyme by using magnetic iron oxide nanoparticles modified with citric acid (CA-Fe_3O_4 NPs) in the presence of nucleotide-hybrid metal coordination polymers (NMCPs). Hence, the activity as well as stability of the CA-Fe_3O_4@Zn/AMP nanofibres-immobilized CRL enzyme gets increased (Li et al. 2017). Their separation behaviour is shown in Fig. 8.5.

8.3.2 Immobilized Enzymes in Biotechnology

Enzymes show one of the attracting roles, that is, they act as natural biocatalysts due to the ability of enzyme to hasten the rate of all biological reactions, but they themselves are not consumed during the reactions. They also do not affect the equilibrium between the reactants and the products. However, the immobilized enzymes show a number of advantages such as they show ability of easy reusability and lower degradation in comparison to free enzymes available in solution. In addition to these advantages, the rate of reaction can be controlled and also prevent contamination of the substrate with enzyme or protein. Stability of enzymes can also be improved by expanding their half-life through immobilizing the enzymes. Immobilized enzymes can work in a broader range of environments that have the ability to increase the stability of enzymes against temperature, pH, contaminants and impurities. Enzymes after their immobilization show improved biocatalytic activity and efficiency that make them highly attractive and therefore show high applicability in numerous varieties of growing biotechnologies (Gerday et al. 2000; Bradley and Wang 2015; Kell et al. 2015; Joshi and Satyanarayana 2015; Tavano 2013; Hyeon et al. 2013; Yu et al. 2015; Wendisch 2015).

8.3.3 Immobilized Enzymes in Biomedicine

Immobilized enzymes or proteins are widely used for the detection and treatment of several diseases, and these show their applications in the field of medicine. Several immobilized antibodies, receptors and enzymes are used in the form of biosensors for analysing different biologically active compounds in diagnosis. Other use of immobilized enzymes in the field of biomedicine includes synthetic cells and the manufacture of some systems that are used in drug delivery for dosing of proteins or enzymes and are well controlled. The development of some enzyme-based electrodes is another important application of immobilized enzymes in the field of biomedicine. Those enzymes that are highly specific and reactive towards its substrate are widely used in biosensor. Hence, the biosensor developed has high reliability, sensitivity and accuracy, showing ease in their handling, and is of low cost in comparison to the usual analytical methods.

Some important applications of bioreactors that are broadly used in the field of medicine for human are:

(a) In the poisoning of organophosphate, degradation of organophosphate can be done by the use phosphotriesterases (Chatterjee et al. 2014; Petrikovics et al. 1999).
(b) Alcohol dehydrogenase (ADH) and acetaldehyde dehydrogenase are used for the conversion of alcohol to its acetate form in the alcohol poisoning (Pei et al. 1995; Magnani et al. 1993).
(c) Repairing of damaged DNA in aging of skin and cancer by the use of DNA repair enzyme (liposome) (Yarosh et al. 1999; Yarosh et al. 1996).
(d) In hypercholesterolemia, hydrolysis of phospholipids can be done by the use of phospholipase A_2 (liposome) (Jørgensen et al. 1999).

8.4 Applications of Nanomaterials in Enzyme Immobilization

Nowadays, researchers are using two types of carriers (i.e. microcarriers and nanocarriers) for immobilizing the enzymes so that the immobilized enzymes can be utilized for several biotechnological, biochemical, bioanalytical or biomedical applications. These micro- and nanocarriers are magnetic nanoparticles and cross-linked enzyme aggregates (CLEAs), respectively. However, among these two carriers, the magnetic nanoparticles show a number of applications in biomedical, bioanalytical, biophysical and biomedical field. Magnetic nanoparticles are widely used in these fields because the nanosized structures have smaller size, excellent magnetic properties and large surface area-to-volume ratio (hence, they have higher surface energy). Therefore, they are used as carriers to which different active substances can easily bind. Several enzyme immobilization methods such as adsorption, chelating or metal binding, affinity binding or covalent binding help in the production of some efficient and stable magnetic nanoparticles bounded with enzyme. The development and synthesis of such materials have immense contribution in the present-day research in the form of imaging agents, sensors, drug delivery targets/vehicles and diagnostic tools. The application of nanomaterial in this field has enhanced the scope of studies.

8.4.1 Gold Nanoparticles as Enzyme Immobilization Templates

Gold nanoparticles are widely used in enzyme immobilization. When the aqueous gold nanoparticles are assembled on the surface of polyurethane (PU) spheres, it leads to the formation of [gold nanoparticle shell]-[polyurethane core] structure. The conjugation of gold nanoparticles occurs easily on the microspheres of polymer because nitrogen in the polymer interacts with the nanoparticles. One essential step in the core-shell structure protocols based on the polymer requires performing some additional modification on the surface of the polymer microspheres. Finally, the nanogold-PU material is allowed to make conjugation with enzyme such as pepsin which leads to the creation of a novel category of biocatalyst. In this way, the enzyme-conjugated material shows some enhanced biocatalytic activity and considerably improved stability to withstand broader pH and temperature range in comparison to the free enzyme in solution. One another advantage of this pepsin bioconjugated gold nanoparticle-labelled polyurethane microsphere is that it can be separated out from the reaction medium without any difficulty and it shows higher reusability of the bioconjugate up to six reaction cycles. In DNA immobilization and detection methodologies, the use of such gold nanoparticle shell-PU core structures is currently being envisaged (Phadtare et al. 2003).

8.5 Conclusion

Enzymes show environmental friendliness, high selectivity, high specificity, low cost and mild reaction conditions, and hence, they have the potential to replace several chemical catalysts. Enzymes are used as important biocatalysts, and hence they show wide applications such as in biosensors, pharmaceuticals, chemicals and foods. Due to low operational stability and short lifetime, enzymes have only limited applications. However, these limitations can be overcome by immobilizing the enzymes. Several materials have been tried for immobilizing the enzymes such as carboxyl-functionalized grapheme oxide, polyurethane foam support, SBA-15 and chitosan. However, the nanostructured carriers are attracting much attention of the researchers and are considered as ideal material for immobilizing the enzymes due to their specific surface area, high dispersibility and lower mass transfer resistance. Many nanomaterials have been used for immobilizing the enzymes so that improvement should be achieved in the enzyme activity and stability. However, the recovery of nanomaterials is very difficult, and hence, they cannot be recycled. Therefore, the magnetic nanosupport would be the most excellent alternative so that recovery of nanostructured immobilized enzymes may become easy by using magnet as the separation medium. Therefore , the magnetic nanoparticles such as iron oxide as well as ferric oxide have been tried successfully for immobilizing the enzymes. In this way, the immobilization support has the tendency to increase the enzyme stability, activity, specificity, lifetime, productivity and structural rigidity.

References

Abedini A, Daud AR, Hamid MAA, Othman NK, Saion E (2013) Nanoscale Res Lett 8:1–10
Andrea CD, Neri F, Ossi PM, Santo N, Trusso S (2009) Nanotechnology 20:245606
Ankamwar B, Chaudhary M, Sastry M (2005a) Synth React Inorg Met-Org Nano-Met Chem 35:19–26
Ankamwar B, Damle C, Ahmad A, Sastry M (2005b) J Nanosci Nanotechnol 5:1665–1671
Asanithi P, Chaiyakun S, Limsuwan P (2012) J Nanomater 2012:963609
Avnir D, Coradin T, Lev O, Livage J (2006) J Mater Chem 16:1013–1030
Behnajady MA, Eskandarloo H, Modirshahla N, Shokri M (2011) Desalination 278:10–17
Bouchat V, Moreau N, Colomer JF, Lucas S (2013) J Surf Eng Mater Adv Technol 3:184–189
Bradley RW, Wang B (2015) New Biotechnol 32:635–643
Brady D, Jordaan J (2009) Biotechnol Lett 31:1639–1650
F. C. Cabrera, H. Mohan, R. J. D. Santos, D. L. S. Agostini, R. F. Aroca, M. A. Rodr'ıguez- P'erez and A. E. Job, J Nanomater, 2013, 2013, 1–10
Cao L (2005) Carrier-bound immobilized enzymes, principles, applications and design. Wiley-VCH, Weinheim
Cao G (2006) Nanostructures and nanomaterials. Imperial College Press, London. Chapters 1–6
Chandran SP, Chaudhary M, Pasricha R, Ahmad A, Sastry M (2006) Biotechnol Prog 22:577–583
Chatterjee K, Sarkar S, Jagajjanani Rao K, Paria S (2014) Adv Colloid Interf Sci 209:8–39
Chen Y, Yeh C (2002) Colloids Surf A Physicochem Eng Asp 197:133–139
Chiang CL, Sung CS (2006) J Magn Magn Mater 302:7–13
Chibata I (1978) Pure Appl Chem 50:667–675
Coccia F, Tonucci L, Bosco D, Bressand M, Alessandro N (2012) Green Chem 14:1073–1078
de Carvalho JF, de Medeiros SN, Morales MA, Dantas AL, Carrico AS (2013) Appl Surf Sci 275:84–87
Dhand C, Dwivedi N, Loh XJ, Ying ANJ, Verma NK, Beuerman RW, Lakshminarayanan R, Ramakrishna S (2015) RSC Adv 5:105003–105037
Dmitrieva O, Acet M, Dumpich G, Kastner J, Antoniak C, Farle M, Fauth K (2006) J Phys D Appl Phys 39:4741–4745
Du JJ, Chen C, Gan YL, Zhang RH, Yang CY, Zhou XW (2014) Int J Hydrog Energy 39:17634–17637
Dyal A, Loos K, Noto M, Chang SW, Spagnoli C, Shafi KVPM, Ulman A, Cowman M, Gross RA (2003) J Am Chem Soc 125:1684–1685
Feng W, Ji P (2011) Biotechnol Adv 29:889–895
Franssen MCR, Steunenberg P, Scott EL, Zuihofac H, Sanders JPM (2013) Chem Soc Rev 42:6491
Gao X, Yu KMK, Tam KY, Tsang SC (2003) Chem Commun:2998–2999
Garcia-Galan C, Berenguer-Murcia A, Fernandez-Lafuente R, Rodrigues RC (2011) Adv Synth Catal 353:2885–2904
Gardimalla HMR, Mandal D, Stevens PD, Yenb M, Gao Y (2005) Chem Commun 35:4432–4434

Gerday C, Aittaleb M, Bentahir M, Chessa JP, Claverie P, Collins T et al (2000) Trends Biotechnol 18:103–107
Ghosh PK, Ahmed SF, Jana S, Chattopadhyay KK (2007) Opt Mater 29:1584–1590
Goldberg RL, Kolibas LM (1990) Connect Tissue Res 24:265–275
Govan J, Gunko YK (2014) Nano 4:222–241
Gupta MN, Kaloti M, Kapoor M, Solanki K (2011) Artif Cells Blood Substit Biotechnol 39:98–109
Hanefeld U, Gardossi L, Magner E (2009) Chem Soc Rev 38:453–468
Hanefeld U, Cao L, Magner E (2013) Chem Soc Rev 42:6211
Hartner S, Ali M, Schulz C, Winterer M, Wiggers H (2009) Nanotechnology 20:445701
Hatakeyama Y, Onishi K, Nishikawa K (2011) RSC Adv 1:1815–1821
Hayashi H, Hakuta Y (2010) Materials 3:3794–3817
Hsieh T, Chuang C, Chou Y, Shu C (2010) Mater Des:31
Hyeon JE, Jeon SD, Han SO (2013) Biotechnol Adv 31:936–944
Ichida D, Uchida G, Seo H, Kamataki K, Itagaki N, Koga K, Shiratani M (2014) J Phys Conf Ser 518:012002
Inès M, Dhouha G (2015) Carbohydr Res 416:59–69
Iravani S (2011) Green Chem 13:2638–2650
Iyer PV, Ananthanarayan L (2008) Process Biochem 43:1019–1032
Jin W, Lee I, Kompch A, Dorfler U, Winterer M (2007) J Eur Ceram Soc 27:4333–4337
Jing Y, Wang H, Chen X, Wang X, Wei H, Guo Z (2014) Appl Surf Sci 316:66–71
Jørgensen K, Kiebler T, Hylander I, Vermehren C (1999) Int J Pharm 183:21–24
Joshi S, Satyanarayana T (2015) Bioresour Technol 176:273–283
Karak N (2009) Fundamentals of polymers: raw materials to finish products, 1st edn. Prentice-Hall Of India Pvt Ltd, New Delhi
Kell DB, Swainston N, Pir P, Oliver SG (2015) Trends Biotechnol 33:237–246
Kim DK, Mikhaylova M, Zhang Y, Muhammed M (2003) Chem Mater 15:1617–1627
Kundu S (2013) Phys Chem Chem Phys 15:14107–14119
Kundu S, Nithiyanantham U (2014) Ind Eng Chem Res 53:13667–13679
Kundu S, Maheshwari V, Saraf RF (2008) Langmuir 24:551–555
Kuthati Y, Kankala RK (2015) Appl Clay Sci 112–113:100–116
Lahde A, Kokkonen N, Karttunen AJ, Jaaskelainen S, Tapper U, Pakkanen TA, Jokiniemi J (2011) J Nanopart Res 13:3591–3598
Lee HS, Zhu L, Weiss RA (2005) Polymer 46:10841–10853, 1684–1687
Lee J, Lee Y, Youn JK, Na HB, Yu T, Kim H, Lee SM, Koo YM, Kwak JH, Park HG, Chang HN, Hwang M, Park JG, Kim J, Hyeon T (2008) Small 4:143–152
Lee D, Tolochko OV, Turaev FR, Kim D, Kim B (2009) J Nanosci Nanotechnol 9:1–6
Li C, Jiang S, Zhao X, Liang H (2017) Molecules 22:179
Lin B, Sundararaj U, Potschke P (2006) Macromol Mater Eng 291:227–238
Liu XD, Chen H, Liu SS, Ye LQ, Li YP (2015) Mater Res Bull 62:217–221
Löhr M, Müller P, Karle P, Stange J, Mitzner S, Jesnowski R et al (1998) Gene Ther 5:1070–1078
Magnani M, Fazi A, Mangani F, Rossi L, Mancini U (1993) Biotechnol Appl Biochem 18:217–226
Martin CR (1994) Science 266:1961
Mateo C, Palomo JM, Lorente GF, Guisan JM, Lafuente RF (2007) Enzym Microb Technol 40:1451–1463
Mikhaylova M, Kim DK, Berry CC, Zagorodni A, Toprak M, Curtis ASG, Muhammed M (2004) Chem Mater 16:2344–2354
Misson M, Zhang H, Jin B (2014) Interface 12:1–20
Mukherjee P, Ahmad A, Mandal D, Senapati S, Sainkar SR, Khan MI, Parishcha R, Ajaykumar PV, Alam M, Kumar R, Sastry M (2001) Nano Lett 1:515–519
Naessens K, van Daele P, Baets R (2001) SPIE-Int Soc Opt Eng Proc 4426:124–127
Nanda A, Saravanan M (2009) Nanomedicine 5:452–456
Netto CGCM, Andrade LH, Toma HE (2009) Tetrahedron Asymmetry 20:2299–2304

Okuyama K, Lenggoro IW (2003) Chem Eng Sci 58:537–547
Ong PL, Mahmood S, Zhang T, Lin JJ, Ramanujan RV, Lee P, Rawat RS (2008) Appl Surf Sci 254:1909–1914
Ozin GA, Arsenault AC (2006) Nanochemistry. RSC Publishing, Cambridge. Chapters 1–6
Pan ZW, Dai ZR, Wang ZL (2001) Science 291:1947
Pei L, Petrikovics I, Way JL (1995) Fundam Appl Toxicol 28:209–214
Petrikovics I, Hong K, Omburo G, Hu QZ, Pei L, McGuinn WD et al (1999) Toxicol Appl Pharmacol 156:56–63
Phadtare S, Kumar A, Vinod VP, Dash C, Palaskar DV, Rao M, Shukla PG, Sivaram S, Sastry M (2003) Chem Mater 15:1944–1949
Ramalingam B, Mukherjee S, Mathai CJ, Gangopadhyay K, Gangopadhyay S (2013) Nanotechnology 24:205602
Roy JJ, Abraham TE (2004) Chem Rev 104:3705–3721
Sathishkumar M, Sneha K, Kwak IS, Mao J, Tripathy SJ, Yun YS (2009) J Hazard Mater 171:400–404
Sevil B, Zuhal K (2010) Macromol Symp 295:59–64
Shankar SS, Rai A, Ahmad A, Sastry M (2004) J Colloid Interface Sci 275:496–502
Shankar SS, Rai A, Ankamwar B, Singh A, Ahmad A, Sastry M (2014) Synth React Inorg Met-Org Nano-Met Chem 3:482–488
Sheldon RA (2007) Adv Synth Catal 349:1289–1307
Sheldon RA (2011) Appl Microbiol Biotechnol 92:467–477; Sheldon RA (2011) Org Process Res Dev 15:213–223
Sheldon RA, van Pelt S (2013) Chem Soc Rev 42:6223
Shibata S, Aoki K, Yano T, Yamane M (1998) J Sol-Gel Sci Technol 11:279
Singh SC, Gopal R (2007) Bull Mater Sci 30:291–293
Solanki JN, Murthy ZVP (2011) Ind Eng Chem Res 50:12311–12323
Brinker CJ, Scherer GW (eds) (1990) Sol-gel science: the physics and chemistry of sol-gel processing. Academic, New York. e-book, http://depts.washington.edu/solgel/documents/class_docs/MSE502/SolGel_Science_The physics_and_chemistry_of_sol- gel_processing_-_Brinker_1990.pdf
Stellacci F, Bauer CA, Meyer-Friedrichsen T, Wenseleers W, Alain V, Kuebler SM, Pond SJK, Zhang Y, Marder SR, Perry JW (2002) Adv Mater 14:175–198
Stepankova V, Bidmanova S, Koudelakova T, Prokop Z, Chaloupkova R, Damborsky J (2013) ACS Catal 3:2823–2836
Sue K, Kawasaki S, Suzuki M, Hakuta Y, Hayashi H, Arai K, Takebayashi Y, Yoda S, Furuya T (2011) Chem Eng J 166:947–953
Suffner J, Agoston P, Kling J, Hahn H (2010) J Nanopart Res 12:2579–2588
Swihart MT (2003) Curr Opin Colloid Interface Sci 8:127–133
Tadic M, Panjan M, Damnjanovic V, Milosevic I (2014) Appl Surf Sci 320:183–187
Takahashi YK, Koyama T, Ohnuma M, Ohkubo T, Hono K (2004) J Appl Phys 95:2690–2696
Tavano OL (2013) J Mol Catal B Enzym 90:1–11
Thurn-Albrecht T, Schotter J, Kästle GA, Emley N, Shibauchi T, Krusin- Elbaum L, Guarini K, Black CT, Tuominen MT, Russell TP (2000) Science 290:2126
Tran DN, Balkus KJ Jr (2011) ACS Catal 1:956–968
Veith GM, Lupini AR, Pennycook SJ, Ownby GW, Dudney NJ (2005) J Catal 231:151–158
Veith GM, Lupini AR, Pennycook SJ, Villa A, Prati L, Dudney NJ (2007) Catal Today 122:248–253
Wang P (2006) Curr Opin Biotechnol 17:574–579
Wang W, Xu Y, Wang DIC, Li Z (2009) J Am Chem Soc 139:12892–12893
Wang C, Huang L, Song S, Saif B, Zhou Y, Dong C et al (2015) Appl Surf Sci 357,. Part B:2077–2086
Wendisch VF (2015) J Biotechnol 201:1

Xing T, Sunarso J, Yang W, Yin Y, Glushenkov AM, Li LH, Howlett PC, Chen Y (2013) Nanoscale 5:7970–7976
Yarosh D, Klein J, Kibitel J, Alas L, O'Connor A, Cummings B et al (1996) Photodermatol Photoimmunol Photomed 12:122–130
Yarosh DB, O'Connor A, Alas L, Potten C, Wolf P (1999) Photochem Photobiol 69:136–140
Yu G, Rosenberg JN, Betenbaugh MJ, Oyler GA (2015) Biotechnol 36:199–204
Zhang Y, Wan J, Skumryev V, Stoyanov S, Huang Y, Hadjipanayis GC, Weller D (2004) Appl Phys Lett 85:5343–5345
Zhang W, Qia X, Chen J, Wang H (2006) J Colloid Interface Sci 302:370–373
Zhang Y, Li J, Han D, Zhang H, Liu P, Li C (2008) Biochem Biophys Res Commun 365:609–613
Zhang X, Yan S, Tyagi RD, Surampalli RY (2011) Chemosphere 82:489–494
Zinchenko A, Miwa Y, Lopatina LI, Sergeyev VG, Murata S (2014) ACS Appl Mater Interfaces 6:3226–3232

Nanomaterial-Immobilized Biocatalysts for Biofuel Production from Lignocellulose Biomass

9

Richa Salwan, Anu Sharma, and Vivek Sharma

Abstract

Enzymes are employed in several fields of basic and applied research as biocatalysts in green chemistry, biosensor, nanobioelectronics, biofuel, and pharmaceutical, agricultural, and biotechnological industries. In present scenario, the diminution of fossil fuels gained the attention of researchers for substitute and sustainable renewable energy resources for biofuel production to combat worldwide energy consumption. The enzyme immobilization as biocatalysts for biofuel applications from lignocellulosic biomasses is found to produce highest percentage of bioethanol. The enzyme immobilization is a fundamental tool to reduce the cost and harness their benefits. The stabilization of enzymes using immobilization helps in efficient recovery from the reaction conditions after biocatalysis and hence makes laborious separation steps easy and permits repetitive use of enzymes. Besides this, it offers several other advantages such as stabilization against harsh reaction conditions, thermodynamic and kinetic stability, surface- and volume-confined enzyme environments, ability to design multi-step reaction, and reduced formation of undesired products which makes easy separation of soluble end products than free enzymes. The different methods of enzyme immobilization either involve adsorption or covalent bonding or encapsulation or a combination of different methods. Several types of nanoparticles and nanocomposites are being used for the stabilization of enzymes which retain the enzyme activity even after immobilization. This book chapter will cover the developments in coupled strategies and the deeper knowledge in stabilization of enzymes with special emphasis on the possibilities of nanomaterial coupled immobilization for operational stabilities in biofuel application.

R. Salwan
College of Horticulture and Forestry, Hamirpur, Himachal Pradesh, India

A. Sharma · V. Sharma (✉)
University Centre for Research and Development, Chandigarh University, Chandigarh, Punjab, India

M. Srivastava et al. (eds.), *Nanomaterials in Biofuels Research*,
Clean Energy Production Technologies, https://doi.org/10.1007/978-981-13-9333-4_9

Keywords
Enzymes · Stabilization · Enhanced activity · Cost · Applications · Nanomaterials

9.1 Introduction

Enzymes perform their activity selectively under different physiological conditions. Enzymes are explored in the areas of food, feed, and pharmaceutical, energy production such as biodiesel and bioethanol, and developing green chemical processes. However, quite often these proteins do not fulfill necessary prerequisite for an effective industrial biocatalyst (Iyer and Ananthnarayan 2008; Sheldon 2007a, b). The applications of enzymes are technically feasible only if the enzymes are stable under extreme conditions such as temperature and pH and occurrence of salts, surfactants, and alkalis. Overall, the enzyme immobilization with any strategy aims to produce a better and stable biocatalyst using chemical alteration (Rodrigues et al. 2011) or exploration of enzymes from the thermophiles (Cowan and Fernandez-Lafuente 2011) or genetic manipulation (Hernandez and Fernandez-Lafuente 2016; Rodrigues et al. 2013). Therefore, there is a need to stabilize the enzyme against adverse environmental conditions such as thermal or alkaline deactivation (Iyer and Ananthnarayan 2008). On a broad note, the applications of enzymes for a vast range of industrial aspects would not be economically viable using conventional chemical methods. Under nonliving conditions, these biomolecules need to be recovered, stored, and recycled for developing sustainable processes. The recent substantial developments in microbiology and molecular biology on the exportation of enzymes from extremophiles such as thermophilic organisms for improving enzyme properties due to directed evolution (Cowan et al. 2005; Rubin-Pitel and Zhao 2006; Schmeisser et al. 2007; Turner 2009) and biochemical engineering (Avnir et al. 2006; Cabana et al. 2009; Figallo et al. 2007; Richter et al. 2008; Sheelu et al. 2008) or immobilization supports and protocols (Avnir et al. 2006; Rusmini et al. 2007); Betancor and Luckarift 2008; Siqueira Jr et al. 2010; Zhang et al. 2005) have enabled to fulfill the enzyme deficiencies as industrial biocatalysts. These tools permit increased efficiency of enzymes as biocatalysts in chemical and industrial sectors (Karel Hernandez 1980).

Several technologies such as cross-linking enzymes, immobilization using microwave, use of mesoporous materials, and nanoparticle-based immobilization are currently used to upsurge the enzyme activity and to minimize the cost for industrial application (Ansari and Husain 2012). The immobilization of enzymes allows the repeated use of enzymes which cut down the cost and allow easy recovery of soluble end product and perform multiple reaction with ease and simplifies the overall design and performance. To address the limitations associated with enzyme applications, efforts to immobilize enzymes by different methods enable cutting down the cost of enzyme as well as purification of the soluble end product (Tischer and Wedekind 1999; Lalonde and Margolin 2002). Since the initial report on immobilization of *Aspergillus oryzae* aminoacylase, for racemic amino acid resolution by Chibata in 1967, several efforts have been made in the past few decades (Brena and Batista-Viera 2006; End and Schoning 2004; Sheldon 2007a, b; Wong et al. 2009). In England, two other non-immobilized enzymes were commercialized on large scale. The first immobilization of penicillin acylase was used for preparing 6-aminopenicillanic acid from penicillin G, whereas glucose isomerase was immobilized as a second enzyme in the USA to convert glucose into fructose. Thereafter, extensive studies on enzyme technology led to an increased number of industrial methods based on immobilized enzymes (Katchalski-Katzir 1993). The term "immobilized enzyme" was coined in 1971 with the suggestions of Ephraim Katchalski-Katzir at Enzyme Engineering Conference at Henniker, New Hampshire, USA, for representing enzymes which are restricted physically or localized for repeated and continuous use with retention of enzyme activities (Katchalski-Katzir 1993). Even in recent times, efforts to stabilize the enzyme of different nature are gaining attentions for converting them into powerful and efficient factories. For example, stabilization of monomeric and oligomeric enzymes via covalent linkage and immobilization of enzyme subunits prevent subunit dissociation.

At present, the immobilized enzymes are the center of research, due to their multiple benefits over soluble enzymes such as stabilization against harsh reaction conditions, thermodynamic and kinetic stability, selectivity, surface- and volume-confined enzyme environments, ability to design multi-step reaction, and reduced formation of undesired products, where soluble end products are easy to collect and reduce the buildup of inhibitory or reactive intermediates, compared to free enzyme (Fernandez-Lafuente 2009; Iyer and Ananthnarayan 2008; Karel Hernandez 1980; Mateo et al. 2007a, b; Polizzi et al. 2007; Tischer and Kasche 1999). In some cases, unexpected significant decrease or enhancement in enzyme activity has been reported, compared with soluble preparation of enzyme. For example, the immobilization of subtilisin was 27 times less active than the soluble enzyme (Schmitke et al. 1996). On the other side, immobilization of lipase as lipoprotein in ester synthesis showed 40-fold enhancement in enzyme efficiency, compared to enzyme powders (Ottolina et al. 1992; Tischer and Kasche 1999). At present, the major emphasis in enzyme immobilization aims at the development of new supporting materials which includes use of mesoporous materials with uniform pore size (2–50 nm) with enormous surface area for the efficient biotransformation and improved stability (Drechsler et al. 2004; Yiu and Wright 2005; Palocci et al. 2007). Moreover, immobilization of enzymes onto macro/micro surfaces enables their usage for other applications including biosensors, developing environmental remediation materials, and other applied fields. The recent developments of site-selective protein immobilization and the use invention of synthetic biology are emerging as a powerful and effective site-specific covalent coupling method. Additionally, several other methods of anchoring or immobilizing such as the fabrication of functional protein microarrays have emerged (Bilitewski 2006; Drager et al. 2007; English et al. 2006; Kingsmore 2006; Wong et al. 2009; Zheng et al. 2005). In parallel, efforts to develop continuous flow systems such as the PASSflow system to streamline production and immobilization of recombinant His6-tagged proteins might be explored as efficient biocatalyst (Drager et al. 2007; Wong et al. 2009).

With increased attention on enzyme immobilization, their massive usage is inevitable in future applications. The routine use of enzymes as soluble homogeneous solution makes the recovery of enzymes difficult; hence their recycling and renewal is impossible. Therefore, the enzyme immobilization could be the probable solution since it can significantly increase the stability of enzymes. Nanobiotechnology has gained attention from enzymologists and has potential to provide biocompatible environment for the enzyme immobilization with high stability and reuse (Ahmad and Sardar 2015; Bosio et al. 2016; Rostro-Alanis et al. 2016). However, the immobilization of enzymes such as amylolytic on the surface via adsorption or covalent linkage or by entrapment is difficult. In addition, the immobilization of enzymes such as amylolytic on the surface via adsorption or covalent linkage or by entrapment is difficult. The large surface area of nanomaterials, high rate of immobilization, high stability (Husain et al. 2011; Husain 2017), and comparatively low steric hindrances offer unique properties for enzyme immobilization. Moreover, nanomaterials with magnetic properties are easy to reuse (Husain 2017).

Biodiesel offers an attractive alternate and renewal source of energy, compared to conventional fossil fuel, due to low level of sulfur and carbon monoxide (Sharp et al. 2000; Srivastava and Prasad 2000). For biofuel production from lignocellulosic biomass, the major associated bottleneck is its structural composition, owing to the complexity of cellulose and hemicellulose to lignin. The production of biofuel is mostly carried by transesterification using corrosive chemicals (Moser 2011). The environmental issues and downstream processing led to a shift in focus for alternative eco-friendly production methods (Bisen et al. 2010; Hama et al. 2007; Kim et al. 2006c; Verma et al. 2013). It demands economically viable and appropriate pretreatment technology (Naresh et al. 2018). To achieve this, several collective conventional pretreatment processes such as acid-alkali, uncatalyzed or acid-catalyzed steam explosion, and ultrasound-assisted ionic liquids have been explored. The limitation of conventional pretreatment processes is they are energy-intensive. Moreover, a series of steps for biomass conversion and diverse nature of lignocellulosic biomass, followed by the formation of inhibitors for fermentation, impede the process. Several attempts have been made for overcoming the technical hurdles by using alternative biological methods of pretreatment (Naresh Kumar et al. 2018). The biological methods based on enzymes have several advantages such as reduction in the formation of fermentation inhibitors, environmentally friendly, and low-cost production. Still, the low hydrolysis rate, cost, and need of longer pretreatment time need to be addressed. Therefore, focusing on coupling of multi-enzyme systems for uninterrupted conversion of plant biomass to final end product is gaining increased attention. For example, coupling of cellulase, laccase, and α-glucosidase through immobilization for efficient biocatalytic pretreatment has been tested to achieve the goal (Betancor and Luckarift 2008).

This book chapter will cover the recent developments in coupling strategies and deep knowledge in stabilization of enzymes with special emphasis on nanomaterial coupled immobilization for operational stabilities in biofuel application.

9.2 Enzyme Immobilization

Enzyme immobilization discovered in 1916 (Nelson and Griffin 1916) was used for the immobilization of invertase enzyme onto a firm carrier such as charcoal or $Al(OH)_3$. Initially, the enzyme immobilization has very low enzyme loading capacity which led to the more refined and better methods of enzyme immobilization. In late 1990s, a number of enzyme immobilization methods were discovered. Still, due to the commercial importance, attentions for enzyme immobilization have increased (Ahmad and Sardar 2015; Brady and Jordaan 2009; Cantone et al. 2013). The immobilization of enzymes was initially done by the use of polymer-based matrices and linkage onto carrier materials. Considerable efforts have been made for cross-linking of enzymes either using protein-based methods or using inactive materials (Silman and Katchalski 1966). The history of confining enzymes is dated back to the 1950s. Use of prefabricated carrier materials for binding is a method of choice, whereas cross-linking enzyme crystals also appears as an interesting alternative (Tischer and Kasche 1999). The method for immobilization is categorized based on the kind of chemical reaction used for binding (Buchholz and Kasche 1997; Kennedy and Melo 1990). Due to the low hydrodynamic and mechanical stability of materials obtained by enzyme cross-linking, researchers opted other methods. The cross-linking-based immobilization is typically support-free and includes formation of cross-linkages usually by the addition of cross-linking reagents such as glutaraldehyde to achieve three-dimensional arrangement (Murty et al. 2002; Öztürk 2001).

Currently, immobilization of enzymes can be done by binding to carrier or encapsulation or cross-linking. The linkage to carrier can be of physical nature through hydrophobic, covalent, van der Waals, and ionic forces. However, physical linkage is weak that it cannot hold the enzyme to the supporting carrier during applications. Here, the carrier can be of synthetic origin such as resin/biopolymer, or inorganic in nature such as (mesoporous) silica/zeolite. The immobilization of enzymes to different materials such as nano-size supports including spheres, tubes, and fibers (Kim et al. 2006a, b; Martin and Kohli 2003; Yim et al. 2003), mesoporous materials (Lei et al. 2006), silica-based nanotubes (Mitchell et al. 2002), inside of phospholipid bilayers (Hamachi et al. 1994), corridors of R-zirconium phosphate (Kumar and McLendon 1997), in Langmuir–Blodgett flicks (Troitsky et al. 2003), on polystyrene particle (Caruso and Mohwald 1999), on self-assembled monolayers (Hung et al. 2006), inside a polymer matrix (Kim et al. 2005a, b), gold nanoparticles accumulated on polymers and zeolite (Bhat et al. 2003; Phadtare et al. 2003) and thermal evaporated fatty lipid layers (Sastry et al. 2002) have been industrialized with their own distinctive advantages and limitations. Encapsulation of enzymes via inclusion in a polymer system such as an organic polymer or a silica sol-gel or a hollow fiber can be done. Here, the physical restraints are weak and cannot avoid enzyme leakage entirely. Therefore, often extra covalent attachment is done. The support binding and entrapment is not always clear. The binding to a prefabricated support regardless of enzyme immobilization on the external or internal surface is referred to as carrier binding. On the other hand, encapsulation involves the synthesis of polymer network in the presence of enzyme. For example, enzyme immobilization in a prefabricated mesoporous silica is carrier binding, whereas during the preparation of silica sol-gel, the enzyme is entrapped.

Conceptually, the immobilization of enzymes can be done on any other protein or on solid support (Hanefeld et al. 2009; Mateo et al. 2007a, b; Palazzo et al. 2014; Sheldon and van Pelt 2013; Vashist et al. 2014). Besides simple adsorption, the carrier is modified using an organic linker which contains combative functional groups. Then, the linkers are directly attached by covalent attachment. The coating can be made up of adsorbed or covalently linked self-assembled monolayer or bilayer of amphiphilic molecules including proteins such as bovine serum albumin and avidin (Vashist et al. 2014) or polymer (Fornera et al. 2012a, b). The organic layer coating may lead to enzyme inactivation upon direct interaction with the solid support. Sometimes, polymer–enzyme conjugates prepared previously in solution are adsorbed (Küchler et al. 2015a, b). Enzymes immobilized on a surface, in some cases, are more stable than their counterpart free enzymes (Küchler et al. 2016). A brief overview of different characteristic features of enzyme immobilization are mentioned in Fig. 9.1.

The enzyme crystals are one of the examples where the crystallized protein acts both as a carrier and catalyst. Here, multiple forces of non-covalent nature stabilize the crystals, compared to cross-linked assemblies of amorphous proteins. In addition, the chemical bonding using cross-linker further stabilizes the crystals in an aqueous solution (Persichetti et al. 1995). Immobilization using non-covalent-based adsorption is beneficial in non-aqueous solutions, where desorption may be ignored due to the little solubility of enzyme in the aqueous environments. This process is the choice for the preparation of lipase in water-immiscible solvent (Balcao et al. 1996; Malcata et al. 1997). The adsorbed enzymes (Antrim and Auterinen 1986) and proteins (Tischer 1992) are widely used in industrial applications. Immobilization

- **Enzyme properties;**
  Molecular mass, prosthetic groups, functional groups on protein surface, purity, kinetic properties such as specific activity, pH and temperature, inhibition, solvents.
- **Carrier chemical characteristics;**
  Chemical composition and behavior, functional groups, accessible surface of matrix, pore size, stability of carrier, particle diameter, flow resistance (for fixed-bed applications), sedimentation velocity (for fluidized beds), abrasion (for stirred tanks) etc.
- **Immobilization method;**
  Bound protein, yield of active enzyme, intrinsic kinetic properties (i.e. properties free of mass- transfer effects).

a

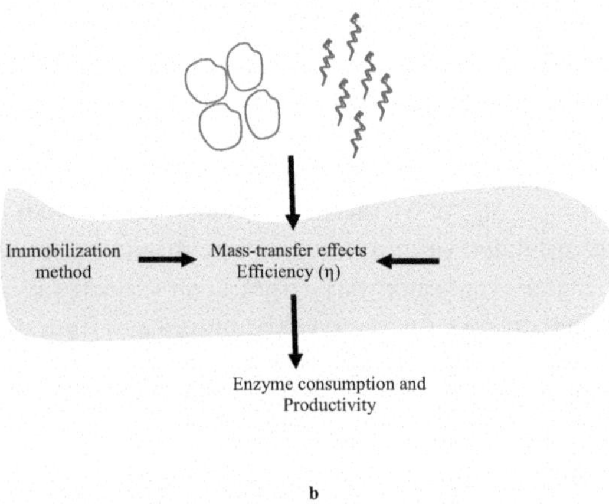

Fig. 9.1 Overview of enzyme immobilization and characteristic features of enzyme immobilization (**a** and **b**). (Adopted from Tischer and Kasche)

of enzymes can be attained by the diffusion of dried and water-soluble preparation of enzyme in water-immiscible organic solvents. Lipases have been immobilized using this method. In non-aqueous conditions, this type of preparation catalyzes hydrolytic reactions in the opposite way by carefully controlling the water content (Bell et al. 1995).

Immobilization by inclusion includes retaining of the enzyme within a membrane such as polymeric networks, hollow fibers, or microcapsules. For simple inclusion, derivatization of the enzyme is not required, and harmful effects of binding forces are not required to be considered. However, protein–carrier interactions lack additional stabilization. Use of polymeric networks is one of the most effective methods when used with other additional binding methods like the linkage of vinyl groups to the enzyme surface, and its incorporation into a rising polymer cable is another method. Such modified enzyme preparations can be combined into "biocatalytic plastics" which are active in aqueous as well as organic media (Wang et al. 1997). There are certain dissimilarities between binding on carriers, cross-linking, and crystallization. The last method primes particles of high enzyme concentration, frequently with an equal distribution. However, immobilization onto preassembled supporting material allows binding onto surfaces with comparative low enzyme concentration but with special binding on the outer covering of the carrier materials (Tischer and Kasche 1999).

9.3 Basic of Enzyme Immobilization

The enzyme immobilization using standard protocol is referred to as "random immobilization." However, in general, the enzyme immobilization is never a complete random process and often leads to protein immobilized in a determined and very similar orientation. The selective enzyme immobilization on the surface using a particular immobilization method can lead to immobilized enzyme molecules with a specific unreachable surface area (Fuentes et al. 2006). Therefore, the challenge associated with different immobilization protocols, changes in enzyme properties, and even inversion of enzyme's specific action (Fernández-Lorente et al. 2007; Mateo et al. 2007a, b; Palomo et al. 2002, 2007; Palomo 2009; Pišvejcová et al. 2006; Terreni et al. 2001) would be difficult to explain. For example, the trypsin-based lysis of penicillin G acylase to Eupergit (Mateo et al. 2000) or glyoxyl agarose (Mateo et al. 2005; Mateo et al. 2006), reveals dissimilarity in the orientation of protein in two methods (Karel Hernandez 1980; Temporini et al. 2010). In fact, a random immobilization of enzyme on a carrier is quite difficult to attain. Researchers have advocated the use of various immobilization methods in a coupled way to achieve the truly random immobilization (Fuentes et al. 2006; Karel Hernandez 1980). The adsorption can be done using H-bridges or hydrophobic or π–π interactions. The covalent binding is more robust and long-lasting. Initially, carrier interacts covalently with protein at one point via major amino group in the protein such as Lys amino acid. However, the non-ionized amino groups are considered better nucleophiles in the reaction (Diána 2015).

9.4 Methods of Immobilization

The immobilization of enzymes can be performed by physical or chemical methods. Physical procedures involve the weak interaction amid carrier and enzyme, whereas the chemical approaches involve covalent bond between the both. The selection of enzyme immobilization method is of vital importance to avoid the damage of any enzyme activity by maintaining the binding site of enzymes. In recent time, the applications of site-selective enzyme immobilization have gained significant attention. The prior knowledge of the enzyme active site and how it can be protected using protective groups which can be detached later on without any loss of enzyme activity are really helpful. Use of adsorbents, covalent linkage, entrapment, and cross-linking are the most common methods of enzyme immobilzation (Ahmad and Sardar 2015; Brady and Jordaan 2009). Even though the basic procedures of enzyme immobilization are classified into four types as mentioned above; several variants, based on combination of these basic procedures, have been industrialized for enzyme immobilization. Each of these method owns their merits and demerits for enzyme immobilization.

Further, based on interactions between enzymes and supporting carrier, different methods of enzyme immobilization are categorized into irreversible and reversible types (Brena and Batista-Viera 2006). In irreversible immobilization, enzyme immobilized onto supporting material cannot be detached by modulating the properties of both, whereas in reversible immobilization, it is possible to detach the enzyme from the supporting material using congenial environment. The covalent bonding, cross-linking, and entrapment are one of the most common methods, for irreversible enzyme immobilization, whereas physical adsorption and non-covalent binding through affinity and chelation binding are reversible type of immobilization methods (Zhao et al. 2015).

9.5 Adsorption of Enzymes

One of the oldest and simple methods of immobilization, the enzyme is adsorbed onto an insoluble support. This method is widely used and offers high capability of enzyme loading. Here, enzyme is mixed with suitable carrier adsorbent under suitable pH and ionic conditions. After that, unbound enzyme is removed by washing, whereas the immobilized enzyme is available for use. The interaction between adsorbent and enzymes are van der Waals forces, hydrogen, and ionic and hydrophobic types. These forces are very weak but operate in large scale. The adsorption usually includes multipoint adsorption between proteins and binding locations on the immobilization surface which may lead to alteration in the protein microenvironment (Johnson et al. 1996). The nature of adsorbent determines the amount of enzyme leakage. To avoid any chemical alterations and side effects to the enzyme activity, the surface property of support is also vital. The easy desorption of enzyme is a major disadvantage of this method (Rao et al. 1998) (Fig. 9.2a).

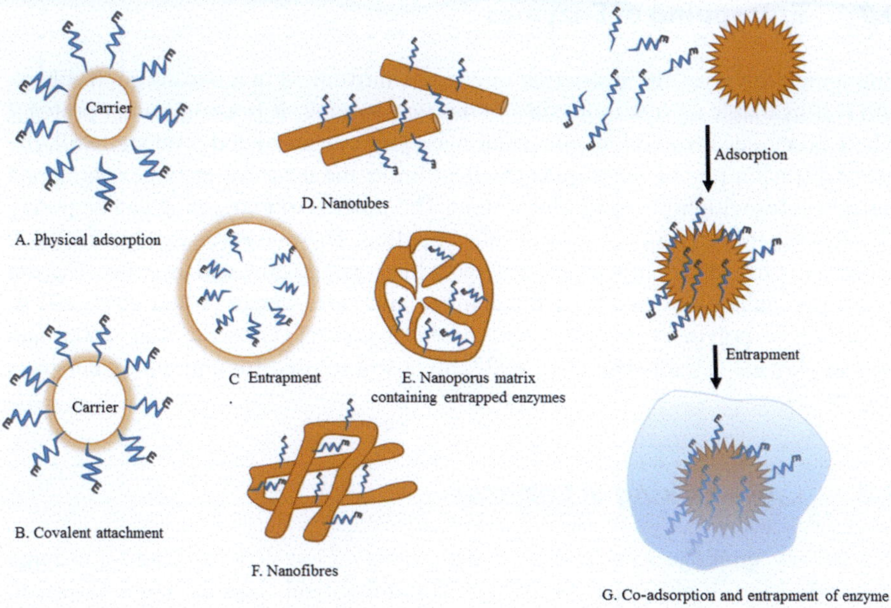

Fig. 9.2 Overview of enzyme immobilization. Basic structures: (**a**) physical adsorption. (**b**) Covalent attachment. (**c**) Entrapment. (**d–f**) Nanosupports for enzyme immobilization. (Adopted from Wang 2006). (**g**) Immobilization of enzymes using co-adsorption method)

9.6 Covalent Binding of Enzymes

For covalent immobilization, the enzyme is conjugated to support matrix via the formation of covalent bonds. Similar to the adsorption of enzymes, the immobilization must be carried in an environment which does not lose enzymatic activity. The covalent linkage of protein to support involves exploration of amino acids on side chain like arginine, histidine, and aspartic acid which are also determined by the reactivity of imidazole, phenolic hydroxyl, and indolyl (D'Souza 1999). Use of peptide-adapted carriers for enzyme linkage leads to high specific activity as well as stability (Fu et al. 2011). The interaction between the supporting material and enzyme can be obtained either by direct bonding or by intercalated linkers known as spacers. The spacer molecule provides better flexibility to the coupled enzyme; therefore the enzyme activity has enhanced as compared to directly coupled enzyme. The activation of functional groups on support matrix is found to facilitate the covalent linkage. Chemicals such as glutaraldehyde, cyanogen bromide agarose, and cyanogen bromide-activated sepharose holding carbohydrate moiety are found to enhance the thermal stability of covalently bound enzymes (Hartmann and Kostrov 2013; Singh et al. 2013) (Fig. 9.2b).

9.7 Entrapping of Enzymes

Entrapment involves the restriction of enzyme movement in a porous gel or fibers; still it is available as free molecules in reaction volume. It is a convenient protocol where processes involve low molecular mass substrates and end products. Still, the lower diffusion rate or mass transfer effect limits the usage of enzymes entrapped for substrates with high molecular weight. The process of entrapping can be purely physical separation using cage or covalent binding. Thermal reverse polymerization is deployed for entrapping natural polymers like agar, gelatin, and agarose, whereas ionotropic gelation is preferred for alginate and carrageenan (Datta et al. 2013). Besides this, polymers like polyvinyl alcohol hydrogel (Grosová et al. 2008) and polyacrylamide (Deshpande et al. 1987) are also explored for entrapping enzymes (Fig. 9.2c).

9.8 Cross-Linking of Enzymes

A relatively simple process of cross-linking involves covalent bonding for enzyme immobilization. It also leads to frequent conformational changes which results in loss of activity; therefore, it is not a preferred immobilization method. Still, it is explored in enzyme immobilization methods particularly for minimizing the leakage of enzymes occurring by adsorption. Glutaraldehyde is one of the most common bifunctional agents extensively used, owing to its little cost, higher efficiency, and stability. The aldehyde groups of glutaraldehyde at the both ends link unrestricted amino groups of enzymes. Here, enzyme is generally immobilized in the presence of a protein like gelatin, collagen, and albumin. The undesirable activity losses probably occur due to the participation of catalytic groups for the immobilization. Another problem associated with cross-linking is lack of control, and therefore, it is challenging to get large-scale enzyme aggregation simultaneously with high activity. Moreover, the gelatinous behavior of the immobilized enzyme is another hurdle which limits the use of immobilized enzymes for several applications. In recent times, enzyme precipitation from aqueous solution in the form of physical aggregates either using salts or water-miscible organic solvents or non-ionic polymers is being explored for developing new immobilized enzymes known as cross-linked enzyme aggregates (CLEAs) (Sheldon 2007a). The physical aggregates are bound, organized by non-covalent interaction. Subsequently, the physical aggregation makes them permanently insoluble without disturbing their native structure and the catalytic activity. CLEAs are very effective, exhibit satisfactory stability as well as performance, and can be formed using economical and effective methods. The methodology can be deployed across the enzymes, even including for cofactor-dependent oxidoreductases (Sheldon 2007b).

9.9 Nature of Supporting Material for Enzyme Immobilization

The carriers of different nature have been explored over a period of time for enzyme immobilization (Datta et al. 2013; White and Kennedy 1980) (Fig. 9.2d–f). The nature of carrier plays decisive role in determining the performance of the enzyme immobilization. An ideally perfect support should be inert, have resistance to physical compression, microbial attack, biocompatible and available at low cost (Brena and Batista-Viera 2006). Polymers including alginate, cellulose, chitin, chitosan, starch, collagen, sepharose, pectin, and other natural polymers are common supporting materials (Datta et al. 2013). Additionally, synthetic polymer and inorganic materials are found to have good mechanical stability and can be modified easily (Bryjak and Trochimczuk 2006; Yong et al. 2008). The inorganic materials including alumina, silica, zeolites, and mesoporous silica (Datta et al. 2013; Hartmann and Kostrov 2013; Hudson et al. 2008) and silica-based materials are one of the most suitable and explored matrices for enzyme immobilization for commercial applications (Hartmann and Kostrov 2013; Ho et al. 2004) and research purpose (Vianello et al. 2000).

9.10 Nanomaterial-Based Enzyme Immobilization

Diverse types of carrier-based techniques for enzyme immobilization have been used in recent times to enhance the enzyme efficiency and stability and for lowering the cost of enzyme in industrial applications. These include use of aggregation, microwave-assisted immobilization, and mesoporous and nanoparticle-based materials. Enzymes/proteins are encapsulated within polymer silica (Frenkel-Mullerad and Avnir 2005) and gold nanoparticles. With the emergence of nanotechnology, the magnetic nanoparticles have gained a considerable attention. The unique presence of specific physiochemical properties of nanoparticles led to their advent in large number of biotechnological applications with better characteristics. The high surface-to-volume ratio of nanoparticles results in a high density of the immobilized entity. These enzyme preparations are active under broad pH and temperature range than free enzymes. The nanoparticles have well-structured core–shell and are uniform in size and shape, easy to prepare without any surfactants and toxic chemicals, and have particle size as per utility limits. Additionally, it is likely to co-immobilize the multi-enzymes on nanoparticles (Ansari and Husain 2012). The application of nano-immobilized enzymes is considered an exciting and emerging area which provides robust nanostructured materials as enzyme carriers for tailoring enzyme scaffolds (Xin et al. 2010). Recent advancements in nanobiology provide diverse range of nanomaterials and the approaches for immobilizing enzymes on the nanosupports.

The higher surface-to-volume ratio offers dense enzyme loading and hence a high activity of immobilized enzyme compared with conventional materials (Gupta et al. 2011; Hwang and Gu 2013; Verma et al. 2013). For example, protein of up to

204 mg/gm dry support for nanoparticles has been achieved compared to the 40 mg/gm of dry support for macroparticle. The high enzyme density leads to maximum enzyme activity units of nanoparticle-immobilized enzymes as compared to per gram of dry supporting macroparticle-immobilized enzymes (Klein et al. 2012). The nanoparticle-based enzyme immobilization clearly revealed that the carrier area-to-volume ratios rather than the carrier area are a better indicator of assessing supporting material's potential. The estimation of enzyme molecules immobilized on the materials can be assessed based on the physical property of the supporting material and enzymes. For example, presuming the spherical shape and the concentration of Fe_3O_4 which is 5.18 g/cm^3 and molecular mass of esterase which is 33,000 dalton, it is possible to estimate the number of esterase molecules conjugated to Fe_3O_4 nanoparticles (Shaw et al. 2006).

The enzyme activity of nanomaterial-bound enzymes is high due to low mass transfer resistance (Kim et al. 2008, 2006c). The stable mono-dispersion behavior of nanomaterial-immobilized enzymes is largely due to significant "Brownian motion" in aqueous suspension (Verma et al. 2014; Wang 2006). The improved diffusion and mobility of nanoparticles can impact the activity of mobilized enzymes (Jia et al. 2003). Therefore, promoting nanoscale structures for enzyme immobilization aims at maximizing the functional surface area and reduces diffusion limitations (Xie and Ma 2009). Other advantages such as thermal stability and augmented surface area can accelerate their potential applications. In recent times, different kinds of nanoparticles have been deployed for the fabrication of different nanomaterials such as nanotubes, nanorings, nanowires, etc. (Ali and Winterer 2010; Ni et al. 2007). Moreover, conjugation of various polymer microspheres and metal nanoparticles with proteins and enzymes has been done successfully (Schuler and Caruso 2000; Willner et al. 2006). The nanoparticle surface alteration using enzymes and biomembrane provides a distinctive biofunctionality to them (Yamashita 2001). Nanoparticles offer several other unique properties for the conjugation and functionalization which imparts advantages for diverse range of applications including clean energy generation and biodiesel production, biological/chemical sensing, drug delivery, and disease diagnostics (Ding et al. 2015). An overview of different nanoparticle-based enzyme immobilization methods is given below:

9.10.1 Enzyme Immobilization Using Magnetic Nanoparticles

The magnetic nanoparticles offer unique property such as superparamagnetism, large and modifiable surface, low toxicity and compatibility compared to bulk material. In this context, the iron oxide based magnetic nanoparticles are an attractive choice for enzyme immobilization. Since, iron oxide based magnetic nanoparticles are attracted to a magnetic force, this limitaion can be overocome once the force is removed. Nanoscaled carrier helps in achieving the upper limits in effective enzyme loading by harmonizing the key features that regulate the efficiency of biocatalysts such as high surface area-to-volume ratios and mass transfer resistance. The unique behavior of nanomaterial-coupled enzymes distinguishes it from conventional immobilized systems and hence shows potential applications in biocatalyst engineering. Still, nanoparticles often form clumps, and therefore, temporary dispersion is required. Moreover, their separation either by centrifugation or membranes is difficult due to their size. Therefore, use of superparamagnetic particles can solve these problems. Such type of particles can easily disperse in solution, and the suspended superparamagnetic particles in the reaction can be removed using an external magnet (Johnson et al. 2011; Liese and Hilterhaus 2013; Yang et al. 2008; Kopp et al. 2014). Magnetite nanoparticles (Fe_3O_4) are one of the most extensively used superparamagnetic supports (Gupta et al. 2011).

The usage of magnetic materials for enzyme immobilization was established on the basis of magnetic particles sequestered from magnetotactic bacteria (Matsunaga and Kamiya 1987). Thereafter, magnetic (maghemite) nanoparticles were explored for the immobilization of *Candida rugosa* lipase (Dyal et al. 2003). Recent developments are focused on the use of magnetic nanoparticle synthesis (Yiu and Keane 2012). The physiochemical attributes of magnetic nanoparticles enable them for several applications. In the last several years, magnetic nanoparticles have been explored as a potential material for enzyme immobilization (Yang et al. 2008). The parting of an immobilized enzyme using magnet is easy where either the immobilized enzyme is recollected using magnetic field or substrate solution removed or vice versa (Dyal et al. 2003; Gardimalla et al. 2005).

The advantages of nanomaterials in the form of shape, size, and functionalization ensure their suitability for enzyme immobilization as nanobiocatalysts. The nanoparticle-immobilized enzymes may be mobilized, directed, and then isolated by magnetic field. The magnetic nanoparticles can be used for enzyme immobilization provided the nanoparticles have adequate geometry and surface property. The magnetic immobilization of the enzymes is done either by the incorporation of magnetic particles during support polymer synthesis (Robinson et al. 1973) or coating of magnetic particles using supporting material such as agarose or dextran (Rusetski and Ruuge 1990). The magnetic particles/supports such as microspheres of different biomaterials or polymers with magnetic elements have been explored with great success (Dyal et al. 2003; Koneracka et al. 1999; Kouassi et al. 2005; Saiyed et al. 2003). The occurrence of amphoteric hydroxyl (OH) groups can directly facilitate the binding of proteins onto magnetic particles (Mehta et al. 1997). The magnetic particles have been utilized for the immobilization of several proteins including bovine serum albumin (BSA), lipase, and streptokinase (Fernandes et al. 2006; Koneracka et al. 2002), trypsin and chymotrypsin (Jordan et al. 2006), glucose oxidase and cholesterol oxidase using CDI as coupling agents (Koneracka et al. 2002; Kouassi et al. 2005; Kuroiwa et al. 2008; Mehta et al. 1997; Pimentel et al. 2007). Conjugation of proteins onto unmasked magnetic particles offers numerous inherent benefits (Saiyed et al. 2007). Magnetite (Fe_3O_4) is one of the most used magnetic materials, whereas matrices including nickel-impregnated silica paramagnetic particles (NSP) are also used for immobilization (Prakasham et al. 2007). The enzymes immobilized on NSP are found more stable and active on extreme pH and temperature limits which may be due to change in enzyme configuration under an immobilized condition (Prakasham et al. 2007).

Similarly, *Bacillus subtilis* keratinase has been immobilized onto polyethylene glycol-supported Fe_3O_4 superparamagnetic nanomaterial (Konwarh et al. 2009), haloalkane dehalogenase of *Xanthobacter autotrophicus* strain GJ10 immobilized on silica-coated or silica-uncoated Fe_3O_4 superparamagnetic particles (Johnson et al. 2008), lipase from *Candida rugose* immobilized on Fe_2O_3 magnetic nanoparticles, and lipase from *C. rugosa* immobilized on novel nanosupport, and phosphorous-containing polyurethanes are found more stable (Guncheva et al. 2011). The pretreatment of the supporting material with polar or non-polar solvents resulted in an undesirable impact on *C. rugose* lipase immobilization, whereas non-ionic surfactants enhanced the enzyme immobilization to 30–40% (Guncheva et al. 2011).

9.10.2 Novel Nanoparticles for Enzyme Immobilization

Gold nanoparticles are one of the outstanding biocompatible materials for enzyme immobilization. The interaction of gold nanoparticles between amino and cysteine group of proteins is equally robust to that of thiol derivative usually used. The proteins can be immobilized on gold nanomaterial directly without using any alteration (Crespilho et al. 2006; Lia et al. 2010; Xu et al. 2007). On the other side, the gold nanoparticles can be functionalized for protein immobilization using thiolated molecules and carboxylic groups (Jun et al. 2001; Vertegel et al. 2004). The zeolite-gold nanoparticle "core–shell" structures have been used for immobilizing fungal proteases (Phadtare et al. 2004). The gold nanoparticles are assembled on the zeolite surfaces through the amino groups present in 3-amino-propyltrimethoxysilane. The protein linked to "core–shell" assemblies can be easily separated from by centrifugation and then reused. The polyurethane microsphere–gold nanoparticle preparations are found to enhance enzyme activity in extreme pH range, shelf-life, as well as thermal stability (Kumar et al. 2003).

9.10.3 Enzyme Immobilization Using Nonmagnetic Nanoparticles

The enzyme immobilization on actuated supports allows the simplification and repeated use of enzymes and their biochemical properties (Gaberc-Porekar and Menart 2001; Iyer and Ananthnarayan 2008; (Mateo et al. 2007a, b). The numerous nanomaterials including nanoparticles, nanotubes, nanofibers, and nanoporous matrices have been extensively used for biotechnological applications (Wang 2006). Nonmagnetic materials such as cellulase immobilized on nanomaterials of precise polymethyl methacrylate (PMMA) core and cellulase shell have been used (Ho et al. 2008; Kin et al. 2008). Similar to other enzyme preparation, the immobilized cellulase showed high thermal stability and residual activity than free enzyme solution. Therefore, this type of methods can be explored for the immobilization of thermally stable enzyme for commercial applications (Ho et al. 2008). These types of nanoparticles contain thick and consistently dispersed enzyme shells. Such types of preparation can produce solid content up to 18% (w/w) concentration (Ho et al. 2008). The other nonmagnetic materials such as epoxy groups are useful for covalent bonding with amino, thiol, and hydroxyl groups present on the surface of enzyme. However, the interaction amid epoxy clusters and soluble enzyme is a slow process. Here, using initial intermolecular physical or chemical interaction, new functional groups are added to the enzyme surface. The subsequent intramolecular covalent linkage between the epoxy groups of the supporting material and the immobilized enzyme is performed. In this process, the initial immobilization regions are strengthened using multipoint covalent linkage (Mateo et al. 2007a, b).

The nonmagnetic nanofibers are created by electrostatic fiber production, and then enzymes can be immobilized into inner or outer surface. For immobilization on inner side, the enzymes are embedded in situ either between the polymer chains or inside the nanofiber. The enzyme immobilization can be on the surface of nanotubes using covalent and secondary binding. The nature of polymer material and buffered media of enzymes influence the enzyme immobilization. The biocompatible water materials such as soluble polymers or gelatins are considered as the most effective and homogeneous medium. The single-walled nanotube offers highly specific surface area, whereas the multi-walled nanotubes are preferred, due to their easy dispersion property. The immobilization on inner side of nanofibers also provides higher stability and enzyme activity than enzyme immobilized on outer surface of fiber materials. The use of other nanomaterial such as carbon nanotubes of both single-walled and multi-walled nature is explored for the enzyme immobilization (Asuri et al. 2006; Chen et al. 2003; Karajanagi et al. 2004; Shim et al. 2002). Presently, multi-walled nanotubes are the preferred material for nanobiocatalysis. The reports on both loss and increase in enzyme activity after multi-walled nanotube immobilization are presented in the literature (Asuri et al. 2006; Shah et al. 2007). The enzyme immobilization using carbon nanotubes is one of the most widely used carrier supports for developing biofuel cells. Other than SiO_2 and iron oxides, nanomaterials such as polynanofibers and superparamagnetic particle-based nanocrystals are also described in the literature for enzyme immobilization (Herricks et al. 2005; Jia et al. 2003) (Fig. 9.3).

Another novel and synthetic method involving single enzyme molecule based immobilization on porous nanostructure known as "single enzyme nanoparticles (SENs)" is also established (Kim and Grate 2003). Here in this method, the armored nanoparticles and a porous and amalgamated organic/inorganic network surround every enzyme molecule. The thin and porous nature of SENs prevents mass transfer effect on the substrate. The nanohybrid enzyme structure is a potential method to stabilize enzymes and hence can be used for diverse applications (Kim and Grate 2003). The fabrication of SENs involves a multistep process including the surface modification of enzyme and polymer growth on the enzyme surface followed by polymerization by condensation which attach polymer into a network surrounding the enzyme (Yang et al. 2008).

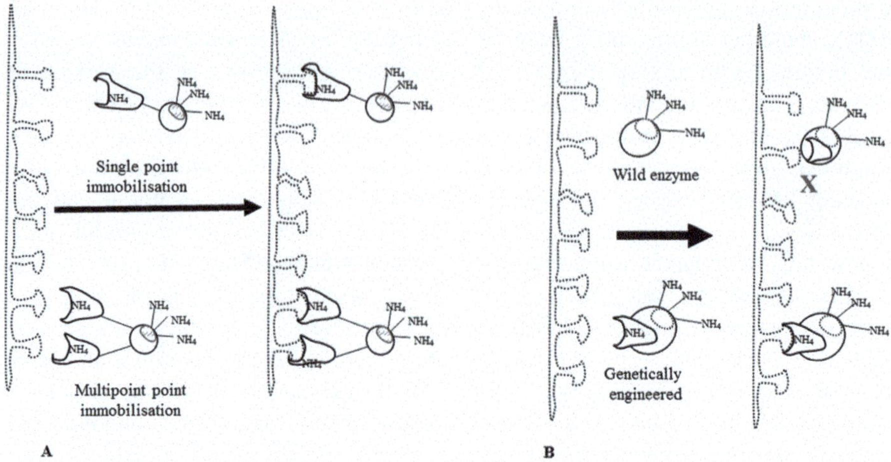

Fig. 9.3 (**a**) Single- and multipoint enzyme immobilization. (**b**) Engineering enzyme for improved tethering on nanoparticle surface

9.11 Methods of Nanomaterial-Based Enzyme Immobilization

The amount of active enzyme present on the carriers is broadly influenced by bound enzyme which is typically 0.1–10% of the total weight of the supporting carrier material (0.1–500 units per mg protein of enzyme activity). There is a need to optimize the reaction volume and amount of enzyme and carrier materials. However, carriers can impart novel properties to enzymes. In crystal-based preparations, the carrier itself is an enzyme protein which determines the crystal weight and has the potential to provide high enzyme density. It can be of special importance for enzymes with low activity. The surface functionalization of nanomaterial offers effective enzyme immobilization. The use of interfacing proteins, agreements the possibilities to discover novel and robust conjugates for different uses (Chronopoulou et al. 2011). In another novel immobilization method based on prefabricated nanophase amphiphilic network, the enzyme is positioned in hydrophilic domains consisting of poly(2-hydroxyethyl acrylate) (PHEA). Here, the hydrophobic polydimethylsiloxane (PDMS) phase-diffusing substrate can contact the enzyme via the bulky interface and allows its isolation and abnormal swelling behavior. In the presence of water, the hydrophilic domain swells, and the polymer conjugated with the enzyme after immersion in an aqueous solution. After drying, the phase shrinks in size and contains trapped enzyme. The immersion of polymer network in organic solvent leads to the hydrophobic PDMS domain swelling and then permits entry of dissolved substrate via the interface (Sheldon 2007a, b).

Enzymes can be adsorbed onto carbon nanomaterial such as nanotubes via covalent or non-covalent methods (Lee et al. 2010). One of the major issues with the adsorption involves leaching of enzyme from nanoporous support. However, it can be regulated using covalent processes and cross-linkers such as glutaraldehyde (Gao et al. 2010). The use of site-specific affinity tag such as his-tagged and single-walled carbon nanotubes, altered with Na,Na-bis(carboxymethyl)-L-lysine hydrate, can overcome the limitations of diffusion and enzyme leaching and achieve proper orientation of active site (Wang and Jiang 2011). Cellulase attached to gold-fixed silica nanoparticles using affinity method is used for the hydrolysis of cellulose to glucose in a single step (Cho et al. 2012). The functionalization of core–shell nanoparticles with amino groups and PEG chains is also explored for the covalent linkage of β-glucosidase (Georgelin et al. 2010). Lipase of *Burkholderia cepacia* is immobilized onto the silica nanowires which is prepared by capping different numbers of alkyl groups. The immobilized lipase exhibited superior performance ascribed to improve mass transfer efficiency of reactants of silica nanowires and the activation of enzyme by capped-octadecyl groups in the aqueous and non-aqueous environmental conditions (Jin et al. 2018). Moreover, nanowires-immobilized lipase can be explored for developing porous biocatalytic membrane for the easy removal of enzyme from end products (Jin et al. 2018). The use of glutaraldehyde as a cross-linker for covalent immobilization to nanoparticle surface is the most commonly explored method (Wang et al. 2009). It offers high enzyme loading of up to 25 milligram of lipase/g particles as well as good activity retention up to 70% (Xie and Ma 2010). Lipase has been immobilized to the carbon nanotubes using carbodiimide-based activation (Ji et al. 2010). The immobilized enzyme also reserved the selectivity of the native enzyme as well as high stability. Use of amidination process for the activation of nitrile groups of the polyacrylonitrile nanofibers is also a preferred method (Li et al. 2011). The activated nanofiber showed high protein loading up to 43 milligram of lipase per gram of fiber and retained enzyme activity of up to 80% than free enzyme. The enzyme immobilization on nanomaterials is executed using free amine present on the enzyme surface with carboxylic acid, created by nanotube oxidation using chemicals such as nitric acid and sulfuric acid. The cross-linkage is also enabled using other chemicals such as EDC (1-ethyl-3-(3-dimethylaminopropyl)carbodiimide) (Saifuddin et al. 2013). The mesoporous silica such as MCF, SBA-15, MCM-41, and MCM-48 has been examined for lipase immobilization (Kim et al. 2007).

9.12 Analytical Tools for Investigating Enzyme–Nanomaterial Interaction

Understanding the interactions of nanocarrier and enzymes at the organizational level is a priority for enhancing the applications of the nanomaterials (Cruz et al. 2010; Pavlidis et al. 2012a, b). The biophysical changes after post immobilization can be observed using scanning electron microscopy (SEM), transmission electron microscopy (TEM), atomic force microscopy (AFM), circular dichroism (CD), UV-vis spectroscopy, Raman spectroscopy, microcalorimetry, time-of-flight secondary-ion mass spectrometry (TOF-SIMS), XPS (X-ray photoelectron) spectroscopy, and Fourier transform infrared spectroscopy (FTIR) that are used to elucidate the structure of enzymes adsorbed on/in the nanomaterials. The conformational stability of immobilized enzymes using these tools led to the finding of novel properties of nanomaterial-immobilized enzymes responsible for enzyme stability and activity (Andrade et al. 2010; Lee et al. 2009). The maintenance of native and active structure of enzyme is determined by the distinctive conformational stability, the nature of the immobilizing carrier, and the intermolecular interaction between the enzyme and carrier (Cruz et al. 2010; Shang et al. 2007).

The possible stabilizing outcome of nanostructured materials is enzyme-specific (Ganesan et al. 2009). The conventional tools for determining enzyme structure which depend on the light scattering phenomena are mostly unproductive for nanomaterial-based protein immobilization studies, due to the background noise of the carrier and possible aggregation. Therefore, the morphology and size of nanosupports prior to immobilization and after the immobilization of enzyme are examined by SEM, TEM, and AFM, respectively. For example, the XRD analysis revealed that there is no change in the crystal phase of magnetic nanoparticles after immobilization to the support (Xie and Ma 2010). FTIR-based comparative analysis of the free and immobilized enzyme revealed the change in the peak pattern for both types of enzyme preparations. CD spectroscopy is a method of preference for defining the impact of carrier addition onto the secondary structure of enzyme (Ganesan et al. 2009). CD spectrum analysis of lipase immobilized to carbon nanotubes revealed that it retained 62% of the α-helix structure of the native lipase (Ji et al. 2010). The sophisticated analytical techniques explain in situ characterization of structural and operational stability of immobilized enzymes on supports (Ganesan et al. 2009).

9.13 Applications of Nanoparticles in Enzymatic Hydrolysis of Lignocellulose in Biofuel/Bioenergy

The continuous depletion of fossil fuels, a growing concern for its contribution to greenhouse gas emissions, global warming, and other political crisis, demands new alternate and sustainable energy resources. Therefore, alternative and renewable energy sources such as lignocellulosic biomass which is mostly unused materials and rich in carbohydrates are a potential substrate for alternative energy source. Energy crops such as perennial grasses, *Miscanthus*, and poplar are one of the main sources of lignocellulosic biomass (Kirupa Sankar et al. 2018). Lipases (EC 3.1.1.3) have been employed for the biodiesel production (Wang et al. 2011a) and cellulases (EC 3.2.1.4) for the saccharification of plant biomass for ethanol production (Abraham et al. 2014; Lee et al. 2010; Margeot et al. 2009; Matano et al. 2013; Puri et al. 2012). Cellulase representing a complex of endoglucanase, exoglucanase, and β-glucosidase works in synergistic way to produce glucose as end product which is afterward fermented into bioethanol (Chandel et al. 2012; Garvey et al. 2013; Lynd et al. 2002). Since, the enzymes immobilized onto nanomaterials are stable and active in extreme environmental conditions (Dordick et al. 2012) therefore these immoblised enzymes can be explored for variety of applications. These molecular machines are extensively used for catalyzing multiple reactions for alternate energy production from renewal energy sources (Johnson et al. 2011; Kim et al. 2008). A list of immobilized enzymes deployed for biofuel production is mentioned in Table 9.1.

Although several processes for immobilization of lipases are available, the major hurdle for the technology transfer is the extraordinary cost of enzyme and limits of repeated use, e.g., the market price of Novozyme 435 is $1000/kg (Stoytcheva et al. 2011). Therefore, to reduce the cost, the techniques to immobilize the enzyme must be available at low prices and high constancy to avoid enzyme leakage or efficiency loss. Among different commercial immobilized lipases, Novozyme 435 is widely explored for biodiesel production. China, for the first time in 2006, started biodiesel production with a capacity of 20,000 tons per year using “Lipozymes TL” and “Novozyms 435.” Thereafter, another biodiesel production plant in 2007 using mebrance immoblised lipase of Candida sp. 99–125 with an annual manufacturing capacity of 10,000 tons was started *Candida* sp. 99–125 has been immobilized on membranes was started. The biodiesel production using immobilized lipase of *Candida* sp. was found to be economical compared to single-time used Novozyme 435 (Zhao et al. 2015).

The lipase and cellulase immobilization has been done on silica, magnetic, gold, cellulose, TiO_2, and other nanomaterials (Cho et al. 2012; Huang et al. 2011; Pavlidis et al. 2012a, b). The use of nanomaterials as carrier materials for enzyme immobilization in biotechnology has been successfully employed for biofuel production (Verma et al. 2014). For biodiesel application, lipases have been immobilized on magnetic Fe_3O_4 nanoparticles at a reactor scale (Wang et al. 2009). Similarly, porcine pancreas lipase, *Candida rugosa* lipase, and *Pseudomonas cepacia* lipase have been immobilized onto the amino-activated Fe_3O_4 nanoparticles. Porcine pancreas lipase with natural resistance to methanol has been used for transesterification enzymatically (Wang et al. 2009). Here, high amount of biodiesel production from soybean oil was recorded, and immobilized enzyme retained 100% of its activity for three cycles (Kaieda et al. 2001). The high amount of protein loading (85% of solution) rendered high enzyme activity to the immobilized enzyme 95% compared to 82% of free enzyme which ultimately led to high biodiesel production. In another study, high biodiesel production was reported for amino-functionalized magnetic nanoparticles covalently immobilized to *Thermomyces lanuginosus* lipase (Xie and Ma 2009, 2010) and nanomaterial’s activated glutaraldehyde cross-linked lipase (Liu et al. 2012). *Burkholderia* sp. lipase immobilized onto hydrophobic magnetic particles without losing any enzyme activity has been successfully used six times for transesterification, and its activity was comparable with Novozyme 435. Lipase obtained from *Pseudomonas cepacia*, immobilized via adsorption onto the nanoporous gold support, achieved biodiesel production of up to 90%, whereas the transformation of soybean oil by free lipase was 74% after 24 h (Wang et al. 2011a, b). In addition, the performance of the enzymes was improved under extreme reaction conditions such as high temperature and pH conditions.

As discussed earlier, lipase immobilized on nanoparticles is most commonly done by covalent or non-covalent methods (Wang et al. 2009) and is found to have

Table 9.1 List of immobilized enzymes currently employed in biofuel production

S. no.	Enzymes	Immobilization method	Material	Properties	Reference
1.	*Burkholderia cepacia* lipase	Ultrasonic emulsification	Soybean oil	High biodiesel yield of 68.6% in 180 min	Murillo et al. (2019)
2.	Lipase	PAN-PEI-SA-$CaCl_2$		High enzyme loading (36.90 mg/g), repeated use (20 cycles), and 2.5-fold increase in enzyme activity (54.47 U/g) compared to Novozyme® 435 were recorded	Li et al. (2019)
3.	Endo-1,4-β-D-glucanase, cellobiohydrolase, β-glucosidase, endo-1,4-β-xylanase, and β-xylosidase	Magnetic nanoparticles, SE-CLEAs, and M-SE-CLEAs	Lignocellulose-based biomass	Single-step saccharification of lignocellulose using nanoparticle-immobilized enzyme produced reducing sugar up to 250 mg/g of biomass.]	
4.	Cellulase and β-glucosidase		Biomass of *Typha angustifolia*, *Arundo donax*, *Saccharum arundinaceum*, and *Ipomoea carnea*	The co-immobilized trienzyme biocatalyst are potential one-pot pretreatment for the production of bioethanol from lignocellulosic biomass	Kirupa Sankar et al. (2018)
5.	Porcine pancreatic lipase	Zirconium-pillared clay using adsorption and cross-linking	Canola oil	Biodiesel production	Colín-Luna et al. (2018)
6.	Hydrolases		*Spirulina platensis* biomass	Bioethanol production from microalgal biomass	Rempel et al. (2019)
7.	Cellulolytic enzymes	Superparamagnetic nanoparticles		The bio-nanocomposites on the membrane prevent direct membrane-foulant interaction for long period, stabilized the enzyme, and hence led to continuous productivity	Gebreyohannes et al. (2018)

8.	Chimera construct of xylanase and lichenase	SpyTag/SpyCatcher		A novel strategy, easy to scale up and, potential to meet the demands of biofuel industry	Lin et al. (2018)
9.	Lipase immobilization	Silica xerogel with protic ionic liquid and bifunctional agents	Olive oil, sunflower, soybean, and colza oils	Application in biodiesel production	Carvalho et al. (2018)
10.	*Pseudomonas* sp. lipase	Activated biochar		Immobilized lipase yield (92.23%) of fatty acid methyl esters, and it can be reused for 3 cycles	Khosla et al. (2017)
11.	*Trichoderma reesei* cellulase	Chitosan-coated magnetic nanoparticles		The immobilized enzyme showed better thermal and storage stability, retained ~80 % of its activity 15 times, and can be easily separated with external magnetic field compared to free enzyme	Sánchez-Ramírez et al. (2017)
12.	*Candida antarctica* lipase	Core–shell magnetic nanoparticles	Waste cooking oil	The immobilized enzyme produced biodiesel by transesterification of waste cooking oil, higher thermal stability, and methanol tolerance compared to the free enzyme	Mehrasbi et al. (2017)
13.	Amylases		Saccharification of microalgal biomass	Immobilized enzyme is a promising alternative toward development of the bioethanol production	Rodrigues et al. (2017)
14.	*Candida* sp. 99–125 lipase	Entrapment of lipase in alginate and support of diatomite		Higher biodiesel yield (92%); repeated use (11 times) was recorded at 40 °C in 24 h; maintained about 70% of its activity after 11 repeated batch reactions	Zhao et al. (2017)
15.	*Burkholderia cepacia* lipase	Nanoporous carbon derived from metal–organic framework	Soybean oil	A promising green and sustainable catalytic system for industrial application	Liu et al. (2017)

(continued)

Table 9.1 (continued)

S. no.	Enzymes	Immobilization method	Material	Properties	Reference
16.	*Streptomyces* sp. strain W007 lipase		Waste cooking oils	High biodiesel yield (~95%); mobilized lipase retained ~70% of its initial activity after four batch cycles	Wang et al. (2017)
17.	Cellulase	Magnetic nanobiocomposite	*Sesbania aculeata* biomass hydrolysate	Maximum bioethanol production of 5.31 g/l under optimal conditions	Baskar et al. (2016)
18.	Lipase	Superparamagnetic Fe_3O_4 hollow submicrospheres	Waste cooking oil	FAME yield exceeded ~93% over a range of temperatures from 10 to 40 °C	Liu et al. (2016)
19.	*Candida parapsilosis* lipase	Synthetic resins (Accurel MP 1000 and Lewatit VP OC 1600)	Jatropha oil	The production of biodiesel by transesterification of jatropha oil with methanol in a lipid/aqueous system	Rodrigues et al. (2016)
20.	*Candida* sp. 99–125 lipase		Waste cooking oil	The immobilized lipase conversion ratio exceeds 70% even after 10 subsequent reactions, in particular which is better than commercial Novozyme 435 and TLIM	Zhao et al. (2014)
21.	Lipases	Silica nanoparticles		The immobilized enzyme can be reused and retained 95% of initial activities after 7 and 15 cycles of the reaction	Babaki et al. (2016)

decent preservation of enzyme activity, extraordinary loading efficiency, as well as higher enzyme activity (Yu et al. 2005; Gao et al. 2010; Georgelin et al. 2010; Ji et al. 2010; Lee et al. 2010; Li et al. 2011; Saifuddin et al. 2013; Wang and Jiang 2011; Xie and Ma 2010). Immobilized lipases have high activity and stability, and the maximum transformation rate up to 75% was documented at 12 h. In comparison, after nine cycles, the transformation rate of only 18.5% was observed in the batch process (Wang et al. 2009). This decrease is probably due to low deactivation rate of the immobilized lipase. Another four-PBR system of immobilized lipase offered higher conversion rate of up to 88% for 192 h and last 75% after 240 h of reaction (Hama et al. 2007; Thanh et al. 2011; Wang et al. 2011a).

Cellulase complex immobilized to gold-doped silica nanoparticles using affinity technique is also used for single-step catalysis of cellulose to glucose (Cho et al. 2012). The nanohybrid-immobilized cellulase loaded at 20 FPU per gm of pretreated bagasse is found to produce reducing sugar of 38.87 gm per liter which resulted in 0.38 gram per gram of ethanol after fermentation. Moreover, the immobilized nanohybrid was recycled up to five times. The co-immobilized laccase, cellulase, and α-glucosidase for the bioethanol production from *Typha angustifolia*, *Arundo donax*, *Ipomoea carnea*, and *Saccharum arundinaceum* lignocellulose biomass are more stable, compared to free enzyme solution. The saccharification of *S. arundinaceum* biomass produced reducing sugar of 205 ± 3.73 milligram per gram of biomass. The high amount of bioethanol up to 63.43 ± 9.35% was obtained with *I. carnea* (Kirupa Sankar et al. 2018).

9.14 Site-Directed Mutagenesis to Enhance the Biochemical Properties of Immobilized Enzymes

The prior information of the protein tertiary structure and basics of enzyme immobilizations can play vital role in developing an efficient system. The usage of site-directed mutagenesis can help to reserve the enzyme role as well as improve immobilization process. An escalation in the quantity of the ionized amino acids on proteins allows the immobilization of relevant enzymes on suitable supports. For example, the introduction of eight Glu residues using site-directed mutagenesis led to decrease in the isoelectric point of the protein from 6.4 to 4.3 (Montes et al. 2007). Therefore, a wild-type enzyme which is difficult to adsorb on anionic exchangers can be adsorbed on DEAE or polyethylenimine-coated supports. Other alternative to site-directed mutagenesis has been explored for enzyme immobilization. For example, the method to increase the formate dehydrogenase residual activity involves entrapping *Candida boidinii* lipase in polyacrylamide gels (Ansorge-Schumacher et al. 2006). Here, the chemical alterations of Lys or the modifications of the hydrophobicity of the medium, due to acrylamide and tetramethylethylenediamine exposure, may lead to inactivation of enzyme (Bordini et al. 2000; Chiari et al. 1992; Dobryszycki et al. 1999). The formate dehydrogenase variants obtained after site-directed mutagenesis reserved high residual activity after entrapment in polyacrylamide gels compared to native enzyme type. Other parameters such as the polymerization method also affect the enzyme variations. Both the immobilization and directed evolution strategies can be explored in coupled fashion for achieving optimized catalyst activity (Brandt et al. 2006; Rai et al. 2016). In general, Cys residues are explored for direct immobilization of the protein on a support containing disulfide groups or gold. Additionally, use of His tagged protein, carrier altered with immobilized metal and other support materials such as glyoxyl or epoxy disulfide or with higher potential and multipoint covalent connection are also explored (Karel Hernandez 1980).

9.15 Conclusion and Future Prospective

The continuously growing industries demand technologies which are more effective, selective, and sustainable than the previous technologies. The commonly used chemical-based methods for the saccharification of plant biomass for biofuel production are promising due to environmental concerns. The biocatalysts for industrial applications are on the edge of evolution, and catalytic processes can be performed under organic or aqueous environments. The substrates can be selectively and effectively modified with enzymes (Schmid et al. 2001). The recent developments in biotechnology and nanobiology for nanobiocatalysts are emerging as attractive choice for the manipulation of nanoscale surroundings of the enzymes by generating nanoscale materials as carrier macroscale reactors. The immobilization on carrier provides specific biochemical and kinetic properties to the enzyme, ease of separation, and even shelf-life from the reaction that can minimize the costs and increase their competitiveness for technical applications (Tischer and Kasche 1999). The benefits of immobilized enzyme can easily compensate the cost of biodiesel production. The option of designing and developing green and efficient operating systems for biodiesel production from waste has been achieved using novel magnetic nanoparticle (Ngo et al. 2013). The properties of immobilized enzymes are administered by the enzyme and the carrier material properties. The enzyme immobilization methods use prefabricated carrier materials for enzyme stabilization through adsorption, conjugation, and entrapments as crystals or powders. The immobilization of enzyme on carrier provides specific biochemical and kinetic properties to the enzyme (Tischer and Kasche 1999).

The immobilization is a valuable tool to cut down the enzyme cost and offers several technological advantages including stability against harsh reaction conditions. It also, permits the repetitive enzyme use which significantly reduces the cost of immobilized enzymes. Moreover, the enzyme can be simply detached from the reaction media. However, the immobilization of enzymes onto planar surfaces may limit their activity due to factors such as alteration in the configuration of native protein, steric interference, and slow diffusion rate. To eliminate the undesirable effects of enzyme immobilization, nanoparticles (NPs) as enzyme carriers are attractive alternative (Ding et al. 2015). Still, the challenges associated with exploring enzymes for biofuel production are enormous including high enzyme costs, inactivation of enzymes by solvents, and difficulties of large-scale production

(Watanabe et al. 2000). The nanomaterial-based efficient methods of enzyme immobilization can cut down the cost to a significant extent. Besides the advantages and limitations of nanomaterials, health and environmental concerns are of paramount importance prior to their usage in biofuel production (Arico et al. 2005; Cui et al. 2005; Krumov et al. 2009). The homogeneous size of nanomaterials and stabilization from aggregation might be addressed by coating with a polymer shell (Rozenberga and Tenne 2008).

The enzymes immobilized by nanostructured materials can provide promising solutions for industrial biotransformation. Use of nanomaterials for fabricating nanobiocatalysts for several enzymes including hydrolase, glucose oxidase, and alcohol dehydrogenase has been achieved with great accomplishments (Kim et al. 2008; Petros and Desimone 2010; Wang et al. 2009; Ginet et al. 2011). The nanomaterial enzymes have lower inactivation and hence can maintain higher activity even after repeated uses (Husain 2017). In recent times, use of nanomaterials for numerous applications has been explored to immobilize enzymes. The development of PBR systems for frequent and efficient use of immobilized lipase (Wang et al. 2011a) and a better understanding of enzyme interactions with nanomaterial surface at the support or in solvent interface can remarkably increase both the assembly (Pavlidis et al. 2012a, b) and efficient nanomaterial-based enzyme immobilization.

In addition, mass transfer effects can reduce the enzyme activity. The resulting enzymes are not only affected by the method adopted for immobilization but also by a decrease in the availability of enzymes or substrate molecules. The mass transfer effects can result in substrate or pH gradient which can decrease the reaction rate and product yield. This also reduces access to substrate and leads to deliberate restriction in the movement of enzyme which affect the access of enzyme for substrate and partition effects of substrate concentrations and decrease the enzyme efficiency as compared to free enzymes. How to avoid and minimize the mass transfer effect and other limitations responsible for reduced enzyme activity during enzyme immobilization needs to be addressed which reduces the enzyme efficiency. Altogether, these factors can offer an accurate measure of enzyme productivity or enzyme consumption (Tischer and Kasche 1999). Therefore, owing to the biochemical properties of enzymes and the varying conditions of reaction technology, there is a need to explore the wealth of methods for developing novel processes of immobilization. Efforts to address the regulated movement and diffusion which primes to reduction in enzyme activity required to be explored using mathematical and other methods. The associated concerns of enzyme immobilization include reduction in enzyme activity or catalytic structures, distortions in the enzyme structure and enzyme inactivation, due to unfolding of the native enzyme structure followed by kinetically irreversible changes through aggregation or covalent changes in the enzyme.

The expansion of methods to stabilize the enzyme structure is an interesting challenge. Compared to enzyme immobilized on micrometric supports, nanobiocatalysts have potential to achieve higher enzyme loading on its surface. Current advances of nanobiocatalytic methods have improved the performance using various nanomaterials, but challenges to cope with mass transfer efficiency and enzyme activity need to be optimized for low cost and efficient processes. The immobilization of enzymes onto nanostructured surfaces has led to numerous benefits compared to the use of free enzymes or immobilization on larger materials. So far nanomaterial-supported enzyme preparations have been explored for biofuel production. The possibility to co-immobilize the multi-enzymes on nanomaterials certainly can enable the efficient use of different enzymes in targeting biomass for biofuel (Alper and Stephanopoulous 2009). However, studies are necessary to know the practical bottlenecks including efficient synthesis techniques for nanomaterials and evaluation of safety guidelines. Regardless of the development in the preparation of nanomaterial–protein conjugates, limited studies have generated thorough evidences on the structure and function of proteins, attached to nanomaterials (Asuri et al. 2007).

Therefore, the biotechnological perspective of enzyme immobilized on nanomaterials is of supreme importance. Keeping in view, the structural and functional properties of nanomaterials for enzyme immobilization and stabilization method, care must be taken for possible modifications in catalytic efficiency. Further, the optimization of stabilizing parameters will prime to a better and efficient processes and hence may upsurge the economic potential in existing enzymatic processes and for other novel areas where enzymes have not been used till now because of their uncertainty.

Acknowledgment Authors are thankful to SEED division, DST, GOI for the award of project under Scheme for Young Scientists and Technologists (SEED-TIASN-023-2018).

References

Abraham RE, Verma ML, Barrow CJ, Puri M (2014) Suitability of magnetic nanoparticle immobilised cellulases in enhancing enzymatic saccharification of pretreated hemp biomass. Biotechnol Biofuels 7:90

Ahmad R, Sardar M (2015) Enzyme immobilization: an overview on nanoparticles as immobilization matrix. Biochem Anal Biochem 4(2):178. https://doi.org/10.4172/2161-1009.1000178

Ali M, Winterer M (2010) ZnO nanocrystals: surprisingly alive. Chem Mater 22:85–91

Alper H, Stephanopoulous G (2009) Engineering for biofuels: exploiting innate microbial capacity or importing biosynthetic potential? Nat Rev Microbiol 7:715–723

Andrade LH, Rebelo LP, Netto CGCM, Toma HE (2010) Kinetic resolution of a drug precursor by Burkholderia cepacia lipase immobilized methodologies on superparamagnetic nanoparticles. J Mol Catal B Enzym 66:55–62

Ansari SA, Husain Q (2012) Potential applications of enzymes immobilized on/in nano materials: a review. Biotechnol Adv 30(3):512–523. https://doi.org/10.1016/j.biotechadv.2011.09.005

Ansorge-Schumacher MB, Slusarczyk H, Schumers J, Hirtz D (2006) Directed evolution of formate dehydrogenase from *Candida boidinii* for improved stability during entrapment in polyacrylamide. FEBS J 273:3938–3945

Antrim RR, Auterinen A-L (1986) A new regenerable immobilized glucose isomerase. Stärke 38:132–137

Arico AS, Bruce P, Scrosati B et al (2005) Nanostructured materials for advanced energy conversion and storage devices. Nat Mater 4:366–377

Asuri P, Karajanagi SS, Sellitto E, Kim DY, Kane RS, Dordick JS (2006) Water-soluble carbon nanotube-enzyme conjugates as functional biocatalytic formulations. Biotechnol Bioeng 95:804–811

Asuri P, Bale SS, Pangule RC, Shah DA, Kane RS, Dordick JS (2007) Structure, function, and stability of enzymes covalently attached to single-walled carbon nanotubes. Langmuir 23(24):12318–12321. https://doi.org/10.1021/la702091c

Avnir D, Lev O, Livage J (2006) Recent bio-applications of sol-gel materials. J Mater Chem 16:1013–1030

Babaki M, Youse M, Habibi Z, Mohammadi M, Youse P, Mohammadi J, Brask J (2016) Enzymatic production of biodiesel using lipases immobilized on silica nanoparticles as highly reusable biocatalysts: effect of water, t -butanol and blue silica gel contents. 91:196–206. https://doi.org/10.1016/j.renene.2016.01.053

Balcao VM, Paiva AL, Malcata FX (1996) Bioreactors with immobilized lipases: state of the art. Enzym Microb Technol 18:392–416

Baskar G, Kumar RN, Melvin XH, Aiswarya R, Soumya S (2016) *Sesbania aculeate* biomass hydrolysis using magnetic nanobiocomposite of cellulase for bioethanol production. Renew Energy 98:23–28. https://doi.org/10.1016/j.renene.2016.04.035

Bell G, Halling PJ, Moore BD, Partridge J, Rees DG (1995) Biocatalyst behaviour in low-water systems. Trends Biotechnol 13:468–473

Betancor L, Luckarift HR (2008) Bioinspired enzyme encapsulation for biocatalysis. Trends Biotechnol 26:566–572

Bhat RR, Genzer J, Chaney BN, Sugg HW, Liebmann-Vinson A (2003) Controlling the assembly of nanoparticles using surface grafted molecular and macromolecular gradients. Nanotechnology 14:1145–1152

Bilitewski U (2006) Protein-sensing assay formats and devices. Anal Chim Acta 568:232

Bisen PS, Sanodiya BS, Thakur GS, Baghed RK, Prasad GB (2010) Biodiesel production with special emphasis on lipase-catalysed transesterification. Biotechnol Lett 32:1019–1030

Bordini E, Hamdan M, Righetti PG (2000) Probing acrylamide alkylation sites in cysteine-free proteins by matrix-assisted laser desorption/ionisation time-of-flight. Rapid Commun Mass SP 14:840–848

Bosio VE, Islan GA, Martínez YN, Durán N, Castro GR (2016) Nanodevices for the immobilization of therapeutic enzymes. Crit Rev Biotechnol 36(3):447–464

Brady D, Jordaan J (2009) Advances in enzyme immobilisation. Biotechnol Lett 31:1639–1650

Brandt B, Hidalgo A, Bornscheuer UT (2006) Immobilization of enzymes in microtiter plate scale. J Biotechnol 1:582–587

Brena BM, Batista-Viera F (2006) Immobilization of enzymes. In: Guisan JM (ed) Methods in biotechnology: immobilization of enzymes and cells. Humana Press Inc, New York, pp 15–30

Bryjak J, Trochimczuk AW (2006) Immobilization of lipase and penicillin acylase on hydrophobic acrylic carriers. Enzym Microb Technol 39:573–578

Buchholz K, Kasche V (1997) Biokatalysatoren und Enzymtechnologie. VCH, Weinheim, pp 166–185

Cabana H, Jones JP, Agathos SN (2009) Utilization of cross-linked laccase aggregates in a perfusion basket reactor for the continuous elimination of endocrine- disrupting chemicals. Biotechnol Bioeng 102:1582–1592

Cantone S, Ferrario V, Corici L, Ebert C, Fattor D et al (2013) Efficient immobilisation of industrial biocatalysts: criteria and constraints for the selection of organic polymeric carriers and immobilisation methods. Chem Soc Rev 42:6262–6276

Caruso F, Mohwald H (1999) Protein multilayer formation on colloids through a stepwise self-assembly technique. J Am Chem Soc 121:6039–6046

Carvalho NB, Vidal BT, Barbosa AS, Pereira MM, Mattedi S, Freitas LDS, Lima ÁS, Soares CMF (2018, Jun 21) Lipase immobilization on silica xerogel treated with protic ionic liquid and its application in biodiesel production from different oils. Int J Mol Sci. 19(7). pii: E1829. https://doi.org/10.3390/ijms19071829

Chandel AK, Chandrasekhar G, Silva MB, Silva SSD (2012) The realm of cellulases in biorefinery development. Crit Rev Biotechnol 32:187–202

Chen RJ, Bangsaruntip S, Drouvalakais KA, Kam NWS, Shim M, Li Y, Kim W, Utz PJ, Dai H (2003) Noncovalent functionalization of carbon nanotubes for highly specific electronic biosensors. Proc Natl Acad Sci 100:4984–4989

Chiari M, Righetti PG, Negri A, Ceciliani F, Ronchi S (1992) Preincubation with cysteine prevents modification of sulfhydryl groups in proteins by unreacted acrylamide in a gel. Electrophoresis 13:882–884

Cho EJ, Jung S, Kim HJ et al (2012) Co-immobilization of three cellulases on Au-doped magnetic silica nanoparticles for the degradation of cellulose. Chem Commun 48:886–888

Chronopoulou L, Kamel G, Sparago C et al (2011) Structure-activity relationships of Candida rugosa lipase immobilized on polylactic acid nanoparticles. Soft Matter 7:2653–2662

Colín-Luna JA, Zamora-Rodea EG, González-Brambila MM, Barrera-Calva E, Rosas-Cedillo R, Medina-Mendoza AK, García-Martínez JC (2018) Biodiesel production using immobilized lipase supported on a zirconium-pillared clay. Effect Immobilization Method Intl J Chem Reactor Eng. https://doi.org/10.1515/ijcre-2017-0260

Cowan DA, Fernandez-Lafuente R (2011) Enhancing the functional properties of thermophilic enzymes by chemical modification and immobilization. Enzym Microb Technol 2011(49):326–346

Cowan D, Meyer Q, Stafford W, Muyanga S, Cameron R, Wittwer P (2005) Metagenomic gene discovery: past, present and future. Trends Biotechnol 23:321–329

Crespilho FN, GhicaME FM, Nart FC, Oliveira ON, Brett CMA (2006) A strategy for enzyme immobilization on layer-by-layer dendrimer-gold nanoparticle electrocatalytic membrane incorporating redox mediator. Electrochem Commun 8:1665–1670

Cruz JC, Pfromm PH, Tomich JM, Rezac ME (2010) Conformational changes and catalytic competency of hydrolases adsorbing on fumed silica nanoparticles: I. tertiary structure. Colloid Surf B 79:97–104

Cui T, Zhang J, Wang J et al (2005) CdS nanoparticles/polymer composite shells on silica nanospheres grown by atom transfer radical polymerization. Adv Funct Mater 15:481–486

D'Souza SF (1999) Immobilized enzymes in bioprocess. Curr Sci 77:69–79

Datta S, Christena LR, Rajaram YRS (2013) Enzyme immobilization: an overview on techniques and support materials. Biotech 3:1–9

Deshpande A, D'souza SF, Nadkarni GB (1987) Co-immobilization of D- amino acid oxidase and catalase by entrapment of *Trigonopsis variabilis* in radiation polymerised Polyacrylamide beads. J Biosci 11:137–144

Diána W (2015) Nanostructured systems for enzyme immobilization. PhD Thesis

Ding S, Cargill AA, Medintz IL, Claussen JC (2015) Increasing the activity of immobilized enzymes with nanoparticle conjugation. Curr Opin Biotechnol 34:242–250. https://doi.org/10.1016/j.copbio.2015.04.005

Dobryszycki P, Rymarczuk M, Bulai G, Kochman M (1999) Effect of acrylamide on aldolase structure 1. Induction of intermediate states. Biochim Biophys Acta 1431:338–350

Dordick JS, Kane RS, Asuri P, et al (2012) Enhanced stability of proteins immobilized on nanoparticles. US Patent, 302870

Drager G, Kiss C, Kunz U, Kirschning A (2007) Enzyme-purification and catalytic transformations in a microstructured PASSflow reactor using a new tyrosine-based Ni-NTA linker system attached to a polyvinylpyrrolidinone-based matrix. Org Biomol Chem 5:3657

Drechsler U, Fischer NO, Frankamp BL, Rotello V (2004) Highly efficient biocatalysts via covalent immobilization of Candida rugosa lipase on ethylene glycol modified gold-silica nanocomposites. Adv Mater 16:271

Dyal A, Loos K, Noto M, Chang SW, Spagnoli C, Shafi KVPM et al (2003) Activity of Candida rugosa lipase immobilized on Fe_2O_3 magnetic nanoparticles. J Am Chem Soc 125:1685–1686

End N, Schoning KU (2004) Immobilized biocatalysts in industrial research and production. Top Curr Chem 242:273

English BP, Min W, van Oijen AM, Lee KT, Luo G, Sun H, Cherayil BJ, Kou SC, Xie XS (2006) Ever-fluctuating single enzyme molecules: Michaelis-Menten equation revisited. Nat Chem Biol 2:87

Fernandes EGR, Queiroz AAAD, Abraham GA, Roman JS (2006) Anti-thrombogenic properties of bioconjugate streptokinase-polyglycerol dendrimers. J Mater Sci Mater Med 17:105–111

Fernandez-Lafuente R (2009) Stabilization of multimeric enzymes: strategies to prevent subunit dissociation. Enzym Microb Technol 45:405–418

Fernández-Lorente G, Palomo JM, Cabrera Z, GuisánJM F-LR (2007) Specificity enhancement towards hydrophobic substrates by immobilization of lipases by interfacial activation on hydrophobic supports. Enzym Microb Technol 41:565–569

Figallo E, Cannizzaro C, Gerecht S, Burdick JA, Langer R, Elvassore N et al (2007) Microbioreactor array for controlling cellular microenvironments. Lab Chip 7:710–719

Fornera S, Bauer T, Schlüter AD, Walde P (2012a) Simple enzyme immobilization inside glass tubes for enzymatic cascade reactions. J Mater Chem 22:502–511

Fornera S et al (2012b) Sequential immobilization of enzymes in microfluidic channels for cascade reactions. ChemPlusChem 77:98–101

Frenkel-Mullerad H, Avnir D (2005) Sol-gel materials as efficient enzyme protectors: preserving the activity of phosphatases under extreme pH conditions. J Am Chem Soc 27(2005):8077–8081

Fu J, Reinhold J, Woodbury NW (2011) Peptide-modified surfaces for enzyme immobilization. PLoS One 6:e18692

Fuentes M, Mateo C, Fernández-Lafuente R, Guisán JM (2006) Detection of polyclonal antibody against any area of the protein-antigen using immobilized protein- antigens: the critical role of the immobilization protocol. Biomacromolecules 7:540–544

Gaberc-Porekar V, Menart V (2001) Perspectives of immobilized-metal affinity chromatography: a review. J Biochem Biophys Methods 49:335–360

Ganesan A, Moore BD, Kelly SM et al (2009) Optical spectroscopic methods for probing the conformational stability of immobilised enzymes. Chem Phys Chem 10:1492–1499

Gao SL, Wang YJ, Diao X et al (2010) Effect of pore diameter and cross-linking method on the immobilization efficiency of *Candida rugosa* lipase in SBA-15. Bioresour Technol 101:3830–3837

Gardimalla HMR, Mandal D, Stevens PD, Yen M, Gao Y (2005) Superparamagnetic nanoparticle-supported enzymatic resolution of racemic carboxylates. Chem Commun 37:4432–4434

Garvey M, Klose H, Fischer R et al (2013) Cellulases for biomass degradation: comparing recombinant cellulose expression platforms. Trends Biotechnol 31:581–589

Gebreyohannes AY, Dharmjeet M, Swusten T, Mertens M, Verspreet J, Verbiest T, Courtin CM, Vankelecom IFJ (2018) Simultaneous glucose production from cellulose and fouling reduction using a magnetic responsive membrane reactor with superparamagnetic nanoparticles carrying cellulolytic enzymes. Bioresour Technol 263:532–540. https://doi.org/10.1016/j.biortech.2018.05.002

Georgelin T, Maurice V, Malezieux B et al (2010) Design of multifunctionalized g-Fe2O3@SiO2 core-shell nanoparticles for enzymes immobilization. J Nanopart Res 12:675–680

Ginet N, Pardoux R, Adryanczyk G, Garcia D, Brutesco C, Pignol D (2011) Single-step production of a recyclable nanobiocatalyst for organophosphate pesticides biodegradation using functionalized bacterial magnetosomes. PLoS One 6(6):e21442. https://doi.org/10.1371/journal.pone.0021442

Grosová Z, Rosenberg M, Rebros M, Sipocz M, Sedlácková B (2008) Entrapment of beta-galactosidase in polyvinylalcohol hydrogel. Biotechnol Lett 30:763–767

Guncheva M, Tashev E, Zhiryakova D, Tosheva T, Tzokova N (2011) Immobilization of lipase from Candida rugosa on novel phosphorous-containing polyurethanes: application in wax ester synthesis. Process Biochem 46:923–930

Gupta MN, Kaloti M, Kapoor M, Solanki K (2011) Nanomaterials as matrices for enzyme immobilization. Artif Cells Blood Substit Biotechnol 39(2):98–109. https://doi.org/10.3109/10731199.2010.516259

Hama S, Yamaji H, Fukumizu T et al (2007) Biodiesel-fuel production in a packed-bed reactor using lipase-producing *Rhizopus oryzae* cells immobilized within biomass support particles. Biochem Eng J 34:273–278

Hamachi I, Fujita A, Kunitake T (1994) Enhanced N-demethylase activity of cytochrome C bound to a phosphate-bearing synthetic bilayer membrane. J Am Chem Soc 116:8811–8812
Hanefeld U, Gardossi L, Magner E (2009) Understanding enzyme immobilisation. Chem Soc Rev 38:453–468
Hartmann M, Kostrov X (2013) Immobilization of enzymes on porous silicas-benefits and challenges. Chem Soc Rev 42:6277–6289
Hernandez K, Fernandez-Lafuente R (2016) LCA studies comparing alkaline and immobilized enzyme catalyst processes for biodiesel production under Brazilian conditions. Enzym Microb Technol 2011(48):107–122
Herricks TE, Kim SH, Kim J, Li D, Kwak JH, Grate JW, Kim SH, Xia Y (2005) Direct fabrication of enzyme-carrying polymer nanofibers by electrospinning. J Mater Chem 14:3241–3245
Ho L-F, Li S-Y, Lin S-C, Hsu W-H (2004) Integrated enzyme purification and immobilization processes with immobilized metal affinity adsorbents. Process Biochem 39:1573–1581
Ho KM, Mao X, Gu L, Li P (2008) Facile route to enzyme immobilization: core–shell nanoenzyme particles consisting of well-defined poly (methyl methacrylate) cores and cellulase shells. Langmuir 24:11036–11042
Huang XJ, Chen PC, Huang F et al (2011) Immobilization of *Candida rugosa* lipase on electrospun cellulose nanofiber membrane. J Mol Catal B Enzym 70:95–100
Hudson S, Cooney J, Magner E (2008) Proteins in mesoporous silicates. Angew Chem Int Ed Eng 47:8582–8594
Hung SW, Hwang JK, Tseng F, Chang JM, Chen CC, Chieng CC (2006) Molecular dynamics simulation of the enhancement of cobra cardiotoxin and E6 protein binding on mixed self-assembled monolayer molecules. Nanotechnology 17:S8–S13
Husain Q (2017) Nanomaterials as novel supports for the immobilization of amylolytic enzymes and their applications: a review. Biocatalysis 3(1):37–53. https://doi.org/10.1515/boca-2017-0004
Husain Q, Ansari SA, Alam F, Azam A (2011) Immobilization of *Aspergillus oryzae* β-galactosidase on zinc oxide nanoparticles via simple adsorption mechanism. Int J Biol Macromol 49:37–43
Hwang ET, Gu MB (2013) Enzyme stabilization by nano/microsized hybrid materials. Eng Life Sci 13:49–61
Igor Alberto Peñarrubia Fernandez, De-Hua Liu, Jinsong Zhao (2016) LCA studies comparing alkaline and immobilized enzyme catalyst processes for biodiesel production under Brazilian conditions. Conserv Recycl 119:117–127. https://doi.org/10.1016/j.resconrec.2016.05.009
Iyer PV, Ananthnarayan L (2008) Enzyme stability and stabilization-aqueous and non-aqueous environment. Process Biochem 43(10):1019–1032. https://doi.org/10.1016/j.procbio.2008.06.004
Ji PJ, Tan HS, Xu X, Feng W (2010) Lipase covalently attached to multiwalled carbon nanotubes as an efficient catalyst in organic solvent. AICHE J 56:3005–3011
Jia H, Zhu G, Wang P (2003) Catalytic behaviors of enzymes attached to nanoparticles: the effect of particle mobility. Biotechnol Bioeng 84:406–414
Jin Q, Li X, Deng C, Zhang Q, Yi D, Wang X et al (2018) Silica nanowires with tunable hydrophobicity for lipase immobilization and biocatalytic membrane assembly. J Colloid Interface Sci 531:555–563. https://doi.org/10.1016/j.jcis.2018.07.035
Johnson RD, Wang ZG, Arnold FH (1996) Surface site heterogeneity and lateral interactions in multipoint protein adsorption. J Phys Chem 100:5134–5139
Johnson AK, Zawadzka AM, Deobald LA, Crawford RL, Paszczynski AJ (2008) Novel method for immobilization of enzymes to magnetic nanoparticles. J Nanopart Res 10:1009–1025
Johnson PA, Park HJ, Driscoll AJ (2011) Enzyme nanoparticle fabrication: magnetic nanoparticle synthesis and enzyme immobilization. Methods Mol Biol 679:183–191
Jordan BJ, Hong R, Gider B, Hill J, Emrick T, Rotello VM (2006) Stabilization ofα-chymotrypsin at air–water interface through surface binding to gold nanoparticle scaffolds. Soft Matter 2:558–560
Jun L, Zhou W, Kumbhar J, Wiemann J, Fang J, Carpentier EE et al (2001) Gold-coated iron (Fe@Au) nanoparticles: synthesis, characterization and magnetic field induced self-assembly. Sol State Chem 159:26–31

Kaieda M, Samukawa T, Kondo A, Fukuda H (2001) Effect of methanol and water contents on production of biodiesel fuel from plant oil catalyzed by various lipases in a solvent-free system. J Biosci Bioeng 91:12–15

Karajanagi SS, Vertegel AA, Kane RS, Dordick JS (2004) Structures and functions of enzymes adsorbed onto single walled carbon nanotubes. Langmuir 20:11594–11599

Karel Hernandez RFL (1980) Control of protein immobilization: coupling immobilization and site-directed mutagenesis. BJOG Int J Obstet Gynaecol 87(11):1015–1021. https://doi.org/10.1016/j.enzmictec.2010.10.003

Katchalski-Katzir E (1993) Immobilized enzymes - learning from past successes and failures. Trends Biotechnol 11(11):471–478. https://doi.org/10.1016/0167-7799(93)90080-S

Kennedy JF, Melo EHM (1990) Immobilised enzymes and cells. Chem Eng Prog 86:81–89

Khosla K, Rathour R, Maurya R, Maheshwari N, Gnansounou E, Larroche C, Thakur IS (2017) Biodiesel production from lipid of carbon dioxide sequestrating bacterium and lipase of psychrotolerant *Pseudomonas* sp. ISTPL3 immobilized on biochar. Bioresour Technol 245(Pt A):743–750. https://doi.org/10.1016/j.biortech.2017.08.194

Kim J, Grate JW (2003) Single-Enzyme nanoparticles armored by a nanometer-scale organic / inorganic network. Nano Lett 3(9):1219–1222. https://doi.org/10.1021/nl034404b

Kim BC, Nair S, Kim J, Kwak JH, Grate JW, Kim SH et al (2005a) Preparation of biocatalytic nanofibres with high activity and stability via enzyme aggregate coating on polymer nanofibres. Nanotechnology 16:S382–S388

Kim BK, Kim YH, Won K, Chang H, Choi Y, Kong K et al (2005b) Electrical properties of polyaniline nanofibre synthesized with biocatalyst. Nanotechnology 16:1177–1181

Kim J, Grate JW, Wang P (2006a) Nanostructures for enzyme stabilization. Chem Eng Sci 61:1017–1026

Kim J, Jia H, Wang P (2006b) Challenges in biocatalysis for enzyme based biofuel cells. Biotechnol Adv 24:296–308

Kim MI, Ham HO, Oh SD, Park HG, Chang HN, Choi SH (2006c) Immobilization of *Mucor javanicus* lipase on effectively functionalized silica nanoparticles. J Mol Catal B Enzym 39:62–68

Kim MI, Kim J, Lee J et al (2007) Crosslinked enzyme aggregates in hierarchically-ordered mesoporous silica: a simple and effective method for enzyme stabilization. Biotechnol Bioeng 96:210–218

Kim J, Grate JW, Wang P (2008) Nanobiocatalysis and its potential applications. Trends Biotechnol 26:639–646

Kin MH, Xuepu M, Lianquan G, Li P (2008) Facile route to enzyme immobilization: core–shell nanoenzyme particles consisting of well-defined poly(methyl methacrylate) cores and cellulase shells. Langmuir 24:11036–11042

Kingsmore SF (2006) Multiplexed protein measurement: technologies and applications of protein and antibody arrays. Nat Rev Drug Discov 5:310–320

Kirupa Sankar M, Ravikumar R, Naresh Kumar M, Sivakumar U (2018) Development of co immobilized tri-enzyme biocatalytic system for one-pot pretreatment of four different perennial lignocellulosic biomass and evaluation of their bioethanol production potential. Bioresour Technol 269:227–236. https://doi.org/10.1016/j.biortech.2018.08.091

Klein MP, Nunes MR, Rodrigues RC et al (2012) Effect of the support size on the properties of b-galactosidase immobilized on chitosan: advantages and disadvantages of macro and nanoparticles. Biomacromolecules 13:2456–2464

Koneracka M, Kopcansky P, Antalmk M, Timko M, Ramchand CN, Lobo D et al (1999) Immobilization of proteins and enzymes to the magnetic particles. J Magn Magn Mater 201:427–430

Koneracka M, Kopcansky P, Timko M, Ramchand CN (2002) Direct binding procedure of proteins and enzymes to fine magnetic particles. J Magn Mater 252:409–411

Konwarh R, Karak N, Rai SK, Mukherjee AK (2009) Polymer-assisted iron oxide magnetic nanoparticle immobilized keratinase. Nanotechnology 20:225–235

Kopp W, da Costa TP, Pereira SC, Jafelicci MJ, Giordano RC, Marques RFC et al (2014) Easily handling penicillin G acylase magnetic cross-linked enzymes aggregates: catalytic and morphological studies. Process Biochem 49:38–46. https://doi.org/10.1016/j.procbio.2013.09.024

Kouassi GK, Irudayaraj J, McCarty G (2005) Examination of cholesterol oxidase attachment to magnetic nanoparticles. J Nanobiotechnol 3:1–9

Krumov N, Perner-Nochta I, Oder S et al (2009) Production of inorganic nanoparticles by microorganisms. Chem Eng Technol 32:1026–1035

Küchler A, Adamcik J, Mezzenga R, Schlüter AD, Walde P (2015a) Enzyme immobilization on silicate glass through simple adsorption of dendronized polymer-enzyme conjugates for localized enzymatic cascade reactions. RSC Adv 5:44530–44544

Küchler A, Bleich JN, Sebastian B, Dittrich PS, Walde P (2015b) Stable and simple immobilization of proteinase K inside glass tubes and microfluidic channels. ACS Appl Mater Interfaces 7:25970–25980

Küchler A, Yoshimoto M, Luginbühl S, Mavelli F, Walde P (2016) Enzymatic reactions in confined environments. Nat Nanotechnol 11(5):409–420. https://doi.org/10.1038/nnano.2016.54

Kumar CV, McLendon GL (1997) Nanoencapsulation of cytochrome c and horseradish peroxidase at the galleries of a-zirconium phosphate. Chem Mater 9:863–870

Kumar A, Mandal S, Selvakannan PR, Pasricha R, Mandale AB, Sastry M (2003) Fractal gold nanostructures produced by the spontaneous reduction of chloroaurate ions in thermally evaporated hexadecylaniline thin films. Langmuir 9:6277–6282

Kuroiwa T, Noguchi Y, Nakajima M, Sato S, Mukataka S, Ichikawa S (2008) Production of chitosan oligosaccharides using chitosanase immobilized on amylase coated magnetic nanoparticles. Process Biochem 43:62–69

Lalonde J, Margolin A (2002) In: Drauz K, Waldmann H (eds) Enzyme catalysis in organic synthesis, vol 1, 2nd edn. Wiley-VCH, Weinheim, pp 163–184

Lee DG, Ponvel KM, Kim M et al (2009) Immobilization of lipase on hydrophobic nano-sized magnetite particles. J Mol Catal B Enzym 57:62–66

Lee SM, Jin LH, Kim JH et al (2010) Beta-glucosidase coating on polymer nanofibers for improved cellulosic ethanol production. Bioprocess Biosyst Eng 33:141–147

Lei CH, Shin YS, Magnuson JK, Fryxell G, Lasure LL, Elliott DC et al (2006) Characterization of functionalized nanoporous supports for protein confinement. Nanotechnology 17:5531–5538

Li Y, Wang H, Lu J, Chu A, Zhang L, Ding Z et al (2019) Preparation of immobilized lipase by modified polyacrylonitrile hollow membrane using nitrile-click chemistry. Bioresour Technol 274:9–17. https://doi.org/10.1016/j.biortech.2018.11.075

Li SF, Fan YH, Hu RF, Wu WT (2011) *Pseudomonas cepacia* lipase immobilized onto the electrospun PAN nanofibrous membranes for biodiesel production from soybean oil. J Mol Catal B Enzym 72:40–45

Lia Y, Schluesenerb HJ, Xu S (2010) Gold nanoparticle-based biosensors. Gold Bull 43:29–41

Liese A, Hilterhaus L (2013) Evaluation of immobilized enzymes for industrial applications. Chem Soc Rev 42:6236–6249. https://doi.org/10.1039/c3cs35511j

Lin Y, Jin W, Wang J, Cai Z, Wu S, Zhang GA (2018) Novel method for simultaneous purification and immobilization of a xylanase-lichenase chimera via SpyTag/SpyCatcher spontaneous reaction. Enzym Microb Technol 115:29–36. https://doi.org/10.1016/j.enzmictec.2018.04.007

Liu CH, Huang CC, Wang YW et al (2012) Biodiesel production by enzymatic transesterification catalyzed by *Burkholderia lipase* immobilized on hydrophobic magnetic particles. Appl Energy 100:41–46

Liu C, Yuan J, Gao H, Liu C (2016) Biodiesel production from waste cooking oil by immobilized lipase on superparamagnetic Fe_3O_4 hollow sub-microspheres. Biocatal Biotransformation 34(6):283–290. https://doi.org/10.1080/10242422.2016.1265948

Liu LH, Shih YH, Liu WL, Lin CH, Huang HY (2017) Enzyme immobilized on nanoporous carbon derived from metal-organic framework: a new support for biodiesel synthesis. 10(7):1364–1369. https://doi.org/10.1002/cssc.201700142

Lynd LR, Weimer PJ, Van Zyl WH, Pretorius IS (2002) Microbial cellulose utilization: fundamentals and biotechnology. Microbiol Mol Biol Rev 66:506–577

Malcata FXH, Reyes R, Garcia HS, Hill CG Jr, Amundson CH (1997) Enzym Microb Technol 14:426–446

Margeot A, Hahn-Hagerdal B, Edlund M et al (2009) New improvements for lignocellulose ethanol. Curr Opin Biotechnol 20:372–380

Martin CR, Kohli P (2003) The emerging field of nanotube biotechnology. Nat Rev Drug Discov 2:29–37

Matano Y, Hasunuma T, Kondo A (2013) Cell recycle batch fermentation of high-solid lignocellulose using a recombinant cellulose- displaying yeast strain for high yield ethanol production in consolidated bioprocessing. Bioresour Technol 135:403–409

Mateo C, Abian O, Fernandez-Lafuente R, Guisan JM (2000) Increase in conformational stability of enzymes immobilized on epoxy-activated supports by favoring additional multipoint covalent attachment. Enzym Microb Technol 26:509–515

Mateo C, Abian O, Bernedo M, Cuenca E, Fuentes M, Fernandez-Lorente G et al (2005) Some special features of glyoxyl supports to immobilize proteins. Enzym Microb Technol 37:456–462

Mateo C, Palomo JM, Fuentes M, Betancor L, Grazu V, López-Gallego F et al (2006) Glyoxyl agarose: a fully inert and hydrophilic support for immobilization and high stabilization of proteins. Enzym Microb Technol 39:274–280

Mateo C, Grazu V, Palomo JM, Lopez-Gallego F, Fernandez-Lafuente R, Guisan JM (2007a) Immobilization of enzymes on heterofunctional epoxy supports. Nat Protoc 2:1022–1033

Mateo C, Palomo JM, Fernandez-Lorente G, Guisan JM, Fernandez-Lafuente R (2007b) Improvement of enzyme activity, stability and selectivity via immobilization techniques. Enzym Microb Technol 40:1451–1463

Matsunaga T, Kamiya S (1987) Use of magnetic particles isolated from magnetotactic bacteria for enzyme immobilization. Appl Microbiol Biotechnol 26:328–332. https://doi.org/10.1007/BF00256663

Mehta RV, Upadhyay RV, Charles SW, Ramchand CN (1997) Direct binding of protein to magnetic particles. Biotechnol Tech 11:493–496

Mehrasbi MR, Mohammadi J, Peyda M, Mohammadi MM (2017) Covalent immobilization of Candida antarctica lipase on core-shell magnetic nanoparticles for production of biodiesel from waste cooking oil. Renew Energy 101:593–602. https://doi.org/10.1016/j.renene.2016.09.022

Mitchell DT, Lee SB, Trofin L, Li N, Nevanen TK, Soderlund H et al (2002) Smart nanotubes for bioseparations and biocatalysis. J Am Chem Soc 124:11864–11865

Montes T, Grazú V, López-Gallego F, Hermoso JA, García JL, Manso I et al (2007) Genetic modification of the penicillin G acylase surface to improve its reversible immobilization on ionic exchangers. Appl Environ Microbiol 73:312–319

Moser BR (2011) Biodiesel production, properties, and feedstocks. In: Tomes D, Lakshmanan P, Sonstad D (eds) Biofuels. Springer, New York, pp 285–347

Murillo G, Ali SS, Sun J, Yan Y, Bartoccie P, El-Zawawy N, Azab M, He Y, Fantozzi F (2019) Ultrasonic emulsification assisted immobilized *Burkholderia cepacia* lipase catalyzed transesterification of soybean oil for biodiesel production in a novel reactor design SC. Renew Energy 135:1025–1034. https://doi.org/10.1016/j.renene.2018.12.080

Murty VR, Bhat J, Muniswaran PKA (2002) Hydrolysis of oils by using immobilized lipase enzyme: a review. Biotechnol Bioprocess Eng 7:57–66

Naresh S, Shuit SH, Kunasundari B, Peng YH, Qi HN, Teoh YP (2018) Immobilization of cellulase from *Bacillus subtilis* UniMAP-KB01 on multi-walled carbon nanotubes for biofuel production. IOP Conf Ser: Mater Sci Eng 318. https://doi.org/10.1088/1757-899X/318/1/012008

Nelson JM, Griffin EG (1916) Adsorption of Invertase. J Am Chem Soc 38:1109–1115

Ngo TPN, Li A, Tiew KW, Li Z (2013) Efficient transformation of grease to biodiesel using highly active and easily recyclable magnetic nanobiocatalyst aggregates. Bioresour Technol 145:233–239

Ni Y, Cao X, Wu G, Hu G, Yang Z, Wei X (2007) Preparation, characterization and property study of zinc oxide nanoparticles via a simple solution-combusting method. Nanotechnology 18:155–161

Ottolina G, Carrea G, Riva S, Sartore L, Veronese FM (1992) Effect of the enzyme form on the activity, stability and enantioselectivity of lipoprotein lipase in toluene. Biotechnol Lett 14:947–952

Öztürk B (2001) Immobilization of lipase from *Candida rugosa* on hydrophobic and hydrophilic supports. (Master thesis). İzmir Institute of Technology, Turkey

Palazzo G, Colafemmina G, Guzzoni Iudice C, Mallardi A (2014) Three immobilized enzymes acting in series in layer by layer assemblies: exploiting the trehalase-glucose oxidase-horseradish peroxidase cascade reactions for the optical determination of trehalose. Sensors Actuators B 202:217–223

Palocci C, Chronopoulou L, Venditti I, Cernia E, Diociaiuti M, Fratoddi I, Russo MV (2007) Lipolytic enzymes with improved activity and selectivity upon adsorption on polymeric nanoparticles. Biomacromolecules 8:3047–3053

Palomo JM (2009) Modulation of enzymes selectivity via immobilization. Curr Org Synth 6:1–14

Palomo JM, Fernandez-Lorente G, Mateo C, Ortiz C, Fernandez-Lafuente R, Guisan JM (2002) Modulation of the enantioselectivity of lipases via controlled immobilization and medium engineering: hydrolytic resolution of mandelic acid esters. Enzym Microb Technol 31:775–783

Palomo JM, Filice M, Fernandez-Lafuente R, Terreni M, Guisan JM (2007) Regioselective hydrolysis of different peracetylated β-monosaccharides by immobilized lipases from different sources. Key role of the immobilization. Adv Synth Catal 349:1969–1976

Pavlidis IV, Vorhaben T, Gournis D et al (2012a) Regulation of catalytic behaviour of hydrolases through interactions with functionalised carbon-based nanomaterials. J Nanopart Res 14:842

Pavlidis IV, Vorhaben T, Tsoufis T et al (2012b) Development of effective nanobiocatalytic systems through the immobilization of hydrolases on functionalized carbon-based nanomaterials. Bioresour Technol 115:164–171

Persichetti RA, Clair NLS, Griffith JP, Navia MA, Margolin AL (1995) Cross-linked enzyme crystals (CLECs) of Thermolysin in the synthesis of peptides. J Am Chem Soc 117:2732–2737

Petros RA, Desimone JM (2010) Strategies in the design of nanoparticles for therapeutic applications. Nat Rev Drug Discov 8:615–627. https://doi.org/10.1038/nrd2591

Phadtare S, Kumar A, Vinod VP, Dash C, Palaskar DV, Rao M et al (2003) Direct assembly of gold nanoparticle "shells" on polyurethane microsphere "cores" and their application as enzyme immobilization templates. Chem Mater 15:1944–1949

Phadtare S, Vinod VP, Mukhopadhyay K, Kumar A, Rao M, Chaudhari RV et al (2004) Immobilization and biocatalytic activity of fungal protease on gold nanoparticle-loaded zeolite microspheres. Biotechnol Bioeng 85:629–637

Pimentel MCB, Leao ABF, Melo EHM, Ledingham WM, Lima-Filho JL, Sivewright M (2007) Immobilization of *Candida rugosa* lipase on magnetized dacron: kinetic study. Art Cell Blood Substit Biotechnol 35:221–235

Pišvejcová A, Rossi C, Hušáková L, Křen V, Riva S, Monti D (2006) β-1,4-Galactosyltransferase-catalyzed glycosylation of sugar derivatives: modulation of the enzyme activity by α-lactalbumin, immobilization and solvent tolerance. J Mol Catal B Enzym 39:98–104

Polizzi KM, Bommarius AS, Broering JM, Chaparro-Riggers JF (2007) Stability of biocatalysts. Curr Opin Chem Biol 11:220–225

Prakasham RS, Sarala DG, Rajya LK, Subba R (2007) Novel synthesis of ferric impregnated silica nanoparticles and their evaluation as a matrix for enzyme immobilization. J Phys Chem C 111:3842–3847

Puri M, Abraham RE, Barrow CJ (2012) Biofuel production: prospects, challenges and feedstock in Australia. Renew Sust Energ Rev 16:6022–6031

Rai M et al (2016) Strategic role of nanotechnology for production of bioethanol and biodiesel. Nanotechnol Rev 5(2):231–250

Rao SV, Anderson KW, Bachas LG (1998) Oriented immobilization of proteins. Microchim Acta 128:127–143

Rao A, Sathiavelu A, Mythili S (2017) Mini review on nanoimmobilization of lipase and cellulase for biofuel production. Biofuel 7269(July). https://doi.org/10.1080/17597269.2017.1348187

Rempel A, de Souza SF, Margarites AC, Astolfi AL, Steinmetz RLR, Kunz A, Treichel H, Colla LM (2019, Sep) Bioethanol from Spirulina platensis biomass and the use of residuals to produce biomethane: an energy efficient approach. Bioresour Technol 288:121588. https://doi.org/10.1016/j.biortech.2019.121588

Richter O, Hoffmann H, Kraushaar-Czarnetzki B (2008) Effect of the rotor shape on the mixing characteristics of a continuous flow Taylor-vortex reactor. Chem Eng Sci 63:3504–3513

Robinson PJ, Dunnill P, Lilly MD (1973) The properties of magnetic supports in relation to immobilized enzyme reactors. Biotechnol Bioeng 15:603–606

Rodrigues RC, Berenguer-Murcia A, Fernandez- Lafuente R (2011) Coupling chemical modification and immobilization to improve the catalytic performance of enzymes. Adv Synth Catal 353:2216–2238

Rodrigues RC, Ortiz C, Berenguer-Murcia Á, Torres R, Fernández-Lafuente R (2013) Modifying enzyme activity and selectivity by immobilization. Chem Soc Rev 42(15):6290–6307. https://doi.org/10.1039/c2cs35231a

Rodrigues ÉF, Ficanha AMM, Dallago RM, Treichel H, Reinehr CO, Machado TP, Nunes GB, Colla LM (2017) Production and purification of amylolytic enzymes for saccharification of microalgal biomass. Bioresour Technol 225:134–141. https://doi.org/10.1016/j.biortech.2016.11.047

Rostro-Alanis MJ, Mancera-Andrade EI, Patiño MBG, Arrieta-Baez D, Cardenas B, Martinez-Chapa SO, Saldívar RP (2016) Nanobiocatalysis: nanostructured materials -a mini-review. Biocatalysis 2:1–24

Rozenberga BA, Tenne R (2008) Polymer-assisted fabrication of nanoparticles and nanocomposites. Prog Polym Sci 33:40–112

Rubin-Pitel SB, Zhao H (2006) Recent advances in biocatalysis by directed enzyme evolution. Comb Chem High-Throughput Screen 9:247–257

Rusetski AN, Ruuge EK (1990) Magnetic fluid as a possible drug carrier for thrombosis treatment. J Magn Magn Mater 85:299–302

Rusmini F, Zhong Z, Feijen J (2007) Protein immobilization strategies for protein biochips. Biomacromolecules 8:1775–1789

Saifuddin N, Raziah AZ, Junizah AR (2013) Carbon nanotubes: a review on structure and their interaction with proteins. J Chem 18:18. Article ID 676815

Saiyed ZM, Telang SD, Ramchand CN (2003) Application of magnetic techniques in the field of drug discovery and biomedicine. Biomagn Res Technol 1:2–8

Saiyed ZM, Sharma S, Godawat R, Telang SD, Ramchand CN (2007) Activity and stability of alkaline phosphatase (ALP) immobilized onto magnetic nanoparticles (Fe_3O_4). J Biotechnol 131(3):240–244. https://doi.org/10.1016/j.jbiotec.2007.06.017

Sánchez-Ramírez J, Martínez-Hernández JL, Segura-Ceniceros P, López G, Saade H, Medina-Morales MA, Ramos-González R, Aguilar CN, Ilyina A (2017) Cellulases immobilization on chitosan-coated magnetic nanoparticles: application for Agave Atrovirens lignocellulosic biomass hydrolysis. Bioprocess Biosyst Eng 40(1):9–22. https://doi.org/10.1007/s00449-016-1670-1

Sastry M, Rao M, Ganesh KN (2002) Electrostatic assembly of nanoparticles and biomacromolecules. Acc Chem Res 35:847–855

Schmeisser C, Steele H, Streit WR (2007) Metagenomics, biotechnology with non-culturable microbes. Appl Microbiol Biotechnol 75:955–962

Schmid A, Dordick JS, Hauer B, Kiener A, Wubbolt M, Witholt B (2001) Industrial biocatalysis today and tomorrow. Nature 409:258–268. https://doi.org/10.1038/35051736

Schmitke JL, Wescott CR, Klibanov AM (1996) The mechanistic dissection of the plunge in enzymatic activity upon transition from water to anhydrous solvents. J Am Chem Soc 118:3360–3365

Schuler C, Caruso F (2000) Preparation of enzyme multilayers on colloids for biocatalysis. Macromol Rapid Commun 21:750–753

Shah S, Solanki K, Gupta MN (2007) Enhancement of lipase activity in non-aqueous media upon immobilization on multi-walled carbon nanotubes. Chem Cent J 1:30

Shang W, Nuffer JH, Dordick JS, Siegel RW (2007) Unfolding of ribonuclease A on silica nanoparticle surfaces. Nano Lett 7:1991–1995

Sharp CA, Howell SA, Jobe J (2000) The effect of biodiesel fuels on transient emissions from modern diesel engines, Part II, unregulated emissions and chemical characterization. SAE Technical Paper. No. 2000-01-1968

Shaw SY, Chen YJ, Ou JJ, Ho L (2006) Preparation and characterization of *Pseudomonas putida* esterase immobilized on magnetic nanoparticles. Enzym Microb Technol 39:1089–1095

Sheelu G, Kavitha G, Fadnavis NW (2008) Efficient immobilization of Lecitase in Gelatin hydrogel and degumming of rice bran oil using a spinning basket reactor. J Am Oil Chem Soc 85:739–748

Sheldon RA (2007a) Cross-linked enzyme aggregates (CLEAs): stable and recyclable biocatalysts. Biochem Soc Trans 35:1583–1587

Sheldon RA (2007b) Enzyme immobilization: the quest for optimum performance. Adv Synth Catal 349:1289–1307

Sheldon RA, van Pelt S (2013) Enzyme immobilisation in biocatalysis: why, what and how. Chem Soc Rev 42:6223–6235

Shim M, Kam NWS, Chem RJ, Li Y, Dai H (2002) Functionalization of carbon nanotubes for biocompatibility and biomolecular recognition. Nano Lett 2:285–288

Silman IH, Katchalski E (1966) Water-insoluble derivatives of enzymes, antigens, and antibodies. Annu Rev Biochem 35:873–908

Singh V, Sardar M, Gupta MN (2013) Immobilization of enzymes by bioaffinity layering. In: Immobilization of enzymes and cells, 3rd edn. Springer/Humana Press, Totawa

Siqueira JR Jr, Caseli L, Crespilho FN, Zucolotto V, Oliveira ON Jr (2010) Immobilization of biomolecules on nanostructured films for biosensing. Biosens Bioelectron 25:1254–1263

Srivastava A, Prasad AR (2000) Triglycerides-based diesel fuels. Renew Sust Energ Rev 4:111–133

Stoytcheva M, Monstero G, Toscano L, Gochev V, Valdez B (2011) The immobilized lipases in biodiesel production. In: Stoytcheva M (ed) Biodiesel – feed- stocks and processing technologies. InTech, Rijeka, pp 397–410

Temporini C, Bonomi P, Serra I, Tagliani A, Bavaro T, Ubiali D et al (2010) Characterization and study of the orientation of immobilized enzymes by tryptic digestion and HPLC-MS: design of an efficient catalyst for the synthesis of cephalosporins. Biomacromolecules 11:1623–1632

Terreni M, Pagani G, Ubiali D, Fernández-Lafuente R, Mateo C, Guisán JM (2001) Modulation of penicillin acylase properties via immobilization techniques: one-pot chemo enzymatic synthesis of Cephamandole from Cephalosporin C. Bioorg Med Chem Lett 11:2429–2432

Thanh LT, Oitsu K, Sadanaga Y et al (2011) A two-step continuous ultrasound assisted production of biodiesel fuel from waste cooking oils: a practical and economical approach to produce high quality biodiesel fuel. Bioresour Technol 101:639–645

Tischer W (1992) In: Finn RK et al (eds) Biotechnology focus 3: fundamentals, applications, information. Hanser, Munich, pp 237–259

Tischer W, Kasche V (1999) Immobilized enzymes: crystals or carriers? Trends Biotechnol 17(8):326–335. https://doi.org/10.1016/S0167-7799(99)01322-0

Troitsky VI, Berzina TS, Pastorino L, Bernasconi E, Nicolini C (2003) A new approach to the deposition of nanostructured biocatalytic films. Nanotechnology 14:597–602

Turner NJ (2009) Directed evolution drives the next generation of biocatalysts. Nat Chem Biol 5:567–573

Vashist SK, Lam E, Hrapovic S, Male KB, Luong JHT (2014) Immobilization of antibodies and enzymes on 3-aminopropyltriethoxysilane-functionalized bioanalytical platforms for biosensors and diagnostics. Chem Rev 114:11083–11130

Verma ML, Barrow CJ, Puri M (2013) Nanobiotechnology as a novel paradigm for enzyme immobilisation and stabilisation with potential applications in biodiesel production. Appl Microbiol Biotechnol 97:23–39

Verma ML, Puri M, Barrow CJ (2014) Recent trends in nanomaterials immobilised enzymes for biofuel production, vol 8551, pp 1–12. https://doi.org/10.3109/07388551.2014.928811

Vertegel AA, Siegel RW, Dordick JS (2004) Silica nanoparticle size influences the structure and enzymatic activity of adsorbed lysozyme. Langmuir 20:6800–6807
Vianello F, Zennaro L, Di Paolo ML, Rigo A, Malacarne C et al (2000) Preparation, morphological characterization, and activity of thin films of horseradish peroxidase. Biotechnol Bioeng 68:488–495
Wang P (2006) Nanoscale biocatalyst systems. Curr Opin Biotechnol 17:574–579
Wang L, Jiang R (2011) Reversible His-tagged enzyme immobilization on functionalized carbon nanotubes as nanoscale biocatalyst. Methods Mol Biol 743:95–106
Wang P, Sergeeva MV, Lim L, Dordick JS (1997) Biocatalytic plastics as active and stable materials for biotransformations. Nat Biotechnol 15:789–793
Wang X, Dou P, Zhao P et al (2009) Immobilization of lipases onto magnetic Fe_3O_4 nanoparticles for application in biodiesel production. ChemSusChem 2:947–950
Wang X, Liu X, Zhao C et al (2011a) Biodiesel production in packed-bed reactors using lipase-nanoparticle biocomposite. Bioresor Technol 102:6352–6355
Wang X, Liu X, Yan X et al (2011b) Enzyme-nanoporous gold biocomposite: excellent biocatalyst with improved biocatalytic performance and stability. PLoS One 6:e24207
Wang X, Qin X, Li D, Yang B, Wang Y (2017) One-step synthesis of high-yield biodiesel from waste cooking oils by a novel and highly methanol-tolerant immobilized lipase. Bioresour Technol 235:18–24. https://doi.org/10.1016/j.biortech.2017.03.086
Watanabe Y, Shimmada Y, Sugihara A et al (2000) Continuous production of biodiesel fuel from vegetable oil using immobilized *Candida antarctica* lipase. J Am Oil Chem Soc 77:355–360
White CA, Kennedy JF (1980) Popular matrices for enzyme and other immobilizations. Enzym Microb Technol 2:82–90
Willner I, Baron R, Willer B (2006) Growing metal nanoparticles by enzymes. Adv Mater 18:1109–1120
Wong LS, Khan F, Micklefield J (2009) Selective covalent protein immobilization: strategies and applications. Chem Rev 109(9):4025–4053. https://doi.org/10.1021/cr8004668
Xie W, Ma N (2009) Immobilised lipase on Fe3O4 nanoparticles as biocatalyst for biodiesel production. Energy Fuel 23:1347–1353
Xie W, Ma N (2010) Enzymatic transesterification of soybean oil by using immobilized lipase on magnetic nano-particles. Biomass Bioenergy 34:890–896
Xin BJ, Si SF, Xing GW (2010) Protease immobilization on γ-Fe_2O_3/Fe_3O_4 magnetic nanoparticles for the synthesis of oligopeptides in organic solvents. Chem Asian J 5:1389–1394. https://doi.org/10.1002/asia.200900696
Xu J, Zeng F, Wu S, Liu X, Hou C, Tong Z (2007) Gold nanoparticles bound on microgel particles and their application as an enzyme support. Nanotechnology 18:265–273
Yamashita I (2001) Fabrication of a two-dimensional array of nanoparticles using ferritin molecule. Thin Solid Films 393:12–18
Yang Z, Si S, Zhang C (2008) Magnetic single-enzyme nanoparticles with high activity and stability. Biochem Biophys Res Commun 367:169–175. https://doi.org/10.1016/j.bbrc.2007.12.113
Yim TJ, Kim DY, Karajanagi SS, Lu TM, Kane R, Dordick JS (2003) Silicon nanocolumns as novel nanostructured supports for enzyme immobilization. J Nanosci Nanotechnol 3:479–482
Yiu HHP, Keane MA (2012) Enzyme-magnetic nanoparticle hybrids: new effective catalysts for the production of high value chemicals. J Chem Technol Biotechnol 87:583–594
Yiu HHP, Wright PAJ (2005) Enzymes supported on ordered mesoporous solids: a special case of an inorganic–organic hybrid. Mater Chem 15:3690–3700
Yong Y, Bai Y-X, Li Y-F, Lin L, Cui Y-J et al (2008) Characterization of Candida rugosa lipase immobilized onto magnetic microspheres with hydrophilicity. Process Biochem 43:1179–1185
Yu LT, Banerjee IA, Gao XY et al (2005) Fabrication and application of enzyme-incorporated peptide nanotubes. Bioconjug Chem 16:1484–1487
Zhang K, Diehl MR, Tirrell DA (2005) Artificial polypeptide scaffold for protein immobilization. J Am Chem Soc 127:10136–10137

Zhao B, Liu X, Jiang Y, Zhou L, He Y, Gao J (2014) Immobilized lipase from Candida sp. 99-125 on hydrophobic silicate: characterization and applications. Appl Biochem Biotechnol 173(7):1802–1814. https://doi.org/10.1007/s12010-014-0967-2

Zhao X, Qi F, Yuan C, Du W, Liu D (2015) Lipase-catalyzed process for biodiesel production: enzyme immobilization, process simulation and optimization. Renew Sust Energ Rev 44:182–197. https://doi.org/10.1016/j.rser.2014.12.021

Zhao K, Cao X, Di Q, Wang M, Cao H, Deng L et al (2017) Synthesis, characterization and optimization of a two-step immobilized lipase. Renew Energy 103:383–387. https://doi.org/10.1016/j.renene.2016.11.035

Zheng G, Patolsky F, Cui Y, Wang WU, Lieber CM (2005) Multiplexed electrical detection of cancer markers with nanowire sensor arrays. Nat Biotechnol 23:1294

Carbon Nanotubes Synthesized by Green/Eco-friendly Technique Potential for Bioenergy Applications

10

Kelvii Wei Guo

Abstract

The multi-walled carbon nanotubes (MWCNTs) were synthesized by homemade equipment with the waste alcohol. The synthesized MWCNTs were characterized by scanning electron microscopy (SEM), transmission electron microscope (TEM), and Raman spectroscopy. Moreover, in confirming the synthesized green/eco-friendly technique, MWCNTs were grown on some of the micropunches to show the attractive potential for bioenergy application compellingly and convincingly. Results show that due to the usage of the waste alcohol, not only is this technique definitely green/eco-friendly to the environment, but also MWCNTs as a protective layer on the micropunch can lengthen the service life of the micropunch by up to 35% without directly affecting the size of the punched holes compared with that of without MWCNTs. Moreover, the punched microholes are definitely quite distinct from that of traditional punching. Besides the fracture band is absent, the areas of rollover and burr are also extremely narrow, which implies the attractive punched microhole can be used directly without any posttreatments.

Keywords

Multi walled carbon nanotubes (MWCNTs) · Green/eco-friendly technique · Bioenergy · Biomass power plant · Waste alcohol · Nanotechnology · Microhole · Micropunch · Service life

K. W. Guo (✉)
Department of Mechanical and Biomedical Engineering, City University of Hong Kong, Kowloon, Hong Kong

M. Srivastava et al. (eds.), *Nanomaterials in Biofuels Research*,
Clean Energy Production Technologies, https://doi.org/10.1007/978-981-13-9333-4_10

10.1 Introduction

Carbon nanotubes (CNTs) have attracted intense attention since their discovery (Iijima 1991, 1993). The past two decades have witnessed their spread in a range of applications including building blocks in modern electronics, field emission sources, photodetectors, nonvolatile memory, ultrasensitive chemical and biosensors, and transparent conductive membranes.

It is well known that CNTs are intrinsically composed of pure carbon atoms that arrange and interact with each other by the strong sp^2 carbon-carbon chemical bonds and form the unique geometric structure of a carbon network; this gives CNTs fascinating and attractive properties, such as electronic, mechanical, and thermal properties. These properties have convinced scientists that CNTs and related nanostructures may play crucial but irreplaceable roles in many different kinds of applications by either replacing the traditional materials or exploring novel functions.

To date, CNTs' production capacity exceeds several 1000 tons per year (De Volder et al. 2013). Advances in CNTs' controlled growth, purification, modification, and processing are showing great advantages in many aspects of everyday life. CNTs have been integrated into various commercial products, ranging from rechargeable batteries, mobile screens, conductive additives, sporting goods, and water filters (Sehrawat et al. 2016; Li et al. 2017; Yadav and Singh 2016; Moreno et al. 2017; Jäckel et al. 2016; Ho et al. 2018; Puch and Hopmann 2014; Zhai et al. 2016; Luzardo et al. 2017; Paramo et al. 2018).

Because of its unique structure, excellent electrical conductivity, mechanical property, high tensile and flexural strengths, high elastic modulus, and high aspect ratio, CNTs have been regarded as an attractive contender in tribological applications. Interesting results were reported with significant reduction in friction and wear rates in nanotube polymers (Pöllänen et al. 2011; Green et al. 2009), ceramics (Zhan et al. 2003; Hvizdoš et al. 2012), and metal composites (Guiderdoni et al. 2013; Bakshi et al. 2011). Theoretically speaking, friction coefficient between the walls of MWCNTs can be extremely low (Damnjanović et al. 2002; Dickrell et al. 2005; Dickrell et al. 2006).

Advances in telecommunication, transportation, and medical fields lead to challenges in microfabrication, miniaturization, and multifunctional technologies in manufacturing. Feature sizes are in the range of a few microns up to hundreds of microns. The ability to fabricate microholes in large quantities has potential applications in microscale batteries and fuel cells (Baik and Seo 2018; Fischer 2018; Price et al. 2018; Shin et al. 2018; Christwardana et al. 2018; Kim and Park 2018; Ali et al. 2018; Takahashi et al. 2018; Martinez et al. 2017a; Hashemi et al. 2017; Torelli et al. 2017; Dubois et al. 2017; Shetzline et al. 2017; Pratyusha et al. 2017; Palecki et al. 2017; Cherevko et al. 2016; Belarb et al. 2016; Fukuzumi 2016; Calabriso et al. 2015; Theodorakakos et al. 2014a; Virkar 2012), fluidic microchemical reactors requiring microscale pumps (Baghel and Mali 2018; Haghighi et al. 2018; Kant et al. 2017; Calvo-López et al. 2017; Guler et al. 2017; Hatsuzawa et al. 2017; Madhavi and Hiremath 2016; Derkus 2016; Liu et al. 2016; Hilber and Jakoby

2015), valves and mixing devices (Asim et al. 2019; Pradeep et al. 2018; di Capaci et al. 2018; Zangl et al. 2018; Hsu et al. 2018; Mohan et al. 2018; Ravi et al. 2018; Silaskar and Shinde 2018; Ferraro et al. 2018; Cortes et al. 2017; de Vries et al. 2017; Hayes et al. 2017; Makaryants 2017; Okhotnikov et al. 2017; Sayindla et al. 2017; Tverskoy et al. 2017; Tesař 2017; Willich and White 2017; Banyai et al. 2016; Farzaneh-Gord and Khoshnazar 2016; Gehlot and Tripathi 2016; Lisowski and Filo 2016; Malama et al. 2016; Ramesh et al. 2016; Edvardsen et al. 2015; Kudźma and Stosiak 2015; Muniak 2015; Singh and Tharakan 2015; Amirante et al. 2014; Rogers et al. 2014), microfluidic systems (Dvořák et al. 2018; Hoysall and Garimella 2018; Massarwi et al. 2018; Samih et al. 2018; Trotta et al. 2018; Tsur and Shamir 2018; Abed et al. 2017; Dittmeyer et al. 2017; Martinez et al. 2017b; Munk et al. 2017; Kowsari et al. 2016; Kuriakose et al. 2016; Sarafraz et al. 2016; Chavoshi and Luo 2015; Kenig et al. 2013), biomedical and medical implants (Affes et al. 2018; Bellemère 2018; Block 2018; Choi et al. 2018; Rupp et al. 2018; Taniguchi et al. 2018; Cristofari et al. 2017; Rittel et al. 2017; Baino and Potestio 2016; Ferraris and Spriano 2016; Hakimi et al. 2016; Nyström and Malmsten 2016; Pina et al. 2016; Ramos and Mesnard 2016; Sharma et al. 2014), diagnostic and remediation devices (Tharmavaram et al. 2018; Gomis-Berenguer et al. 2017; Rosace et al. 2017; Tanimu et al. 2017; Ferreira et al. 2016; Monošík and Angnes 2015; Nguyen 2013; Schütze et al. 2013), cooling holes in jet turbine blades (Fraas et al. 2019; Kim and Kim 2018; Lee et al. 2018; McInturff et al. 2018; Enagi et al. 2017; Al-Zurfi and Turan 2016; Mazaheri et al. 2016; Park et al. 2016; Ali and Janajreh 2015; Satta and Tanda 2015; Sciubba 2015; Chung et al. 2014; Krewinkel 2013), fiber optics (Pakarzadeh et al. 2019; Bhattacharya 2018; Kumar et al. 2018; Gaona-Reyes and Bermudez 2017; Tu et al. 2014), micro-nozzles for higher energy efficiency (Do et al. 2017; Krishnamoorthi and Malayalamurthi 2018; Ranjan et al. 2017; Saadati and Roohi 2015; Theodorakakos et al. 2014b; Torre et al. 2011) and less pollutant emissions to the environment (Boncel et al. 2011; Gutierrez et al. 2012; Tofighy and Mohammadi 2011; Upadhyayula and Gadhamshetty 2010; Tiusanen et al. 2012), micro-molds and deep X-ray lithography masks, optical lenses, and micro-components in daily life products such as compact disc players, air bags, and inkjet printers. Currently, with the ever-increasing demand for smaller, higher-quality, and lower-priced products from almost all fields of industry, household equipment, and entertainment electronics, the development of manufacturing methods which are tailor-made for the micro-system technique with higher precision, lower cost, and larger quantities and more eco-friendly to the environment is extremely urgent.

The trend for micromanufacturing is more focused on miniaturizing or down-scaling both conventional and nonconventional methods to produce microproducts. Manufacturing processes can be categorized according to the type of energy used in the process itself, such as mechanical, chemical, electrochemical, electrical, and laser processes. The working principles behind each process include consideration of mechanical forces, thermal effects, ablation, dissolution, solidification, recomposition, polymerization/lamination, and sintering. According to the way in which components/products are to be made, general manufacturing processes can also be classified into subtractive, additive, forming, joining, and hybrid processes. The classification is equally applicable to micromanufacturing. For subtractive processes, there are micromechanical cutting (milling, turning, grinding, polishing, etc.), micro-EDM, micro-ECM, laser beam machining, electron beam machining, and photochemical machining. For additive processes, the relevant techniques are surface coating (CVD, PVD), direct writing (inkjet, laser-guided), microcasting, microinjection molding, sintering, photoelectron forming, chemical deposition, polymer deposition, and stereolithography. For deforming processes, there are microforming (stamping, extrusion, forging, bending, deep drawing, incremental forming, superplastic forming, hydroforming, punching), hot embossing, and micro-/nano-imprinting. For joining processes, the related techniques are micromechanical assembly, resistance, laser, vacuum soldering, bonding/welding, and gluing. For hybrid processes, there are microlaser ECM, LIGA, and LIGA combined with laser machining, micro-EDM and laser assembly, shape deposition and laser machining, laser-assisted microforming, micro-assembly injection molding, combined micromachining, and casting (Mamalis 2005; Janssen et al. 2008; Uhlmann et al. 2016; Korkmaz et al. 2017; Cole et al. 2017; Biasetti et al. 2017; Singh et al. 2018; Saxena et al. 2018).

Punching is the process of forcing a punch through the material and into a die to create a hole in the workpiece. It is often an economic way of creating shaped holes in mass production. Micropunching was demonstrated in the punching of circular and noncircular holes as small as 5 microns in size. Being a mechanical process that does not involve the use of chemicals, punching is environment-friendly. Therefore, micropunching can be an economic and environment-friendly way of fabricating shaped microholes in mass production.

It is well known that punching is the process of forcing a punch through the material and into a die to create a hole in the work piece. Tool wear is an important issue in micropunching. Even with punches made of hard and tough materials like tungsten carbide and cobalt (WC/Co), the quality of the punched holes declines rapidly under repeated punching (Dornfeld et al. 2006; Mark 2006; Bhattacharyya 2015; Luttge 2016; Knobloch and Kaminorz 2004; Mohamed 2006; Edwards 2006; Oliver et al. 2004; Johnstone and Parameswaran 2004; Brousseau et al. 2010).

The higher quality of the micropunches is a key to the successful application of micropunching. When the punching number exceeds 1000, the quality of the micropunch decreases obviously due to the serious wear of the micropunch. During the micropunching, the friction between the micropunch and the substrate deteriorates the micropunch. As a result, the quality of the punched holes decreases simultaneously. In addition, the price of WC/Co micropunches is costly (Guo and Tam 2011).

Because of the unique property of CNTs, the friction between the micropunch and the substrate can be decreased by forming a CNT coating on the surface of the micropunches to increase the wear property of the micropunches and improve the tool life in the end. Therefore, aiming to form an effective CNT coating on WC/Co micropunches, the multi-walled carbon nanotubes (MWCNTs) were synthesized by homemade equipment with the waste alcohol, which is green/eco-friendly to the environment. Meanwhile, the effect of synthesized MWCNT coating on the tool life of WC/Co micropunches has been fully investigated by confocal laser, SEM, digital balance, etc. Results show that the effect of MWCNTs synthesized by the waste alcohol is significantly efficient. The wear of micropunch coated with MWCNTs decreases obviously, even in the severe wear period. Moreover, the wear loss is definitely less than that of micropunch without MWCNT coating. This promising result is due to the formation of a transfer film at the contact region by rubbing the MWCNT forest. MWCNTs adhere to the surface of the micropunch avoiding direct contact with the substrate and providing lubricant properties to the interface by virtue of their graphitic nature.

10.2 Experimental Materials and Procedures

10.2.1 Experimental Materials

Micropunch made by Ultrahardness tools company in Japan with 75% volume fraction WC particle and 25% volume fraction Co particle of 50 μm mean size, 150 μm in diameter, is shown in Fig. 10.1. Specimens for punching experiments were pure titanium sheets 200 μm in thickness.

Figure 10.2 shows the surface texture of micropunch and its elements' (carbon, tungsten, cobalt) distribution. Ploughs induced by the ultraprecise machining can be obviously detected, and some debris are distributed on the surface of the micropunch. Moreover, the distribution of tungsten element is denser than that of both carbon element and cobalt element, which indicates that the hardness of the surface of the micropunch is hard and suitable for the micropunching process.

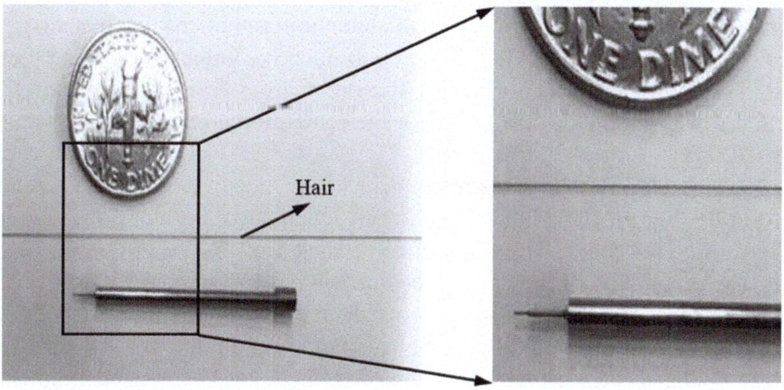

Fig. 10.1 Profile of Ø150 μm micropunch

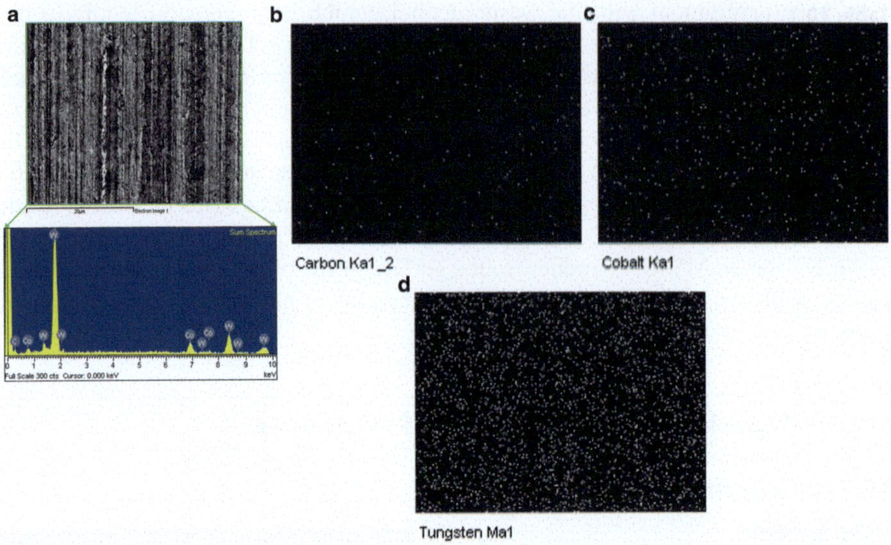

Fig. 10.2 Surface texture of WC/Co micropunch and elements' distribution. (**a**) Surface texture and mapping of elements. (**b**) Carbon distribution. (**c**) Cobalt distribution. (**d**) Tungsten distribution

10.2.2 Experimental Procedures

10.2.2.1 CNT Growth

Micropunches were cleaned by acetone and pure ethyl alcohol for contaminant removal prior to CNT growth/deposition.

In this research, Fe was taken as the catalyst for CNT forming. In the initial, the catalyst was coated on the required area of the micropunch by electron cyclotron resonance (ECR). The relevant processing parameters are listed in Table 10.1. It is well known that the distribution of cobalt element in the micropunch is also a catalyst for CNT forming and CNTs will be consequently synthesized during the CNT growth. Therefore, in order to protect the other areas from synthesizing CNTs, aluminum foil is adopted as the cover for the remaining areas of the micropunches (Dupuis 2005).

For Fe coating process, the micropunches were set in a vacuum chamber for CNT forming. Figure 10.3 shows the relevant schematic diagram of the homemade CNT forming equipment. The micropunch was set on the heating area for CNT growth. The dimension of the vacuum chamber was Ø100 mm × 150 mm.

The waste alcohol, which was directly taken from the alcohol after the specimens' cleaning, met with the GB-HW06 standard, was put into a ceramic container under the heater to provide carbon source for chemical vapor deposition (CVD). Conditions for CNT growth were summarized in Table 10.2.

Table 10.1 ECR processing parameters for Fe deposition

Substrate	WC/Co micropunch
Catalyst	Fe
Irradiation time	60 s
Accelerated voltage	2500 V
Ion current density	12.0 mA/cm^2
Gas	Ar
Gas flow rate	0.6 SCCM
Vacuum	1.5×10^{-4} pa

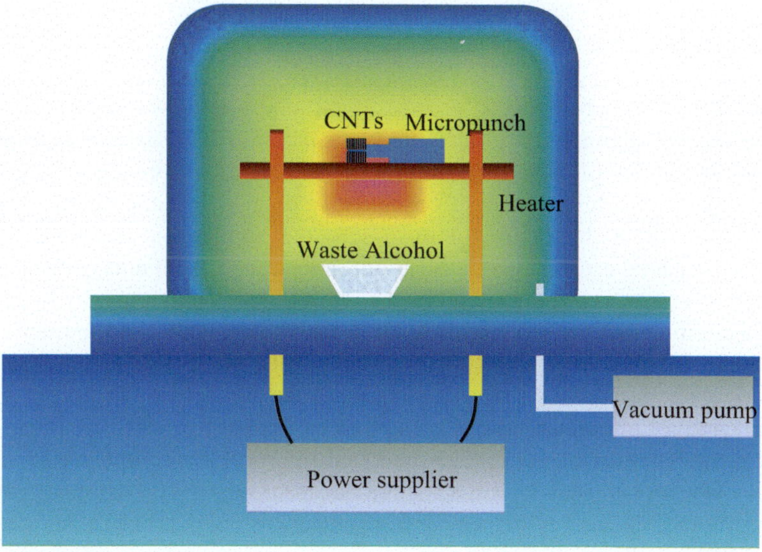

Fig. 10.3 Setup for CNT growth

Table 10.2 Parameters for CNT growth

Heating rate	3~4 A/s
Cooling rate	5 A/s
Reaction direct current	38~40 A
Reaction time	10 s
Vacuum	1×10^{-2} pa

10.2.2.2 Micropunching

The acetone and pure ethyl alcohol were taken as cleaner to make the substrate-pure titanium sheet clean the contaminants. After that, the substrate is carefully put into the microdie. When all is set properly, the micropunching is in process using the microprocessing machine MP50 (made in Japan), where the federate of micropunching was 2 mm together with 20 pulses in 1 min.

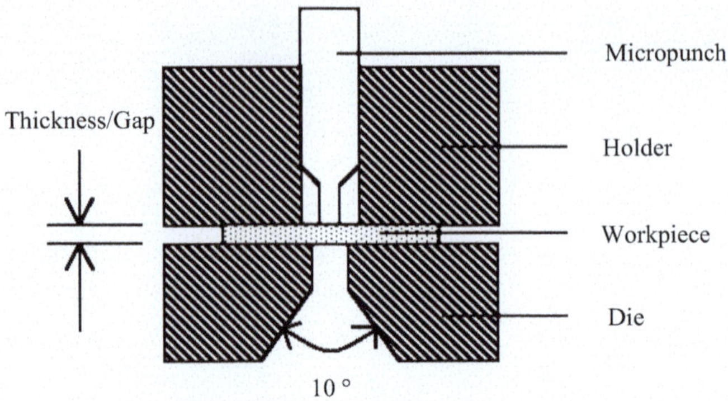

Fig. 10.4 Schematic diagram of micropunching

Along with the micropunching process, the in-house setup was used to make the accurate vertical alignment. The accurate vertical alignment is extremely important to the micropunching. Without it, the micropunch is easily broken and seriously worn during the punching process. Consequently, the punched microholes lack high quality, resulting in the failure of the microhole application afterward. In this research, the lateral error is less than 1.5 μm in the range of 20 mm vertical stroke. A *z*-stepper is taken as the stroke and load with the 20 N in maximum by a push rod. In the meantime, the bush was fabricated by the same material as the micropunch (tungsten carbide/cobalt) with a 200 μm through-hole in 10° taper. Both faces of the bushes were carefully polished to ensure flatness and parallelism. The relevant schematic diagram of micropunching is shown in Fig. 10.4.

The characteristics of microholes punched by WC/Co micropunch in various periods were investigated by LEXT confocal laser-OLS3000, scanning electron microscopy (SEM), energy dispersive X-ray analysis (EDX), optical surface profiler (OSP)-Veeco/Wyko NT9300, etc.

10.3 Results and Discussion

10.3.1 Micropunch with MWCNT Coating

An image of a CNT-coated micropunch is shown in Fig. 10.5, and its higher magnification image is expressed in Fig. 10.6. It illustrates that the synthesized CNTs coat on the surface of the micropunch successfully and the height of CNTs is about 15 μm. Meanwhile, CNTs vertically align with high density. Moreover, CNTs appear wiggly because they were tightly packed during growth. TEM image of CNTs is shown in Fig. 10.7. It indicates that CNTs are multi-walled nanotubes (MWCNTs) with a diameter of about 3–5 nm.

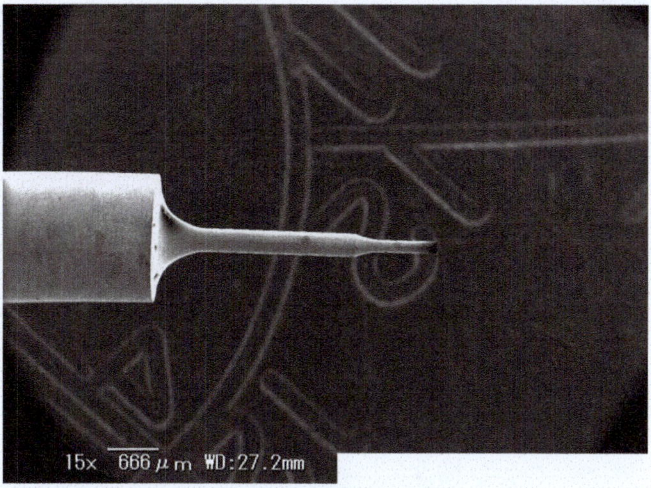

Fig. 10.5 A micropunch with MWCNT coating

10.3.2 Raman Spectroscopy Analysis

Raman spectroscopy analysis is taken for testing the property of the synthesized vertically aligned MWCNTs. The parameters of Raman spectroscopy analysis are listed in Table 10.3. It is well known that the *G*-band, the *D*-band, and the RBM are included in the typical Raman shifts (cm^{-1}) and the corresponding peak of the *G*-band is at about 1590 cm^{-1} and the *D*-band 1350 cm^{-1}. The characteristic of the synthesized MWCNTs is evaluated according to the ratio between the *G*-band and the *D*-band.

Figure 10.8 expresses the result of Raman spectroscopy analysis of the synthesized vertically aligned MWCNTs. It shows the ratio of IG/ID is about 0.75. Meanwhile, the peak of *D*-band is higher than that of *G*-band. It reveals that the synthesized vertically aligned MWCNTs are highly disordered. As the media in the tribology, the disordered MWCNTs will be more inclined to show their attractive lubricant characteristics more easily. Therefore, the micropunch with such MWCNTs must be more effective for the practical applications than that of the micropunch without MWCNT coating.

10.3.3 Wear Loss of Micropunches

During the micropunching, the micropunches vary during the various periods due to the wear of the micropunch along with the weight changes of the micropunches. The variation of the micropunches was measured by the intermittent weight calibration by digital balance during punching.

Fig. 10.6 MWCNTs coated on the surface of micropunch

Results clearly show that there are three stages during the micropunching process; they are initial stage, quasi-stable stage, and severe wear stage. The initial stage is also known as the run-in stage. In this stage, for the micropunch without CNT coating (cycle no. <500), the wear loss of the micropunch is definitely obvious and relatively high. With the effect of friction between the micropunch and the substrate, the micropunch wears and the surface is smoothened. At the same time, the crests on the surface of the micropunch are lowered. When the punching number is in the range of 500–1200, the wear characteristic of the micropunch is quasi-stable. In this stage, the wear loss is very little, and the variation of the surface texture of the micropunch is also hardly detected. However, when the punching number exceeds 1200, due to the severe cobalt loss, which is the bonding material in the micropunch, tungsten carbide is more easily peeled off from the micropunch seriously resulting in a distinctive weight loss. Moreover, the micropunch is more inclined to break in this stage.

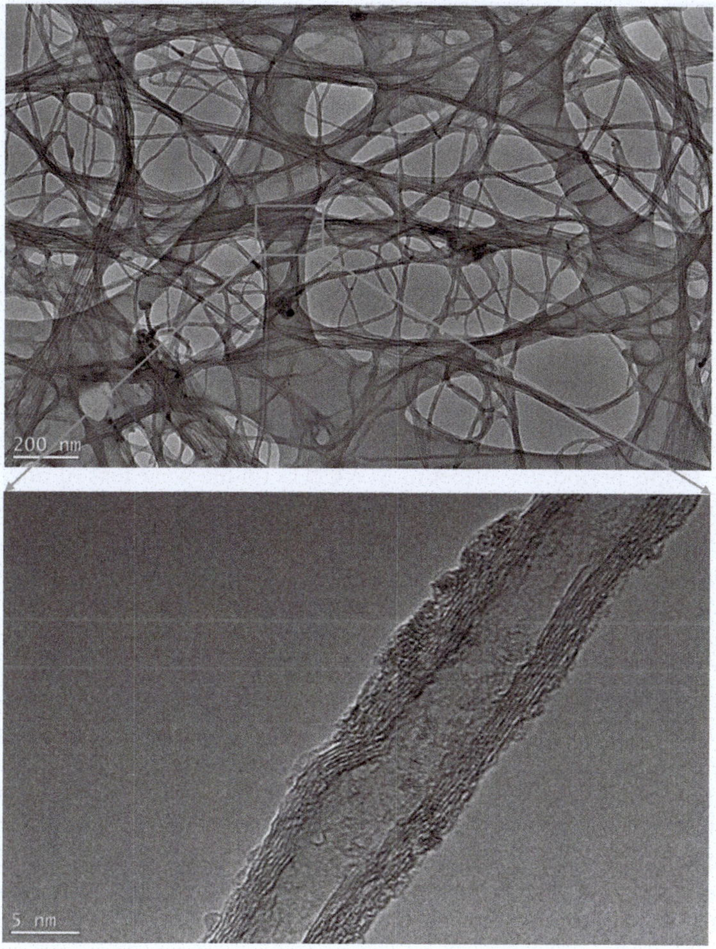

Fig. 10.7 TEM image of synthesized MWCNTs

Table 10.3 Conditions for Raman spectroscopy analysis

Laser wavelength	532 nm
Laser power	30 mW × 0.5%
Irradiation spot	5 μm
Integration times	30

It also expresses that the wear characteristics of the micropunches with CNT coating are nearly the same as that without CNT coating. The attractive result of the micropunches with CNT coating is that the quasi-stable stage forwards and begins at 450. At the same time, the quasi-stable stage delays and ends at 1400. Moreover, the wear loss of non-coated/coated micropunches in the quasi-stable

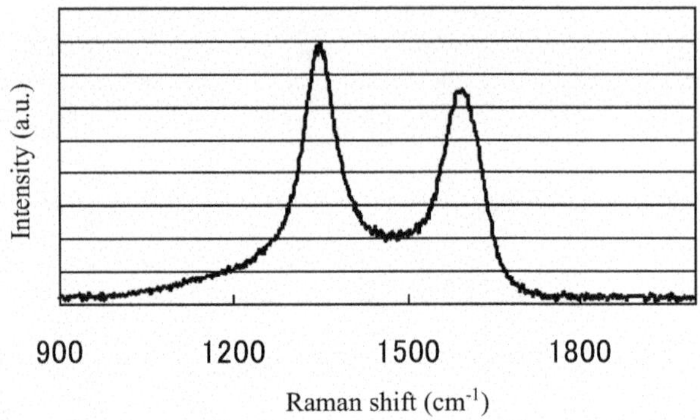

Fig. 10.8 Raman spectra of synthesized MWCNTs

period is definitely attractive, especially for punching number from 500 to 1200 for non-coated micropunches and from 450 to 1400 for MWCNT-coated micropunches. In this period, the effect of MWCNTs on the wear loss of micropunch sounds obvious. In addition, the comparison between the micropunch with and without CNT coating is affirmatively distinctive. For the severe wear stage, the wear loss of the micropunches (both with CNT coating and without CNT coating) is certainly obvious. When the punching number exceeds 1200 for the micropunches without CNT coating and the punching number exceeds 1400 for the micropunches with CNT coating, the weight loss of the micropunches at both conditions is all serious, and the beneficial effect of the synthesized MWCNTs on the wear loss disappears, especially with the punching number increasing further.

It should be noted that with the favorable effect of MWCNTs synthesized on the surface of the micropunches, the start of the quasi-stable period is pushed forward, and the end of the quasi-stable period is postponed. It elucidates that the effective quasi-stable period of the micropunches with CNT coating is longer than that of without CNT coating on the surface of micropunches. For MWCNT-coated micropunches, it is 1400–450 = 950. By comparison, for non-coated micropunches, it is 1200–500 = 700. Furthermore, the total wear loss of MWCNT-coated micropunches is less than that of non-coated micropunches. It demonstrates that the life of micropunch is improved or prolonged evidently.

10.3.4 Profile of Microhole Punched by WC/Co Micropunch

The diameter of the punched microhole by the micropunch was measured by LEXT confocal laser-OLS3000.

It illustrates that in the different period, with the weight variation of the micropunch, the diameter of microhole changes correspondingly. In the initial period, the diameter of microhole decreases obviously with the increment of punching number. While the punching number is from 500 to 1200, the diameter of microhole relatively keeps stable due to the unvariation of the micropunch which is indicated by the near-zero weight loss of the micropunch in the quasi-stable period. With the punching number increasing further, especially when the punching number exceeds 1525, the diameter of microhole decreases remarkably because of the serious loss of cobalt. Consequently, the serious wear of micropunch mainly relies on tungsten carbide particles which are easily peeled off without cobalt as bonding material in the severe period of the micropunching. Meanwhile, due to the temperature increment with punching number increasing, tungsten carbide particles are more easily peeled off, resulting in more intensive wear loss. Moreover, if the pulses per minute increases, the wear loss of the micropunch will be more drastically.

Moreover, it also illustrates that for both cases (non-MWCNT coated and MWCNT coated on the surface of the micropunches), in the different wear conditions, the diameter of the punched microhole changes correspondingly. In the initial condition, the diameter variation of the two kinds of micropunches (with and without CNT coating) essentially overlaps, which is the same as the trend of the weight change. There is a distinctive diameter decrease of the punched microholes when the weight loss of the micropunch is fast in the initial stage. At this period, the beneficial effect of MWCNTs on the size of the punched microhole is not clear. At a result, the evolution of the size of the punched microholes relies mainly on the weight loss of the micropunches. The attractive results of the diameter variation of the punched microholes at both conditions (with and without CNT coating) in the quasi-stable stage are very small along with the very little weight loss of the micropunches. For the micropunches without CNT coating synthesized on the surface, the diameter of the micropunches decreases slightly faster consistent with the slightly faster weight reduction of the corresponding punch. The diameter of the punched microhole for the micropunches with CNT coating on the surfaces reduces from 174.2 μm to 172.9 μm and from 173.1 μm to 170.9 μm for the micropunches without CNT coating on the surfaces. The holes achieved by the CNT coated on the surface of the micropunches are slightly bigger than that of without CNT coated on the surface of the micropunches which reveals the definitely beneficial effect of CNT coating compellingly and convincingly.

The diameter of the punched microholes reduces rapidly in the severe wear stage. After 1600 shots, the diameter of the punched microholes reduces to 163.8 μm for punching with CNT coating on the surface of the micropunches. For the micropunches without CNT coating on the surface, the diameter of the punched microholes reduces to 161.0 μm. These scenarios indicate about 6% reduction from the nominal diameter in the quasi-stable stage. More seriously, near the end the reduction rate is about 6.5 μm per 100 shots for both micropunches. For the surface of the micropunches coated with MWCNTs, due to the lack of attached MWCNTs on the surface of the micropunches, the relevant wear characteristic is same as that of non-MWCNTs coated the surface of the micropunches. Consequently, the serious weight loss of the micropunches is the obvious characteristic in the severe wear stage. In the meantime, the relevant diameter of the punched microholes reduces remarkably.

Meanwhile, the variation of the diameter of the punched microholes meets very well with the wear loss of the micropunches.

10.3.5 Surface Texture of MWCNT-Coated Micropunch

Figure 10.6 shows the initial surface texture of MWCNTs coated micropunch are distributed as MWCNTs forest, MWCNTs densely populate, mutually tangle and distribute uniformly on the surface of the micropunch.

It can be seen that MWCNTs are distributed non-uniformly on the surface of the micropunch, and under the rubbing effect of interaction between the micropunch and the substrate, a bundle or bulk of MWCNTs attaches on the micropunch surface. The microtextures of the micropunch surface in the quasi-stable period show that the uniformly distributed MWCNTs attached on the surface of the micropunch form a favorable transfer film as a lubricant layer between the micropunch and the substrate in the micropunching procedure. Moreover, owing to this transfer film formation at the interaction area under the effect of CNT forest stripping, MWCNTs or the formed debris adhere to the surface of the micropunch (or the mating surfaces) averting the direct contact between the micropunch and the substrate in the micropunching process and are taken as the lubricant to the interface by virtue of their graphitic nature. Results definitely provide a promising prospect for the future prolonging the life of micropunches in the following practical applications.

When the punching number increases, it illustrates that MWCNTs distribute sparsely and disappear at the last stage. It reveals that the beneficial effect of MWCNTs is distributed initially on the surface of the micropunch lost. As a result, the severe wear of the micropunch appears.

10.4 Features of Punched Microhole

Actually, material shear and fracture in the micropunching with foil (thickness less than 200 μm) are different from those of punching in traditional punching. Figure 10.9 shows the distinguishing features of traditional punching, which includes four sections: (1) rollover, (2) shear/burnish zone, (3) fractured zone, and (4) burr. Most attractively, the features of microhole punched by micropunching are definitely quite distinct from that of traditional punching. The fracture band is absent in the punching of a 150 μm hole in titanium with the thickness of 200 μm as shown in Fig. 10.10. In the relevant fractrography of inner micropunched microhole, it indicates that the areas of rollover and burr are extremely narrow besides the absence of the fracture band. The most part of the inner microhole is the shear/burnish band. Figure 10.11 expresses its three-dimensional topography observed by optical surface profiler (OSP)-Veeco/Wyko NT9300, which agrees well with the results shown in Fig. 10.10.

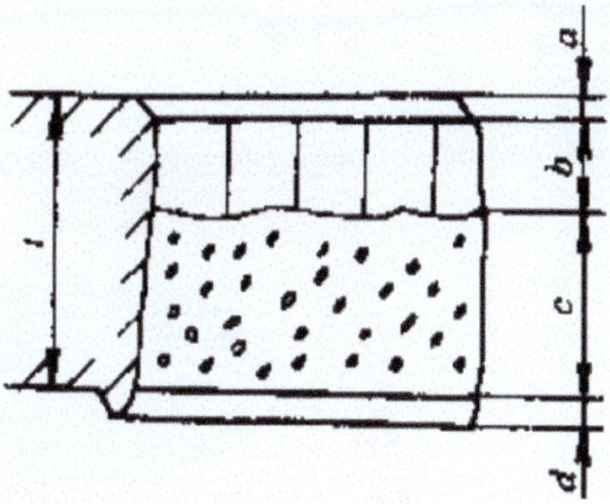

Fig. 10.9 Traditional punching: (*a*) rollover, (*b*) shear/burnish zone, (*c*) fractured zone, and (*d*) burr

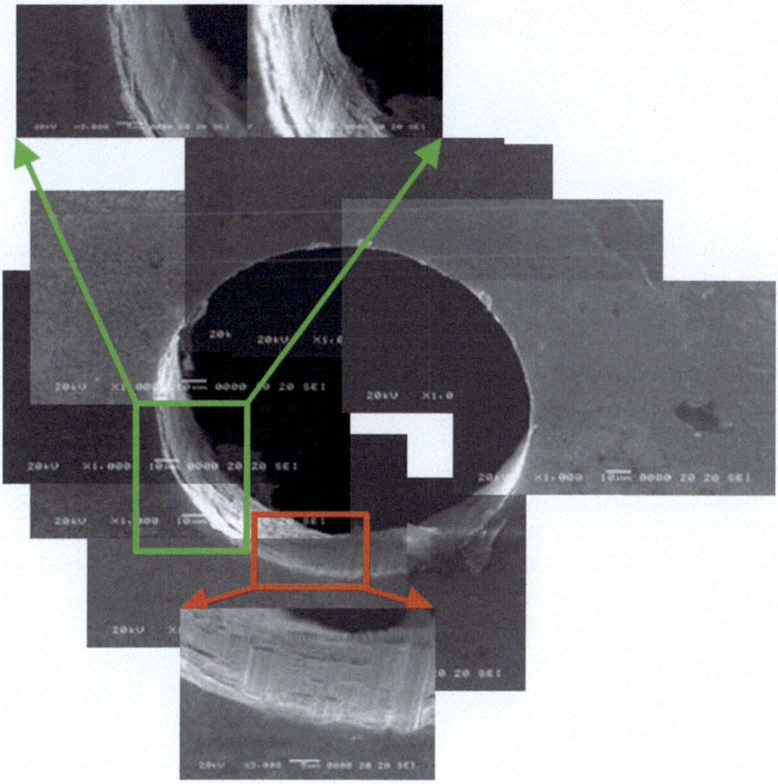

Fig. 10.10 Fractrography of the inner microhole formed by micropunching

It indicates that how the punch and foil interact during punching at microscale must be quite different from that at millimeter-scale punching. Therefore, future investigation shall be pinpointed to look into better understanding of the related mechanism of punch-foil interaction in micropunching, which can

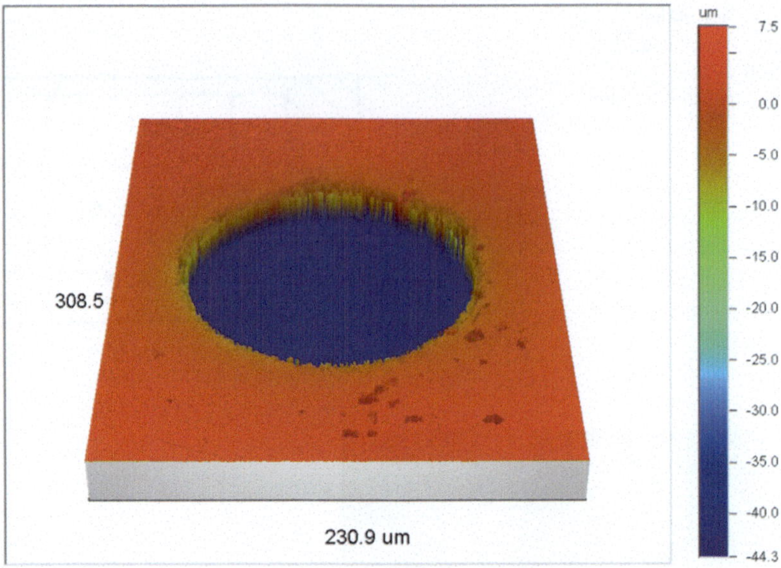

Fig. 10.11 3D topography of the punched microhole

ultimately lead to the improved micropunches and the quality of the punched microholes simultaneously.

10.5 Microhole Potential for Bioenergy

Bioenergy known as the promising renewable energy is generated from materials derived from biological sources. Among bioenergy, as a kind of renewable energy source, biomass is attained from the biological materials of the living including both animals and plants. Up to now, the energy such as heat or electricity is generated by burning biomass to produce the steam or convert to other useful things such as ethanol, methane, etc. At present, there are a lot of biomass power plants all over the world with the increasing share of usage of electricity. The relevant power plant performance is evaluated through energetic performance criteria: electrical power and thermal efficiency. As an effective method in the power plants, microholes can beneficially enhance fluid flow and heat exchange, especially for the turbines. Moreover, materials with microholes of special physicochemical properties can be applied in energy conversion and storage devices to distinctively improve the related efficiency, especially for the electrochemical areas, along with the minimal environmental impact.

10.6 Conclusion

A dense layer of packed and tangled MWCNTs was grown on the working regions of WC/Co micropunches by the waste alcohol eco-friendly to the environment, and the height of the MWCNTs is about 15 μm with the diameter of about 3–5 nm. In the micropunching process, the weight loss of MWCNT-coated micropunches could be divided into the initial run-in stage, quasi-stable stage, and severe wear stage, just like the case of micropunches without MWCNT coating. MWCNT coating helps to lubricate the micropunch and specimen interface. Moreover, MWCNT coating could extend the length of the quasi-stable stage by about 250 shots (or 35%). In addition, the attractive features of punched microhole are definitely quite distinct from that of traditional punching. Besides the fracture band is absent in the punching of a 150 μm hole in titanium with the thickness of 200 μm, the areas of rollover and burr are also extremely narrow. Meanwhile, the diameter of the front side of the punched microhole is almost the same as that of the back side even in the severe period, which implies the punched microhole can be directly used in biomass power plants without any posttreatments.

References

Abed WM, Domingues AF, Poole RJ, Dennis DJC (2017) Heat transfer enhancement in a cross-slot micro-geometry. Int J Therm Sci 121:249–265

Affes F, Ketata H, Kharrat M, Dammak M (2018) How a pilot hole size affects osteosynthesis at the screw-bone interface under immediate loading. Med Eng Phys 60:14–22

Ali ARA, Janajreh I (2015) Numerical simulation of turbine blade cooling via jet impingement. Energy Procedia 75:3220–3229

Ali A, Khan ZS, Jamil M, Khan Y, Ahmad N, Ahmed S (2018) Simultaneous reduction and sulfonation of graphene oxide for efficient hole selectivity in polymer solar cells. Curr Appl Phys 18(5):599–610

Al-Zurfi N, Turan A (2016) LES of rotational effects on film cooling effectiveness and heat transfer coefficient in a gas turbine blade with one row of air film injection. Int J Therm Sci 99:96–112

Amirante R, Distaso E, Tamburrano P (2014) Experimental and numerical analysis of cavitation in hydraulic proportional directional valves. Energy Convers Manag 87:208–219

Asim T, Mishra R, Oliveira A, Charlton M (2019) Effects of the geometrical features of flow paths on the flow capacity of a control valve trim. J Pet Sci Eng 172:124–138

Baghel R, Mali HS (2018) A study on effects of discharge energy on geometric characteristics of high aspect ratio micro-holes on TiN-Al_2O_3 ceramics. Mater Today Proc 5(9, Part 3):17828–17837

Baik KD, Seo IS (2018) Metallic bipolar plate with a multi-hole structure in the rib regions for polymer electrolyte membrane fuel cells. Appl Energy 212:333–339

Baino F, Potestio I (2016) Orbital implants: state-of-the-art review with emphasis on biomaterials and recent advances. Mater Sci Eng C 69:1410–1428

Bakshi SR, Keshri AK, Agarwal A (2011) A comparison of mechanical and wear properties of plasma sprayed carbon nanotube reinforced aluminum composites at nano and macro scale. Mater Sci Eng A 528:3375–3384

Banyai D, Sfarlea I, Opruta D (2016) Experimental research on variable hydraulic resistors of servo-hydraulic valves. Energy Procedia 85:44–51

Belarb E, Blas-Ferrando VM, Haro M, Maghraoui-Meherzi H, Gimenez S (2016) Electropolymerized polyaniline: a promising hole selective contact in organic photoelectrochemical cells. Chem Eng Sci 154:143–149
Bellemère P (2018) Pyrocarbon implants for the hand and wrist. Hand Surg Rehab 37(3):129–154
Bhattacharya R (2018) Generation of phase singular optical beams in microstructure optical fibers. Opt Commun 428:15–21
Bhattacharyya B (2015) Chapter 12. Recent advancements in EMM for micro and nanofabrication. In: Electrochemical micromachining for nanofabrication, MEMS and nanotechnology, pp 219–240
Biasetti DA, Di Liscia EJ, Torchia GA (2017) Optical waveguides fabricated in Cr:LiSAF by femtosecond laser micromachining. Opt Mater 73:25–32
Block MS (2018) Dental implants: the last 100 years. J Oral Maxillofac Surg 76(1):11–26
Boncel S, Müller KH, Skepper JN, Walczak KZ, Koziol KKK (2011) Tunable chemistry and morphology of multi-wall carbon nanotubes as a route to non-toxic, theranostic systems. Biomaterials 32:7677–7686
Brousseau EB, Dimov SS, Pham DT (2010) Some recent advances in multi-material micro- and nano-manufacturing. Int J Adv Manuf Technol 47:161–180
Calabriso A, Cedola L, Zotto LD, Rispoli F, Santori SG (2015) Performance investigation of passive direct methanol fuel cell in different structural configurations. J Clean Prod 88:23–28
Calvo-López A, Puyol M, Casalta JM, Alonso-Chamarro J (2017) Multi-parametric polymer-based potentiometric analytical microsystem for future manned space missions. Anal Chim Acta 995:77–84
Chavoshi SZ, Luo XC (2015) Hybrid micro-machining processes: a review. Precis Eng 41:1–23
Cherevko S, Kulyk N, Mayrhofer KJJ (2016) Durability of platinum-based fuel cell electrocatalysts: dissolution of bulk and nanoscale platinum. Nano Energy 29:275–298
Choi YJ, Oh EK, Lee MJ, Kim NJ, Choung HY, Khwarg SI (2018) Atypical mycobacterial infection in anophthalmic sockets with porous orbital implant exposure. Am J Ophthalmol 195:131–142
Christwardana M, Frattini D, Accardo G, Yoon SP, Kwon YC (2018) Early-stage performance evaluation of flowing microbial fuel cells using chemically treated carbon felt and yeast biocatalyst. Appl Energy 222:369–382
Chung H, Park JS, Sohn HS, Rhee DH, Cho HH (2014) Trailing edge cooling of a gas turbine blade with perforated blockages with inclined holes. Int J Heat Mass Transf 73:9–20
Cole KM, Kirk DW, Singh CV, Thorpe SJ (2017) Optimizing electrochemical micromachining parameters for Zr-based bulk metallic glass. J Manuf Process 25:227–234
Cortes DF, Tang TX, Capelluto DGS, Lazar IM (2017) Nanoflow valve for the removal of trapped air in microfluidic structures. Sensors Actuators B Chem 243:650–657
Cristofari F, Piotrowski B, Pesci R (2017) Mechanical properties of a nanoporous membrane used in implantable medical devices. Correlation between experimental characterization and 2D numerical simulation. J Mech Behav Biomed Mater 74:43–53
Damnjanović M, Vuković T, Milosević I (2002) Super-slippery carbon nanotubes: symmetry breaking breaks friction. Eur Phys J B 25:131–134
De Volder MF, Tawfick SH, Baughman RH, Hart AJ (2013) Carbon nanotubes: present and future commercial applications. Science 339:535–539
de Vries SF, Florea D, Homburg FGA, Frijns AJH (2017) Design and operation of a tesla-type valve for pulsating heat pipes. Int J Heat Mass Transf 105:1–11
Derkus B (2016) Applying the miniaturization technologies for biosensor design. Biosens Bioelectron 79:901–913
di Capaci RB, Vaccari M, Pannocchia G (2018) Model predictive control design for multivariable processes in the presence of valve stiction. J Process Control 71:25–34
Dickrell PL, Sinnott SB, Hahn DW, Raravikar NR, Schadler LS, Ajayan PM, Sawyer WG (2005) Frictional anisotropy of oriented carbon nanotube surfaces. Tribol Lett 18(1):59–62. https://doi.org/10.1007/s11249-004-1752-0

Dickrell PL, Pal SK, Bourne GR, Muratore C, Voevodin AA, Ajayan PM, Schadler LS, Sawyer WG (2006) Tunable friction behavior of oriented carbon nanotube films. Tribol Lett 24(1):85–90
Dittmeyer R, Boeltken T, Piermartini P, Selinsek M, Loewert M, Dallmann F, Kreuder H, Cholewa M, Wunsch A, Belimov M, Farsi S, Pfeifer P (2017) Micro and micro membrane reactors for advanced applications in chemical energy conversion. Curr Opin Chem Eng 17:108–125
Do KH, Kim T, Han YS, Choi BI, Kim M (2017) Investigation on flow distribution of the fuel supply nozzle in the annular combustor of a micro gas turbine. Energy 126:361–373
Dornfeld D, Min S, Takeuchi Y (2006) Recent advances in mechanical micromachining. CIRP Ann Manuf Technol 55(2):745–768
Dubois A, Ricote S, Braun RJ (2017) Benchmarking the expected stack manufacturing cost of next generation, intermediate-temperature protonic ceramic fuel cells with solid oxide fuel cell technology. J Power Sources 369:65–77
Dupuis AC (2005) The catalyst in the CCVD of carbon nanotubes – a review. Prog Mater Sci 50:929–961
Dvořák M, Seip KF, Pedersen-Bjergaard S, Kubáň P (2018) Semi-automated set-up for exhaustive micro-electromembrane extractions of basic drugs from biological fluids. Anal Chim Acta 1005:34–42
Edvardsen S, Dorao CA, Nydal OJ (2015) Multiphase flow in complex valve geometry. Energy Procedia 64:91–100
Edwards SA (2006) The nanotech pioneers. Angew Chem 45(43):7116–7117
Enagi II, Al-attab KA, Zainal ZA (2017) Combustion chamber design and performance for micro gas turbine application. Fuel Process Technol 166:258–268
Farzaneh-Gord M, Khoshnazar H (2016) Valve fault detection for single-stage reciprocating compressors. J Nat Gas Sci Eng 35(Part A):1239–1248
Ferraris S, Spriano S (2016) Antibacterial titanium surfaces for medical implants. Mater Sci Eng C 61:965–978
Ferraro D, Serra M, Ferrante I, Viovy JL, Descroix S (2018) Microfluidic valve with zero dead volume and negligible back-flow for droplets handling. Sensors Actuators B Chem 258:1051–1059
Ferreira ADBL, Nóvoa PRO, Marques AT (2016) Multifunctional material systems: a state-of-the-art review. Compos Struct 151:3–35
Fischer F (2018) Photoelectrode, photovoltaic and photosynthetic microbial fuel cells. Renew Sust Energ Rev 90:16–27
Fraas M, Glasenapp T, Schulz A, Bauer HJ (2019) Optimized inlet geometry of a laidback fan-shaped film cooling hole - experimental study of film cooling performance. Int J Heat Mass Transf 128:980–990
Fukuzumi S (2016) Artificial photosynthesis for production of hydrogen peroxide and its fuel cells. Biochimica et Biophysica Acta (BBA) – Bioenergetics 1857(5):604–611
Gaona-Reyes JL, Bermudez D (2017) The theory of optical black hole lasers. Ann Phys 380:41–58
Gehlot R, Tripathi B (2016) Thermal analysis of holes created on ceramic coating for diesel engine piston. Case Stud Thermal Eng 8:291–299
Gomis-Berenguer A, Velasco LF, Velo-Gala I, Ania CO (2017) Photochemistry of nanoporous carbons: perspectives in energy conversion and environmental remediation. J Colloid Interface Sci 490:879–901
Green MJ, Behabtu N, Pasquali M, Adams WW (2009) Nanotubes as polymers. Polymer 50:4979–4997
Guiderdoni C, Pavlenko E, Turq V, Weibel A, Puech P, Estournès C, Peigney A, Bacsa W, Laurent C (2013) The preparation of carbon nanotube (CNT)/copper composites and the effect of the number of CNT walls on their hardness, friction and wear properties. Carbon 58:185–197
Guler MT, Beyazkilic P, Elbuken C (2017) A versatile plug microvalve for microfluidic applications. Sensors Actuators A Phys 265:224–230
Guo W, Tam HY (2011) Effects of extended punching on wear of the WC/Co micropunch and the punched microholes. Int J Adv Manuf Technol 59:955–960. https://doi.org/10.1007/s00170-011-3567-0

Gutierrez F, Rubianes MD, Rivas GA (2012) Dispersion of multi-wall carbon nanotubes in glucose oxidase: characterization and analytical applications for glucose biosensing. Sensors Actuators B Chem 012(161):191–197
Haghighi F, Talebpour Z, Nezhad AS (2018) Towards fully integrated liquid chromatography on a chip: evolution and evaluation. TrAC Trends Anal Chem 105:302–337
Hakimi O, Ventura Y, Goldman J, Vago R, Aghion E (2016) Porous biodegradable EW62 medical implants resist tumor cell growth. Mater Sci Eng C 61:516–525
Hashemi SMH, Neuenschwander M, Hadikhani P, Modestino MA, Psaltis D (2017) Membraneless micro fuel cell based on two-phase flow. J Power Sources 348:212–218
Hatsuzawa T, Ito D, Nisisako T, Yanagida Y (2017) Micro-rotary ratchets driven by migratory phytoplankton with phototactic stimulus. Precis Eng 48:107–113
Hayes SA, Nix AC, Nestor CM, Billups DT, Haught SM (2017) Experimental investigation of the influence of freestream turbulence on an anti-vortex film cooling hole. Exp Thermal Fluid Sci 81:314–326
Hilber W, Jakoby B (2015) Ethanol fermentation as the basis for autonomous, long-term and high-pressure fluid transport in microfluidics. Proc Eng 120:100–105
Ho KC, Teow YH, Mohammad AW, Ang WL, Lee PH (2018) Development of graphene oxide (GO)/multi-walled carbon nanotubes (MWCNTs) nanocomposite conductive membranes for electrically enhanced fouling mitigation. J Membr Sci 552:189–201
Hoysall DC, Garimella S (2018) Investigation of a serpentine micro-pin fin heat and mass exchanger for absorption systems. Int J Refrig 93:108–121
Hsu MC, Alfadhel A, Forouzandeh F, Borkholder DA (2018) Biocompatible magnetic nanocomposite microcapsules as microfluidic one-way diffusion blocking valves with ultra-low opening pressure. Mater Des 150:86–93
Hvizdoš P, Puchý V, Duszová A, Dusza J, Balázsi C (2012) Tribological and electrical properties of ceramic matrix composites with carbon nanotubes. Ceram Int 38:5669–5676
Iijima S (1991) Helical microtubules of graphitic carbon. Nature 354:56–58
Iijima S (1993) Single-shell carbon nanotubes of 1-nm diameter. Nature 363:603–605
Jäckel N, Weingarth D, Schreiber A, Krüner B, Zeiger M, Tolosa A, Aslan M, Presser V (2016) Performance evaluation of conductive additives for activated carbon supercapacitors in organic electrolyte. Electrochim Acta 191:284–298
Janssen PJM, Hoefnagels JPM, de Keijser TH, Geers MGD (2008) Processing induced size effects in plastic yielding upon miniaturization. J Mech Phys Solids 56:2687–2706
Johnstone RW, Parameswaran M (2004) An introduction to surface-micromachining. Kluwer Academic, Dordrecht
Kant R, Singh D, Bhattacharya S (2017) Digitally controlled portable micropump for transport of live micro-organisms. Sensors Actuators A Phys 265:138–151
Kenig EY, Su YH, Lautenschleger A, Chasanis P, Grünewald M (2013) Micro-separation of fluid systems: a state-of-the-art review. Sep Purif Technol 120:245–264
Kim JH, Kim KY (2018) Film-cooling performance of converged-inlet hole shapes. Int J Therm Sci 124:196–211
Kim T, Park S (2018) Modeling flash boiling breakup phenomena of fuel spray from multi-hole type direct-injection spark-ignition injector for various fuel components. Energy Convers Manag 160:165–175
Knobloch H, Kaminorz Y (2004) MicroNano integration. Springer-Verlag, Berlin/Heidlberg
Korkmaz E, Gozen BA, Bediz B, Ozdoganlar OB (2017) Accurate measurement of micromachining forces through dynamic compensation of dynamometers. Precis Eng 49:365–376
Kowsari K, Amini MH, Papini M, Spelt JK (2016) The effects of fluid vapor pressure and viscosity on the shapes of abrasive slurry-jet micro-machined holes and channels. Int J Mach Tools Manuf 110:80–91
Krewinkel R (2013) A review of gas turbine effusion cooling studies. Int J Heat Mass Transf 66:706–722

Krishnamoorthi M, Malayalamurthi R (2018) Engine characteristics analysis of chaulmoogra oil blends and corrosion analysis of injector nozzle using scanning electron microscopy/energy dispersive spectroscopy. Energy 165,. Part B:1292–1319

Kudźma Z, Stosiak M (2015) Studies of flow and cavitation in hydraulic lift valve. Arch Civil Mech Eng 15(4):951–961

Kumar R, Kumar A, Singh I (2018) Electric discharge drilling of micro holes in CFRP laminates. J Mater Process Technol 259:150–158

Kuriakose S, Mangalan AV, Namboothiri B, Ray A (2016) Micro machining process selection: an integrated theory. Proc Technol 25:862–868

Lee S, Hwang W, Yee KJ (2018) Robust design optimization of a turbine blade film cooling hole affected by roughness and blockage. Int J Therm Sci 133:216–229

Li PC, Hu CC, You TH, Chen PY (2017) Development and characterization of bi-functional air electrodes for rechargeable zinc-air batteries: effects of carbons. Carbon 111:813–821

Lisowski E, Filo G (2016) CFD analysis of the characteristics of a proportional flow control valve with an innovative opening shape. Energy Convers Manag 123:15–28

Liu YH, Chang TH, Wang CC (2016) Heat transfer enhancement of an impinging synthetic air jet using diffusion-shaped orifice. Appl Therm Eng 94:178–185

Luttge R (2016) Chapter 2. Basic technologies for microsystems. In: Nano- and microfabrication for industrial and biomedical applications, 2nd edn. pp 11–54

Luzardo FHM, Velasco FG, Correia IKS, Silva PMS, Salay LC (2017) Removal of lead ions from water using a resin of mimosa tannin and carbon nanotubes. Environ Technol Innovat 7:219–228

Madhavi JB, Hiremath SS (2016) Investigation on machining of holes and channels on borosilicate and sodalime glass using μ-ECDM setup. Proc Technol 25:1257–1264

Makaryants GM (2017) Fatigue failure mechanisms of a pressure relief valve. J Loss Prev Process Ind 48:1–13

Malama B, Kuhlman KL, Brauchler R, Bayer P (2016) Modeling cross-hole slug tests in an unconfined aquifer. J Hydrol 540:784–796

Mamalis AG (2005) Advanced manufacturing engineering. J Mater Process Technol 161:1–9

Mark JJ (2006) Microfabrication and nanomanufacturing. CRC/Taylor & Francis, Boca Raton

Martinez S, Michaux G, Salagnac P, Bouvier JL (2017a) Micro-combined heat and power systems (micro-CHP) based on renewable energy sources. Energy Convers Manag 154:262–285

Martinez S, Michaux G, Salagnac P, Bouvier JL (2017b) Micro-combined heat and power systems (micro-CHP) based on renewable energy sources. Energy Convers Manag 154:262–285

Massarwi F, Machchhar J, Antolin P, Elber G (2018) Hierarchical, random and bifurcation tiling with heterogeneity in micro-structures construction via functional composition. Comput Aided Des 102:148–159

Mazaheri K, Zeinalpour M, Bokaei HR (2016) Turbine blade cooling passages optimization using reduced conjugate heat transfer methodology. Appl Therm Eng 103:1228–1236

McInturff P, Suzuki M, Ligrani P, Nakamata C, Lee DH (2018) Effects of hole shape on impingement jet array heat transfer with small-scale, target surface triangle roughness. Int J Heat Mass Transf 127(Part A):585–597

Mohamed GH (2006) The MEMS handbook. CRC/Taylor & Francis, Boca Raton

Mohan B, Du JG, Sim J, Roberts WL (2018) Hydraulic characterization of high-pressure gasoline multi-hole injector. Flow Meas Instrum 64:133–141

Monošík R, Angnes L (2015) Utilisation of micro- and nanoscaled materials in microfluidic analytical devices. Microchem J 119:159–168

Moreno M, Arribas AS, Bermejo E, Zapardiel A, Chicharro M (2017) Carbon nanotubes as analytical tools in capillary electromigration methods. Appl Mater Today 9:456–481

Muniak DP (2015) A proposal for a new methodology to determine inner authority of the control valve in the heating system. Appl Energy 155:421–433

Munk DJ, Kipouros T, Vio GA, Steven GP, Parks GT (2017) Topology optimisation of micro fluidic mixers considering fluid-structure interactions with a coupled Lattice Boltzmann algorithm. J Comput Phys 349:11–32

Nguyen TD (2013) Portraits of colloidal hybrid nanostructures: controlled synthesis and potential applications. Colloids Surf B: Biointerfaces 103:326–344

Nyström L, Malmsten M (2016) Surface-bound microgels-from physicochemical properties to biomedical applications. Adv Colloid Interf Sci 238:88–104

Okhotnikov I, Noroozi S, Sewell P, Godfrey P (2017) Evaluation of steady flow torques and pressure losses in a rotary flow control valve by means of computational fluid dynamics. Int J Heat Fluid Flow 64:89–102

Oliver G, Henning K, Pieter T (2004) Microsystem engineering of lab-on-a-chip devices. Angew Chem 43(34):4382–4390

Pakarzadeh H, Rezaei SM, Namroodi L (2019) Hollow-core photonic crystal fibers for efficient terahertz transmission. Opt Commun 433:81–88

Palecki S, Gorelkov S, Wartmann J, Heinzel A (2017) Frost induced damages within porous materials - from concrete technology to fuel cells technique. J Power Sources 372:204–211

Paramo E, Palmero S, Heras A, Colina A (2018) Carbon nanostructured films modified by metal nanoparticles supported on filtering membranes for electroanalysis. Talanta 178:736–742

Park S, Jung EY, Kim SH, Sohn HS, Cho HH (2016) Enhancement of film cooling effectiveness using backward injection holes. Int J Therm Sci 110:314–324

Pina VG, Amigó V, Muñoz AI (2016) Microstructural, electrochemical and tribo-electrochemical characterisation of titanium-copper biomedical alloys. Corros Sci 109:115–125

Pöllänen M, Pirinen S, Suvanto M, Pakkanen TT (2011) Influence of carbon nanotube–polymeric compatibilizer masterbatches on morphological, thermal, mechanical, and tribological properties of polyethylene. Compos Sci Technol 71:1353–1360

Pradeep A, Raj SV, Stanley J, Nair BG, Babu TGS (2018) Automated and programmable electromagnetically actuated valves for microfluidic applications. Sensors Actuators A Phys 283:79–86

Pratyusha T, Sivakumar G, Gupta D, Yella A (2017) Unique hole transport layer of PCDTBT doped with PCPDTBT for perovskite solar cells. Mater Today Proc 4(7, Part 1):6820–6826

Price C, Hamzehloo A, Aleiferis P, Richardson D (2018) Numerical modelling of fuel spray formation and collapse from multi-hole injectors under flash-boiling conditions. Fuel 221:518–541

Puch F, Hopmann C (2014) Morphology and tensile properties of unreinforced and short carbon fibre reinforced nylon 6/multiwalled carbon nanotube-composites. Polymer 55(13):3015–3025

Ramesh S, Ramirez DG, Ekkad SV, Alvin MA (2016) Analysis of film cooling performance of advanced tripod hole geometries with and without manufacturing features. Int J Heat Mass Transf 94:9–19

Ramos A, Mesnard M (2016) A new condyle implant design concept for an alloplastic temporomandibular joint in bone resorption cases. J Cranio-Maxillofac Surg 44(10):1670–1677

Ranjan R, Chou SK, Riaz F, Karthikeyan K (2017) Cold gas micro propulsion development for satellite application. Energy Procedia 143:754–761

Ravi D, Yohannes S, Koteswararao B, Satish K, Kishorebabu KS (2018) Structural analysis of down the hole button bit with different materials. Mater Today Proc 5(2), Part 1):4711–4719

Rittel D, Dorogoy A, Shemtov-Yona K (2017) Modelling dental implant extraction by pullout and torque procedures. J Mech Behav Biomed Mater 71:416–427

Rogers CI, Oxborrow JB, Anderson RR, Tsai LF, Nordin GP, Woolley AT (2014) Microfluidic valves made from polymerized polyethylene glycol diacrylate. Sensors Actuators B Chem 191:438–444

Rosace G, Trovato V, Colleoni C, Caldara M, Re V, Brucale M, Piperopoulos E, Mastronardo E, Milone C, De Luca G, Plutino MR (2017) Structural and morphological characterizations of MWCNTs hybrid coating onto cotton fabric as potential humidity and temperature wearable sensor. Sensors Actuators B Chem 252:428–439

Rupp F, Liang L, Geis-Gerstorfer J, Scheideler L, Hüttig F (2018) Surface characteristics of dental implants: a review. Dent Mater 34(1):40–57

Saadati SA, Roohi E (2015) Detailed investigation of flow and thermal field in micro/nano nozzles using Simplified Bernoulli Trial (SBT) collision scheme in DSMC. Aerosp Sci Technol 46:236–255

Samih S, Latifi M, Farag S, Leclerc P, Chaouki J (2018) From complex feedstocks to new processes: the role of the newly developed micro-reactors. Chem Eng Process Process Intensif 131:92–105

Sarafraz MM, Hormozi F, Peyghambarzadeh SM (2016) Pool boiling heat transfer to aqueous alumina nano-fluids on the plain and concentric circular micro-structured (CCM) surfaces. Exp Thermal Fluid Sci 72:125–139

Satta F, Tanda G (2015) Effect of discrete-hole arrangement on film-cooling effectiveness for the endwall of a turbine blade cascade. Appl Therm Eng 91:507–514

Saxena KK, Qian J, Reynaerts D (2018) A review on process capabilities of electrochemical micromachining and its hybrid variants. Int J Mach Tools Manuf 127:28–56

Sayindla S, Lund B, Ytrehus JD, Saasen A (2017) Hole-cleaning performance comparison of oil-based and water-based drilling fluids. J Pet Sci Eng 159:49–57

Schütze C, Dietrich P, Sauer U (2013) Diagnostic monitoring to identify preferential near-surface structures for CO_2 degassing into the atmosphere: tools for investigations at different spatial scales validated at a natural analogue site. Int J Greenhouse Gas Control 18:285–295

Sciubba E (2015) Air-cooled gas turbine cycles – part 1: an analytical method for the preliminary assessment of blade cooling flow rates. Energy 83:104–114

Sehrawat P, Julien C, Islam SS (2016) Carbon nanotubes in Li-ion batteries: a review. Mater Sci Eng B 213:12–40

Sharma DS, Patel NP, Trivedi RR (2014) Optimum design of laminates containing an elliptical hole. Int J Mech Sci 85:76–87

Shetzline JA, Bukola S, Creager SE (2017) A convenient miniature test platform for polyelectrolyte membrane fuel-cell research. J Electroanal Chem 797:8–15

Shin HS, Kwon OJ, Oh BS (2018) Correlation between performance of polymer electrolyte membrane fuel cell and degradation of the carbon support in the membrane electrode assembly using image processing method. Int J Hydrog Energy 43(45):20921–20930

Silaskar SM, Shinde VB (2018) Weight optimization of valve for cost effectiveness: using value analysis. Proc Manuf 20:329–337

Singh VK, Tharakan TJ (2015) Numerical simulations for multi-hole orifice flow meter. Flow Meas Instrum 45:375–383

Singh A, Manikandan NA, Sankar MR, Pakshirajan K, Roy L (2018) Experimental investigations and surface morphology of bio-micromachining on copper. Mater Today Proc 5(2, Part 1):4225–4234

Takahashi H, Nakai A, Shimoyama I (2018) Waterproof airflow sensor for seabird bio-logging using a highly sensitive differential pressure sensor and nano-hole array. Sensors Actuators A Phys 281:243–249

Taniguchi N, Jinno T, Takada R, Koga D, Ando T, Okawa A, Haro H (2018) Do screws and screw holes affect osteolysis in cementless cups using highly crosslinked polyethylene? A 7 to 10-year follow-up case-control study. Orthop Traumatol Surg Res 104(3):307–315

Tanimu A, Jaenicke S, Alhooshani K (2017) Heterogeneous catalysis in continuous flow microreactors: a review of methods and applications. Chem Eng J 327:792–821

Tesař V (2017) Fluid mechanics of needle valves with rounded components part I: configurations and models. Sensors Actuators A Phys 254:101–108

Tharmavaram M, Pandey G, Rawtani D (2018) Surface modified halloysite nanotubes: a flexible interface for biological, environmental and catalytic applications. Adv Colloid Interf Sci 261:82–101

Theodorakakos A, Strotos G, Mitroglou N, Atkin C, Gavaises M (2014a) Friction-induced heating in nozzle hole micro-channels under extreme fuel pressurization. Fuel 123:143–150

Theodorakakos A, Strotos G, Mitroglou N, Atkin C, Gavaises M (2014b) Friction-induced heating in nozzle hole micro-channels under extreme fuel pressurization. Fuel 123:143–150

Tiusanen J, Vlasveld D, Vuorinen J (2012) Review on the effects of injection moulding parameters on the electrical resistivity of carbon nanotube filled polymer parts. Compos Sci Technol 72:1741–1752

Tofighy MA, Mohammadi T (2011) Adsorption of divalent heavy metal ions from water using carbon nanotube sheets. J Hazard Mater 185:140–147

Torelli R, Som S, Pei YJ, Zhang Y, Traver M (2017) Influence of fuel properties on internal nozzle flow development in a multi-hole diesel injector. Fuel 204:171–184

Torre FL, Kenjereš S, Moerel JL, Kleijn CR (2011) Hybrid simulations of rarefied supersonic gas flows in micro-nozzles. Comput Fluids 49(1):312–322

Trotta G, Volpe A, Ancona A, Fassi I (2018) Flexible micro manufacturing platform for the fabrication of PMMA microfluidic devices. J Manuf Process 35:107–117

Tsur EE, Shamir A (2018) Computer-aided design of resistance micro-fluidic circuits for 3D printing. Comput Aided Des 98:12–23

Tu J, Paleocrassas AG, Reeves N, Rajule N (2014) Experimental characterization of a micro-hole drilling process with short micro-second pulses by a CW single-mode fiber laser. Opt Lasers Eng 55:275–283

Tverskoy MM, Andrianov VN, Sokolov AV (2017) Creating new generation of actuators for shut-off and control ball valves with double-gate. Proc Eng 206:1303–1308

Uhlmann E, Mullany B, Biermann D, Rajurkar KP, Hausotte T, Brinksmeier E (2016) Process chains for high-precision components with micro-scale features. CIRP Ann Manuf Technol 65:549–572

Upadhyayula VKK, Gadhamshetty V (2010) Appreciating the role of carbon nanotube composites in preventing biofouling and promoting biofilms on material surfaces in environmental engineering: a review. Biotechnol Adv 28:802–816

Virkar AV (2012) Transport through mixed proton, oxygen ion and electron/hole conductors: analysis of fuel cells and electrolyzer cells using Onsager equations. Int J Hydrog Energy 37(17):12609–12628

Willich C, White AJ (2017) Heat transfer losses in reciprocating compressors with valve actuation for energy storage applications. J Energy Storage 14., Part(2):322–328

Yadav SP, Singh S (2016) Carbon nanotube dispersion in nematic liquid crystals: an overview. Prog Mater Sci 80:38–76

Zangl K, Danzl R, Helmli F, Prantl M (2018) Highly accurate optical μCMM for measurement of micro holes. Proc CIRP 75:397–402

Zhai P, Isaacs JA, Eckelman MJ (2016) Net energy benefits of carbon nanotube applications. Appl Energy 173:624–634

Zhan GD, Kuntz JD, Wan JL, Mukherjee AK (2003) Single-wall carbon nanotubes as attractive toughening agents in alumina-based nanocomposites. Nat Mater 2(1):38–42

Synthesis of Iron Oxide Nanomaterials for Biofuel Applications

11

Misbah Ghazanfar and Muhammad Irfan

Abstract

The world is paying attention on exploring and establishing alternative modes of energy production instead of utilizing fossil fuels, because of the truth that conventional fuels are being squandered and the liberated chemicals of these fuels have become detrimental to the environment. In this context, production of biofuels can solve the issue, and nanotechnology can play a significant role for biofuel industry. Iron oxide nanomaterials are helpful and have great potential for enhanced production of biofuels. Many investigations have been conducted on the synthesis and use of iron oxide nanomaterials. There are different methods to synthesize iron oxide nanomaterials, and they are efficiently used in biofuel synthesis.

Keywords

Nanotechnology · Iron oxide · Biofuel · Nanomaterials · Nanoparticles · Bioethanol · Biodiesel · Renewable · Production

11.1 Introduction

Almost every country today is facing the issue of current energy requirements and future supplies (Waqas et al. 2018). Recently, more than around 75% of this energy is formed from nonrenewable sources. We are disregarding the outcomes of overuse of natural resources on our environment and the necessities of our coming generations (Nizami et al. 2017). Environmental concerns and the rapid exploitation of the fossil fuels lead to exploration of a substitute fuel (Balat and Balat 2010). Raising need for fossil fuel from the last several years consequences in the decline

M. Ghazanfar · M. Irfan (✉)
Department of Biotechnology, University of Sargodha, Sargodha, Pakistan

M. Srivastava et al. (eds.), *Nanomaterials in Biofuels Research*, Clean Energy Production Technologies, https://doi.org/10.1007/978-981-13-9333-4_11

of natural resource availability. Moreover, fossil fuel has a major issue like the discharge of carbon monoxide, which causes global warming. So, there is a greater need for alternate fuel (Ghazanfar et al. 2018).

Environmental, economic, and energy safety worries developing from imprudent dependence on petroleum products are compelling almost all the countries on the earth to find alternative fuel in the form of biodiesel and bioethanol (Farrell et al. 2006). To make progress toward renewable and sustainable energy systems, there have been rigorous development and research work done in the field of biofuels. It is considered that biofuel industry has a significant potential part in dealing the issues of climate change and atmospheric pollution, along with the renewable energy production as it is quickly developing (Rai and Da Silva 2017).

The establishment of renewable and substitute fuels is a very critical research area in the last several years. Biomass is earning extreme attraction as a renewable energy source, because of the world fuel need and distresses on the climatic influence of production of carbon dioxide. Biomass has innumerable advantages as compared to fossil fuels, e.g., low emissions of oxides of sulfur and nitrogen, carbon dioxide, locally produced, harmless, as well as renewable (Zhang et al. 2011). Advance living standards demand elevated mobility rate, and, as a result, transport is one of the biggest and speedy establishing energy-demanding areas. Increase in competitive agribusiness automatization also leads to a high energy need. However, due to anxiety on the negative impact of fossil fuels on the environment, the use of biofuels appears as an encouraging alternative that is moderately becoming technically and economically feasible. The development of lifestyle forces the chase for continual energy so that world energy demands across the world can be fulfilled (Demirbas 2010), while global problems such as global warming and environmental pollution are also caused by the utilization of fossil fuels because they burnt as the chief means of energy (Kiran et al. 2014). This situation leads to the discovery of eco-favorable, sustainable, and renewable energy by industrial, government, and energy areas (Shafiee and Topal 2009).

Liquid biofuels were preferred over others among renewable energies, as they constitute approximately 40% of the global energy consumption (Tan et al. 2008) because the utilization of liquid biofuels also causes a decline in liberation of greenhouse gas, making opportunities for job, supply security and regional development (Demirbas and Balat 2006).

Solar, wind, geothermal, and biomass are different options which constitute renewable energy. Due to the frequent availability of plant residues that can be employed to produce gaseous and liquid fuels or electricity, the attention in the latter has been raised. Utilization of this green waste can be in the form of direct biomass burning to produce electricity and vapor or after few alterations that permit a better employment of this energy supply (Antunes et al. 2017).

Researchers are making efforts to enhance overall constituents of the biofuel industry such as pretreatment of feedstock/biomass, process conditions and parameters, designs of reactor, yields and quality of product, and the inclusive optimization of process, in addition to the capital investment, approval by public, and availability for market for different biofuels (Hussein 2015).

Nonetheless, the strategies for the handling of biomass linger restricted by economical and technological barricades. Therefore, novel and innovative strategies and perspectives are essential to attain viable operations as well as bigger and quicker alterations of world energy pattern (Antunes et al. 2017). Nanotechnology is a generative field of investigation and research among the new developments, which includes auspicious potential in several application zones of industrial and social benefits.

Nanotechnology along other connecting technologies (information technology, cognitive sciences, and biotechnology) is capable to enhance human living standard having implementations in various fields such as life sciences, electronics, pharmaceutical and material development, etc. (Demetzos 2016; Wolf and Medikonda 2012). Due to their dynamic physicochemical characteristics, nanomaterials have the prospective for their commercial utilization. In the future, it is predicted that the commercial significance of nanoparticles will elevate distinctly and probably it would rise from 2000 tons as in 2004 to around 58000 tons in 2011–2020 (Shi and Ma 2010, 2011). Various fields including industries belonged to biofuel production and environmental remediation; there are vast applications of nanoparticles (Srivastava et al. 2014b, a). Efficacy of green fuels like bioethanol/biohydrogen can be improved by employing nanoparticles; thus nanomaterials can be supportive in the improvement of the production of biofuels (Srivastava et al. 2015a). Studies also revealed that the production of cellulases, their strength, and catalytic ability can also be enhanced employing nanomaterials. Furthermore, the presence of nanoparticles also enhances the pretreatment of lignocellulosic biomass (Srivastava et al. 2015a). Iron oxide nanomaterials have several crystalline phases, relying on their stoichiometry, such as goethite [$FeO(OH)$], ferrihydrite [$Fe_5HO_8(4H_2O)$], magnetite (Fe_3O_4), wüstite (FeO), hematite (a-Fe_2O_3), maghemite (g-Fe_2O_3), etc. Fe_3O_4 and g-Fe_2O_3 are the most investigated iron oxide nanomaterials (Ubale and Belkhedkar 2015), which are used extensively in different applications (Cavas et al. 2013; Haddad et al. 2015; Jia et al. 2005) as well as in pretreatment and hydrolysis of biomass to produce sugars (Srivastava et al. 2017a). Figure 11.1 indicated various applications of nanomaterials in the process of production of biofuels. According to the truth as mentioned prior, the purpose of this chapter is to describe the synthesis of iron oxide and its applications in enhancing biofuel production. In present chapter, we tried to project an overview of the biofuels and synthesis of iron oxide nanomaterials and their implementations in biofuel production.

11.2 Worlds' Fuel Demand and Biofuels

Presently, world energy requirements are executed primarily by hydro, natural gas, coal, nuclear, oil, and inexhaustible means (Brazilian Energy Balance 2015). From 2002 to 2030, the requirement of biofuels are expected to rise up to 57%. In 2005, the average cost of patrol was \$2.56 per gallon, which was \$0.67 greater than the average cost of patrol in the earlier year. Until now the average cost of petrol in the United States increased up to \$4.10 per gallon (Kumar et al. 2009). Investigations

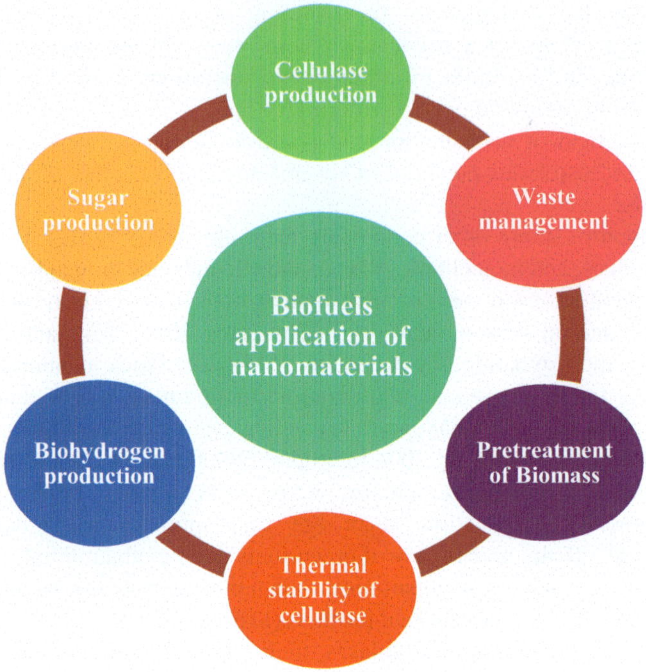

Fig. 11.1 Different applications of nanomaterials in production of biofuels

for substitute energy means, particularly that obtained from inexhaustible matter, e.g., biomass, accelerated as the global energy need and oil reserves shrink significantly (Saxena et al. 2009).

Total world energy chief production has been raised from the last decades, and currently the major producers are (MTOE) Saudi Arabia (630), Russia (1334), Australia (357), China (2555), Brazil (26), Indonesia (457), the United States (1989), Iran (308), Canada (452), and India (571) (Yearbook Enerdata 2015a).

Reviewing various energy sources, energy supply of the world is fulfilled principally by oil, chased by naturally occurring gas and coal. Figure 11.2 illustrates the profile of world energy supply in 1971 and 2013 which distinctly depicted that the planet greatly relies on traditional fuels.

Whereas, from 40 years, renewable energy contributes for only about 10% of total in world energy supply (Brazilian Energy Balance 2015). Regardless of this scenario, green energy is being used in several countries. For instance, electricity production from inexhaustible energy is the foremost source of energy production in Portugal (62.6%), Brazil (73.4%), Chile (42.8%), New Zealand (79%), Sweden (58.5%), Norway (98%), Venezuela (62.8%), Italy (42.1%), Colombia (70%), and Canada (62.5%) (Yearbook Enerdata 2015b). Furthermore, share of solar and wind in electricity production is powerfully dispersed in Romania (9.6), Belgium (10.0),

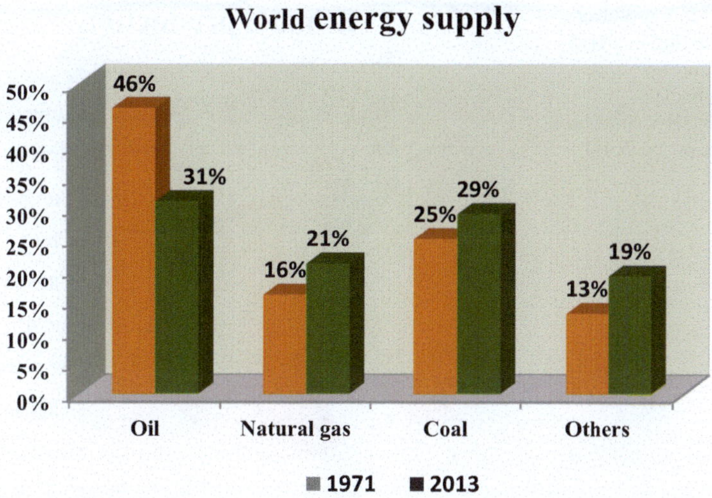

Fig. 11.2 Profile of world energy supply in 1971 and in 2013. (Source: Based on International energy agency from the Key World Energy Statistics © OECD/IEA 2015, www.iea.org/statistics. Licence: www.iea.org/t&c; As redesign by present authors)

Portugal (24.5), Spain (23.9), the Netherlands (6.6), the United Kingdom (10.3), Germany (15.2), Sweden (7.6), Italy (16.7), and New Zealand (21.7) (Yearbook Enerdata 2015c).

Recently biofuels have secured much attention and are being produced with enhanced production in different countries. For instance, the United States applying corn as a crude matter and Brazil applying sugarcane juice as a source of carbon are the bigger producers of bioethanol in the world by producing around 14,000 and 7000 million of gallons per year, respectively. The overall bioethanol production in the world is approximately 25 million gallons annually (Renewable Fuels Association 2015). In 2006, Goldemberg (2006) suggested that carbon liberation can be decreased up to 66 million tons annually if bioethanol formed from juice of sugarcane can be substituted 10% of total utilized patrol around the globe. Another principal biofuel is biodiesel and is commonly synthesized through a transesterification reaction which is catalyzed chemically, between oils and fats from plants or animals and low chain length alcohols (mainly methanol), and the synthesis of this compound is improved too. For instance, currently European Union is producing around 3.35 billion liters of biodiesel, and in 2025, this production can be elevated up to nearly 4.14 billion gallons (Statista, The statistic portal 2016a). The largest producers of biodiesel in the world in 2014 were the United States (4.7 billion liters) chased by Germany (3.4 billion liters), Brazil (3.4 billion liters), Indonesia (3.1 billion liters), and Argentina (2.9 billion liters) (Statista, The statistic portal 2016b). In Fig. 11.3 we can observe an outline of the major producers of renewable fuel in the world in 2015.

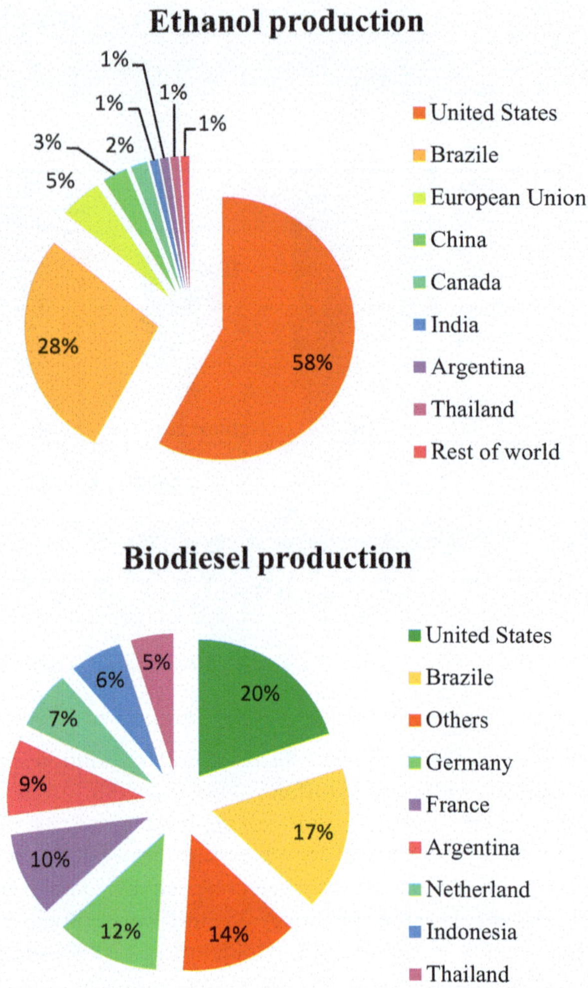

Fig. 11.3 Main producers of renewable fuel in world in 2015. (Source: Renewable Fuels Association 2015; Statista, The statistic portal 2016b)

Lignocellulosic biomass is the most abundant renewable source of carbon on earth as it is frequently found in plant residues, hence accounted as potential material for biofuel production. It is estimated that this green waste is generated from 10–50 billion tons annually and about millions of tons of predominant biomass can be feasibly reutilize (Zhao et al. 2012). For self-assisted formation of energy, biomass is burnt in boilers in different industrial processes. Cellulose and hemicellulose are found in green biomass and constitute its two-thirds polymeric carbohydrates (Antunes et al. 2014) and therefore can be employed in bioprocessing for production of biofuels and other valuable products.

So the green technology along with an advance approach like nanotechnology has immense prospective to contribute various commercial zones with huge need of synthesis within the necessitated conception of feasible development methods.

11.3 Biomass to Biofuels

Although field of synthesis of biofuels receives huge importance for the substitute of fossil fuels, it reinforces to lessen the atmospheric pollution, yet away from the viable implementations due the large formation price. Therefore, uninterrupted attempts are being made for the cheap formation of biofuels all around the globe to balance its economy. So, if we use lignocellulosic biomass as substrate, it can be cost-effective. This substrate is a biological waste; cost-effective, huge source of carbon, inexhaustible in character, and easily found on the earth in large amount. These lignocellulosic biowastes can be easily utilized as the substrate for the production of biofuels (Srivastava et al. 2015a, b). Over the last many years, many scientists have conducted researches on thermochemical pathways, fermentation, and biomethanation for the transformation of biomass to biofuels as origin of energy, which has recently become the area of interest. In general, the lignocellulosic biomass is a complicated mixture of organic compounds, e.g., carbohydrates (cellulose, hemicelluloses, and starch), proteins, lignin, and fats; however, the physiochemical attributes of biowaste differ in distinct pattern according to their origin. For example, the basic components of the biomass from forest or agricultural origins are lignin and carbohydrate which can differ with the variety of plant. The roots, seeds, and seed residue of plant or crop which is profuse in fats and starch are sources of biomass, whereas most of the biomass types are residues or byproducts of crops, human waste, cattle waste, municipal waste, demolition waste, construction waste, and forest remains. As it has been discussed above, the usage of biomass for liquid biofuels is an essential substitute to prevent hazardous influences of direct burning of biomass (as untreated or crude solid fuels) which can lead to air pollution, subsidiary pollution, and hazardous health effects (Verma et al. 2012).

Cellulosic biowaste can be converted into fermentable sugars by cellulase enzymes, and then the product can be additionally transformed into desirable biofuels. While cellulase enzyme performs an important role in the formation of biofuels, the cellulose enzyme production is the most costly step in the whole method of production of biofuels. So, huge expenses on cellulase enzyme formation increase the price of biofuels, and it contributes 40% of the total price (Bhalla et al. 2013; Srivastava et al. 2015a, b). Moreover, one of the major reasons to enhance the cost of formation is lack of effective preprocessing of lignocellulosic biomass (Sun and Cheng 2002). Other than the abovementioned facts, many factors like effectiveness of cellulase enzyme, its stability, and catalytic activity at different pH and medium components influence the entire process of formation of biofuels (Yeoman et al. 2010).

As a substitute origin of energy, the use of biomass fuels has become necessity of the present era to avoid the environmental pollution and the crises of fossil fuels (Barakat et al. 2012; Barnard et al. 2010). The transformation of these biomasses into fuels has enticed huge concentrations because they are inexhaustible in nature and have feasible origin of cellulose and there is broad presence of these lignocellulosic biomasses. Therefore, greater than 20,000 scientific research papers rely on the ISI Web of Knowledge database, and approximately 3000 international

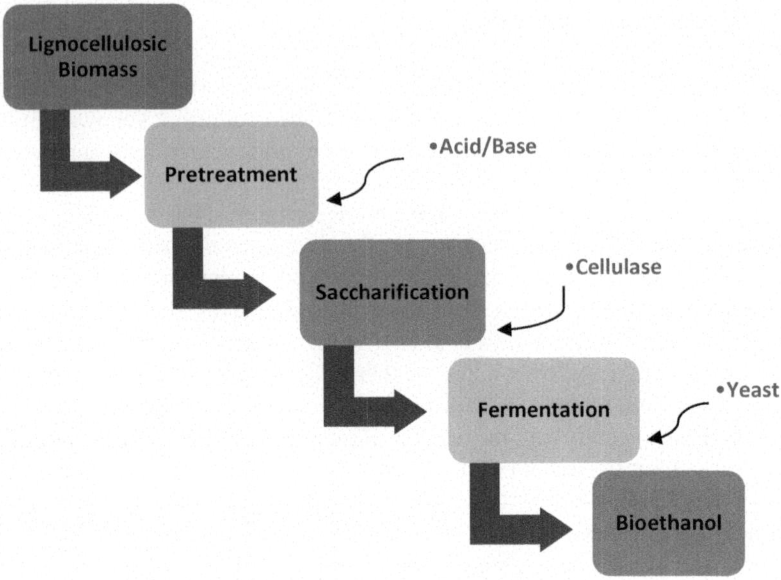

Fig. 11.4 Overview of biofuel productions from lignocellulosic biomass

patents, according to the SpaceNet database, have been published in this field till date (Chaker Ncibi and Sillanpaa 2013). The immense amount of research publications and patent undoubtedly recommend conversion of biowaste into biofuels is among the most provoking research fields to further search its practical application. Figure 11.4 represents flowchart of process of formation of biofuels using the lignocellulosic biomass as the substrate. Cellulose-containing biomass undergoes suitable pretreatment process and then by the action of cellulolytic enzymes converted into fermentable sugar, and then fermentive microorganisms use these sugars and produce bioethanol/biohydrogen through the fermentation process.

In many countries in the world, the production of biodiesel from the transesterification of animal fats, vegetable oils, and discarded frying oils has increased principally, because transesterification is one of the most common procedures employed to decrease viscosity of oil. This is actually the manufacturing of a fuel, biodiesel, comprised of mono-alkyl esters of long-chain fatty acids. Biodiesel, because of its strategic, technical, and environmental advantages, is the most vastly believed substitute fuel for diesel engines (Enweremadu et al. 2011). Furthermore biodiesel is strategically powerful as traditional, petroleum-obtained diesel fuel and needs effectively no changes in the fuel dispensation fundaments. Decline of most dissipated liberation, greater flash point, inherent lubricity, and biodegradability are among the other benefits of biodiesel compared to petro-diesel (Haas et al. 2001), and the truth that it is of local emergence. To avoid pollution and retrench the biodiesel synthesis with the raising requirement of biodiesel, it is indispensable to minimize loads on the eatable oil and to produce biodiesel from discarded non-eatable fatty acids. Thermal transformation processes of biowaste involve direct burning, thermal breakdown, liquefaction, and gasification. Quick thermal breakdown of biomass is supposed to be a favorable thermochemical method in which biomass is transformed to pyrolytic oil, gas, and char (Zhang et al. 2011). With advantages of unburned hydrocarbons, small level of smoke, and carbon monoxide than fossil fuel-derived diesel, biodiesel is accepted as an inexhaustible and atmosphere-friendly fuel compassionate to petro-diesel. With biodiesel it is feasible to attain virtuous burning, consequently reduction in level of waste release by utilizing oxygenates as supplement antioxidants and nanomaterials as additives in green diesel. Currently, several investigators concentrated their focus on production of fuel strategy for gaining good implementation and liberation properties.

Bioethanol is the ethanol merely formed by fermentation of the sugar and starch ingredients of plants, renewable energy source. Agricultural residues of beetroot, sugarcane, rice, corn, potatoes, and currently employed grapes, banana, dates, and other wastes are being used for the production of bioethanol. The production of such biofuel is due to the declining quantity of fossil fuels, substitute energy origins required to be inexhaustible, feasible, powerful, inexpensive, appropriate, and secure (Chum and Overend 2001).

When conventional fuel became costly and environmental issues emerged, advance ethanol industry initiated in the 1970s. In 1975, the Brazilian government launched a pioneer program known as "Pro'alcool" (Pro-Alcohol) with two main objectives: to reduce the impact caused by oil price increases and, at the same time, mitigate the fall of sugar price in the international market. Between 1980 and 2002, over 5 billion dollars was invested on sugarcane agriculture and industry to expand alcohol fuel production. "Pro'alcool" is known worldwide for its positive effect on biofuel promotion. Nowadays, in Brazil, 20–25% of anhydrous ethanol is used as an additive in gasoline. Moreover, since 2003, flex-fuel vehicles, which can use alcohol, gasoline, or gasoline+alcohol, are on the market. Ethanol is the leading liquid biofuel used for transportation (Bolivar-Telleria et al. 2018). Among alternative fuels, bioethanol has an extensive history and vast application in transportation sector. Germany and France began to employ bioethanol as a substitute fuel in internal combustion engines (ICEs) in 1984. In 1925 Brazil started utilization of bioethanol. Until the early 1900s, Europe and the United States widely used bioethanol. Until the fuel crisis in the 1970s, the utilization of bioethanol was ignored after World War II because of its costly synthesis as compared to petroleum fuel. Since the 1980s, the attention toward bioethanol has been growing, and it has been believed as a good replacement in several countries. Global production of ethanol raised from 13.12 billion of gallons in 2007 to 25.68 of gallons in 2015 with a little reduction in 2012 and 2013 (Azhar et al. 2017). In 2015 with the synthesis of around 15 billion gallons, the United States is the largest ethanol producer in world. Eighty-five percent of the world's ethanol is produced by the United States and Brazil. Bioethanol is also called ethyl alcohol and chemically C_2H_5OH or EtOH. It can be utilized as mixed with patrol to form "gasohol" or can be used directly as pure ethanol (Staniszewski et al. 2007). In diesel-bioethanol mixture, bioethanol is used as a gasoline booster and octane improver and also contributes in reduction of the emission of harmful gasses (Pejin et al. 2009). Green ethanol offers various benefits over patrol, e.g., enhanced heats of vaporization, greater flame speeds, greater octane number (108), and wider flammability limits (Balat and Balat 2009). Contrary to gasoline, bioethanol is quickly biodegradable and little toxic and produces lesser atmospheric pollution (John et al. 2011). Production of bioethanol through the first, second, and third generation is carried out by using different types of feedstock. The first-generation production of bioethanol involves raw materials plentiful in sucrose such as sugar beet, sweet sorghum, sugarcane, and fruits and starch such as potato, corn, barley, cassava, wheat, rice, and sweet potato, while lignocellulosic biomass such as straw, wood, grasses, etc. are used for second-generation production of bioethanol. However algal biomass including macroalgae as well as microalgae is employed for production of third-generation bioethanol (Nigam and Singh 2011).

11.4 Nanomaterials and Biofuel Production

In 1959, the term "nanotechnology" was introduced by Feynman which refers to the synthesis and utilization of nanomaterials, e.g., materials which have at least one dimension smaller than 100 nm (Palaniappan 2017). Since then, many researchers and scientists around the world have explored the effects of nanomaterials in a wide variety of fields, ranging from power conversion equipments, integrated circuits, and food products (Wu et al. 2008). Compared to their large-scaled materials, the nanomaterials are said to have distinctive chemical, magnetic, mechanical, electronic, and optical characteristics (Biswas et al. 2012). Nanomaterials are said to perform a critical part in power generation fields because of their special structure, increased energy electrical storage capacity, comparatively great specific area, and relatively good efficacy of heating and lighting (Ansari and Husain 2012). Dimension-wise, they can be compared to biological macromolecules like enzymes or nucleic acids (Verma et al. 2013a). At the outset, it is first essential to understand the production mechanisms and basic characteristics of nanomaterials before using them to enhance biofuel production. Nanomaterials basically include a variety of substances ranging from nanocomposites, nanoparticles, nanotubes, nanosheets metal-based nanomaterials to carbon-based nanomaterials and nanocrystalline materials. There are various ways of nanomaterial production (Palaniappan 2017).

Nanoparticles are described as distinct bodies that have three dimensions of the order of 100 nanometers or less (Scenihr 2007); they own distinctive physical and chemical properties that are dramatically different from large materials (Osaka et al. 2006). By utilizing various kinds of the nanomaterials, nanotechnology has great power to increase the production of bioenergy. Nanoparticles have great chemical, electrical, and physical characteristics, which make them different from the bulk material. They are capable of enhancing the production of energy, hence serving as the solution to the issue of energy production. Nanotechnology can be employed to enhance different kinds of renewable energies such as solar, biogas, water, etc. (Hussein 2015).

Employment of nanoparticles is supposed as one of the major techniques these days to facilitate and improve biofuel formation process from waste biomass (Choedkiatsakul et al. 2011). The convenience to incorporate nanoparticles will particularly give more worth to green fuel formation method to become more sustainable by lowering the price and positive environmental impacts. Two-thirds of the lignocellulosic waste is covered by cellulose and hemicelluloses, apart from lignin which can be efficiently transformed into green fuels following the pretreatment (Alvira et al. 2010; Cai et al. 2011; Chang and Yao 2011; Hamelinck et al. 2005; Rawat et al. 2014; Srivastava et al. 2015a, b; Sun and Cheng 2002). Moreover, using the nanomaterials as the catalyst, the price of reagent used for the pretreatment of biomass may be reduced (Alvira et al. 2010; Srivastava et al. 2015a, b). The sugar production can be improved providing high yield, by using suitable pretreatment technology, which can further be used for biofuel production. That is why, by applying the suitable pretreatment process, more scientists are governing to enhance the biotransformation ability of lignocellulosic biomass. In one of the investigations by Wei et al. (2015), raised sugar formation was observed in the availability of iron oxide nanomaterials utilizing acidic pretreatment method of corn stover, employed as the substrate. This integrated acid nanocarrier-treated substrate as compared to without nanoparticles makes approximately 13 to 19% more xylose and glucose. Furthermore, results from investigations found that at 100 °C, the use of iron oxide nanoparticle for the pretreatment of substrate causes formation of increased sugar over the metallic iron. Likewise, it has been reported that Fe_3O_4–RGO–SO_3H nanocomplex can enhance the sugar formation through enzymatic breakdown (Yang et al. 2015). These researches unlock a new pathway for the cheap formation of biofuels employing the nanomaterials, but still this field is at the very initial phase. So, there should be more feasible applications of biofuels because they are very effective alternative source of energy. Highly effective cellulase system should perform in severe conditions, for the remarkable enzymatic breakdown of discarded biomass. In many researches, utilization of several cofactors like metal ions is reported for upgrading the cellulase effectiveness and its commercial formation (Srivastava et al. 2014b, c). Recently, to enhance the enzyme stability, use of nanoparticles is growing as an advance avenue in the zone of biofuels (Dutta et al. 2014; Jordan et al. 2011; Singh et al. 2016; Srivastava et al. 2014a, b). There are very restricted but promising studies that have been outlined in this research field. For instance, Dutta et al. (2014) observed an enhanced cellulase formation by bacteria in the availability of hydroxyapatite nanocarriers. The enzyme employed in this investigation was extremely thermostable which at 80 °C kept its half-life. Also, Srivastava et al. (2015c) have described an enhancement in the formation of cellulase enzyme and its thermal stability in the availability of iron (II, III) oxide/alginate nanocomplex. Furthermore, an enhanced thermal stability is also noted in the availability of nickel cobaltite ($NiCo_2O_4$) nanoparticle (Srivastava et al. 2014a). In the availability of zinc oxide nanoparticle, increased pH stability of unprocessed cellulases has been reported by Srivastava et al. (Srivastava et al. 2016). Likewise, Verma et al. (2013b) have explored the stability of the β-glucosidase enzyme against temperature, which was enhanced in the existence of iron oxide nanoparticles, displayed its half-life at 70 °C. The aforementioned investigations distinctly demonstrate that the nanomaterials have capabilities to play a significant part to change the whole biotransformation method (Ansari and Husain 2012). At normal temperature, i.e., 45–50 °C, there are various disadvantages of performing enzymatic breakdown, which include incomplete hydrolysis, overall slow reaction, and twitchy to microbial spoilage. Hence, decreased yield of fermentable sugars is generally observed and hence needs greater enzyme stuffing (Lu et al. 2006; Singh et al. 2016; Wang et al. 2010). It is accounted that, to remove the above bottlenecks; the thermostable or thermophilic enzyme is an effective way (Viikari et al. 2007; Yeoman et al. 2010). In early studies, betterment in stability of cellulase by employing nanomaterials is reported (Dutta et al. 2014; Srivastava et al. 2015c). An enhanced stability of cellulase enzyme against temperature at 80 °C was obtained in the availability of calcium hydroxyapatite nanoparticles and afterward enhanced sugar formation (Dutta et al. 2014). Furthermore, 5.71-fold more D-xylose and 15-fold more other reducing sugars have been reported by these authors in availability of the hydroxyapatite nanoparticles. In the current conducted research of Srivastava et al. (2015c), greater stability against temperature and productivity of sugar with iron (II, III) oxide/alginate nanocomposite-treated cellulase have also been observed. Though capability of nanoparticles is investigated by various investigations, their precise mechanism in reaction is however ambiguous and requires even more investigations in this field. Implementations of nanoparticles are much investigated in the field of biohydrogen formation among several subsisting biofuels (Ivanova et al. 2009; Zhang and Shen 2007). It is reported that production of biohydrogen is much complex, and microorganism-assisted method relies on various physical and chemical parameters such as inorganic sustenance comprising metal ions, working condition, and nature of substrates (Wang and Wan 2008). Moreover microorganisms, which perform significant part in the formation of biohydrogen, can receive benefit of nanomaterials in anaerobic environment, because the shifting of electron is much feasible to acceptors in presence of nanomaterials (Beckers et al. 2013). Efficacy of biohydrogen formation was enhanced in the availability of gold nanoparticle in the research by Zhao et al. (2013). In addition to silver and gold, nanocarriers can also enhance the substrate transformation effectiveness in biohydrogen process. Enhanced biohydrogen was observed in the existence of hematite nanoparticles in the investigation by Han et al. (2011). Though struggles are being made to enhance the biohydrogen formation by acquiring various strategies and different types of nanomaterials, this field is yet in grappling stage for smaller- to larger-scale study. Applications of nanomaterials in the field of biofuel formation are summarized in Table 11.1.

Table 11.1 Feasible applications of nanomaterials (Srivastava et al. 2017b)

Biofuel areas	Feasible applications of nanomaterials
Pretreatment of cellulosic biomass	Enhance the liberation of hemicelluloses and lignin structure
Cellulase enzyme	Enhance the synthesis, pH, and thermal stability
Bioethanol production	Enhance the synthesis and resistance of microorganisms
Biohydrogen production	Enhance the synthesis and stability of nitrogenase and hydrogenase

For future scenario, biofuels are supposed to be a better choice to substitute the fossil fuels and are probably an advanced and new research zone to offer a replacement in the form of inexhaustible energy. Currently, the United States and Brazil produce about 80% of the world's biofuels, and both countries are producing principally bioethanol. In future, worlds' requirement of fuel for transport is accounted to rise up to 55% by 2030 as compared to 2004, and according to this estimate, the demand of production of biofuels will increase. Internationally a number of efforts and studies are being done to upgrade the economical production of biofuels (Srivastava et al. 2017a).

The nanomaterials as supplements for biodiesel production have appeared as an advance favorable fuel supplement among the current fuel supplements to green diesel, for acquiring extreme betterment in the production and greatest possible decrease of harmful liberation (Prabu 2017). Dreizin (2000) noticed that the nanoparticle scattered test fuels exhibit good thermal and physical characteristics due to its greater surface to volume ratio and perform as an oxygen buffer regarding NO liberation. Kuo et al. (2003) investigated later that nanomaterials increase the heat shifting estimates because of its greater specific surface area. Idriss (Idriss 2004) conducted a research on the advancement of cerium oxide nanomaterials with ethanol and found that nanoparticles could be applied as a fuel-formed supplement in hydrocarbon liquid fuels. Farfaletti et al. (2005) later applied ceria nanomaterials as supplement in diesel admixture fuel and reveal critical decline of particulate material, carbon monoxide, and unburned hydrocarbon emission.

Singh et al. (2016) stated that in the current years, nanotechnology has attained greater concentration due to extensive limits of technological and engineering implementations. Due to their diverse physical and chemical characteristics, nanomaterials have the power for their industrial use. It is also believed that the commercial value of nanomaterials would elevate in future years, and probably it is will raised from 2000 tons in 2004 to around 58,000 tons in 2011 to 2020 (Shi and Ma 2011). Several sectors covering industries belonged to atmospheric amendatory and biofuel synthesis have vast use of nanoparticles. Production of biofuel can be enhanced by use of nanomaterials; for example, effectiveness of biofuels like bioethanol/biohydrogen can be getting better by applying nanoparticles. Use of nanomaterials can also enhance the production of cellulase enzyme, its hydrolysis efficacy, and stability. Furthermore, nanoparticles also play part in the improvement of pretreatment of lignocellulosic biomass (Srivastava et al. 2017a). The advantages of implementing nanotechnology to catalysis involve lifetime, resistance to poisoning, better activity, and other new efficiencies (Mahmood and Hussain 2010; Mahmood et al. 2010). These enhancements and advances cannot be gained consistently with catalysts synthesized by other processes. Cobalt-based catalysis is an illustration of extending the catalytic efficiencies of regular catalysts by nanoscale modifications. The structure, size, and other supplementary constituents, such as Mg, Si, and Ni, may influence the cobalt nanoparticle catalysts (Hussain et al. 2009; Kesavan et al. 2001; Son et al. 2002; Yang et al. 2001). Particularly being inexpensive, cobalt nanoparticles require gentle reaction environment for greater yields of products in less reaction times relative to the conventional catalysts (Mahmood and

Hussain 2010; Mahmood et al. 2010). In gasification, nickel catalyst is broadly employed (Davidian et al. 2007; Hussain et al. 2009; Svoboda et al. 2007). Guo et al. (1995) reported that the size and structure of nanoparticles influence the melting point and catalytic ability increases on merging nanoparticles of cobalt and nickel. X-ray diffraction, scanning electron microscope, and transmission electron microscope are used to study the characteristics of nanoparticles (Salavati-Niasari et al. 2008). For catalytic reduction of carbonyl and aromatic compounds, nickel nanoparticles can be employed (Kidwai et al. 2006; Mahmood and Hussain 2010; Mahmood et al. 2010). Kidwai et al. (2006) accounted that nickel nanoparticles can also be employed for the selective and effective catalysis of the reduction of aldehydes in the availability of additional functional groups, e.g., –CN, $-NO_2$, and double bonds at α and β places. Catalytically the nickel nanomaterials perform principal part in the aldehydic groups' reduction. In addition to nickel, cobalt too performs a critical character in alteration of the chemistry of fatty acid. It was distinctly observed that cobalt nanoparticles (Wu et al. 2004) extended the band of the carboxyl group of pure liquid and takes to the synthesis of two novel bands with property of the symmetric cis (COO-) and the asymmetric trans (COO-) stretch.

Increasing rates of purified products of crude oil and rates of crude oil itself are a clear indication of the continuing depletion of global nonrenewable fossil fuels, namely, coal, crude oil, and natural gas. One study has indicated that the world's fossil fuel reserves would be diminished by 2050 at this rate of consumption, which is roughly about 105 times nonfaster than that the nature can actually create (Satyanarayana et al. 2011). Apart from the rising prices, fossil fuels are also said to contribute considerably to environmental pollution and ecological devastation, causing depletion of ozone layer and global warming and destroying both the biosphere and the geosphere. The environmental concentration of the main greenhouse gas, carbon dioxide, has been steadily rising to reach a level of about 390 parts per million and is expected to reach 750 parts per million by the end of this century (Ahrens 2011). This carbon dioxide is capable of trapping the sun's infrared radiation and thereby causing a rise in the global temperature. These two factors have led the fuel industry to move toward sustainable sources of alternative energy or, in other terms, renewable energy sources. Sustainable sources of alternative energy can range from solar, wind, tidal, and geothermal energies to energies generated from biomass, termed as biofuels. Currently about 90% of the world's energy demands are met by fossil fuels, which is expected to reduce to around 50% by 2040 with the advent of the abovementioned renewable energy sources (Hussein 2015). Biofuels utilize domestically available renewable resources and are able to decompose biologically and nonpoisonous. They can be formed from numerous raw materials such as vegetable fats and biowastes. Anaerobic digestion of agricultural residues, animal manures, organic food wastes, sewage sludge, and various other energy crops yield yet another renewable source of energy known as biogas.

Nanoparticles are at the frontline of quick establishment in nanotechnology. Their unique size-dependent characteristics of nanoparticles form these materials requisite and predominant in several zones of human affairs (Salata 2004). Iron has the status of pillar in the present infrastructure because of being one of the most contemporaneous transition metals in the crust of Earth (Huber 2005). Although, as compared to other group elements like nickel, cobalt, platinum, gold, and iron oxides are ignored to some extent. There are nearly 16 known iron oxides, formed by chemical combination of iron and oxygen. In the form of rust, iron (III) oxide is available in nature (Cornell and Schwertmann 2003). Commonly, iron oxides are widespread and greatly used because they are cost-effective and perform a mandatory part in several geological and biological operations. Humans also use them widely, e.g., as hemoglobin and catalysts, iron ores in thermite, and long-lasting pigments (colored concretes, paints, and coatings) (Laurent et al. 2008). In nature hematite (α-Fe_2O_3), maghemite (g-Fe_2O_3), and magnetite (Fe_3O_4) are the most usual kinds of iron oxides. In the area of scientific technology, these oxides play very important role (Teja and Koh 2009). Nanoparticles are made up of ferromagnetic matter, and their size range from less than 10–20 nanometer shows an idiosyncratic type of magnetism, particularly superparamagnetism. Alloys, oxides, elemental metals, and other chemical compounds are the ferromagnetic materials that can be magnetized by an exterior magnetic field. Normally, only the nanoparticle systems have the outstanding phenomenon (Ali et al. 2016).

Cellulases are the complex enzymes that comprise cellobiohydrolases, β-glucosidase, and endoglucanases and are the imperative constituent in the hydrolysis of lignocellulosic biomasses, produced by different species of protozoans, bacteria, and fungi. Exploiting the hydrolytic expertise of these enzymes permits for the effective transformation of inexhaustible naturally occurring supplies into fermentable sugars. Regrettably, the feasible implementations of cellulases are restrained by their hydrophilic nature when they undergo the complications of the reactions involved (Ho et al. 2008). For the manufacturing of a variety of functional materials or equipments, the immobilization of biomolecules is a significant tool onto insoluble aids (Tischer and Wedekind 1999). Feasible isolation from reaction mixture, improved stability, possible modulation of the enzymatic characteristics, effortless avoidance of microbial growth, etc. are among its most distinct advantages (Bornscheuer 2003). The employment of magnetic supports for immobilization is of specific interest. Magnetic nanocarriers accomplish two main actions in the way, they have a magnetic matter which grant the required magnetic characteristics to accumulate with the species to be isolated and they can have surface attributes which allow a discriminatory isolation (Jordan et al. 2011).

Nanotechnology has numerous implementations such as amendment in raw materials, establishing much effective catalysts, and many others in biofuels and bioenergy field. Enzymes, for example, are being immensely employed to catalyze synthesis of biodiesel from oils and fats or to convert biomass into biofuels such as biogas and ethanol (Michalska et al. 2015; Verma and Barrow 2015). Nanomaterials, in this scenario, could be employed to immobilize the enzyme or to substitute the enzymes; consequently more productive catalysis is gained or enhanced the recuperation of biocatalysts from media. Furthermore, this strategy involves modifications in which magnetic characteristics can be appended to systems that are immobilized (Rai et al. 2016; Verma and Barrow 2015).

Anaerobic breakdown of organic wastes such as animal, human, plant, and agricultural wastes can produce biogas. The liberation of energy from anaerobic operations relies on the carbon and nitrogen ratio, whereas these organic residues have frequent carbon and nitrogen sources (Feng et al. 2014). It is also reported that the production of energy can be increased by incorporation of small amount of metal ions which act as catalyst and enhancing the activity of methanogenic bacteria. As methanogenic bacteria need little quantity of iron, nickel, and cobalt for the anaerobic digestion, scientists reported that use of nanomaterials is advantageous instead of using bulky or atomic materials (Feng et al. 2010). Yang et al. (2015) reported that due to powerful paramagnetic characteristic and great coercivity, magnetic nanoparticles can be employed in the procedure of methanogenesis. In 2015, Abdelsalam et al. (2015) also observed that nickel and cobalt nanoparticles improved the production of methane gas. They also demonstrated that the action of iron oxide nanoparticles is greater than that of iron nanoparticles. Later Abdelsalam et al. (2016) investigated that the influence of various nanoparticles, e.g., iron, iron oxide, nickel, and cobalt maximum methane and biogas is synthesized by anaerobic digestion of cattle manure. It was also reported that when iron oxide nanocarriers were employed to the hydrocarbon matter in the anaerobic digester, improvement of the disintegration activity and high yield of biogas and methane production were noticed (Casals et al. 2014). Various food feedstocks like sucrose from sugarcane, starch from corn, plant oils, and animal fats are usually used for the synthesis of first generation of biofuels (Naik et al. 2010). Continuous exploitation of these raw materials broaches several worries, and therefore, new concept of second-generation biofuel is obtaining importance in the world scenario for biofuel production by employing non-food feedstock such as lignocellulosic materials, e.g., agricultural remains, wood wastes, and others (Eggert and Greaker 2014). However, second-generation biofuels have both pros and cons as the use of waste material is the advantageous prospect, while the high production cost and technological issues and infrastructure are the demerits of it. Therefore, in this situation, there is a need to establish effective strategies to deal these problems in treatment of mass and enhanced output in biofuels. Pursuing these approaches, implementation of nanotechnology can beat the aforesaid problems by offering the opportunity to amend the properties of feedstock matter for synthesis of biofuel. Because of extraordinary physiochemical characteristics, nanoparticles have deliberated uses in synthesis of biofuel. Many nanomaterials such as titanium dioxide, iron oxides, stannic oxide, zinc oxide, graphene, fullerene, and carbon presenting distinctive characteristics are being utilized for production of biofuel production. Furthermore, some characteristics such as quantum properties, greater surface to volume ratio, and ability to immobilize because of their little size, nanoparticles (magnetic) have broad applications in the area of biofuel production. In addition to these, the easy recovery of nanoparticles from reaction mixture by employing appropriate magnetic field makes them more significant (Ahmed and Douek 2013). Biodiesel fulfils the specific standards and hence can be employed as fuel in diesel engines; it is a mixture of esters, which are generally synthesized by transesterification of animal fats or vegetable oils with small-chain alcohols, e. g., ethanol or methanol. Biodiesel

Table 11.2 Use of different nanocatalysts and feedstock used for biodiesel production

Sr no.	Nanocatalyst	Feedstock	Yield (%)	References
1	Lithium-impregnated calcium oxide (Li-Cao)	Karanja oil and *Jatropha* oil	99.0	(Kaur and Ali 2011)
2	K2O/γ-Al2O3	Rapeseed oil	94.0	(Han and Guan 2009)
3	MgO nanoparticles on TiO2 support	Soybean oil	95.0	(Mguni et al. 2012)
4	KF/γ-Al2O3/honeycomb ceramic (HC) monolithic catalyst	Palm oil	96.0	(Gao et al. 2015)
5	MgO	Sunflower oil and rapeseed oil	98.0	(Verziu et al. 2008)
6	Sulfamic and sulfonic acid-functionalized silica-coated crystalline Fe/Fe3O4 core/shell magnetic nanoparticles	Glyceryl trioleate	95.0	(Wang et al. 2015)
7	KF/CaO	Chinese tallow seed oil	96.8	(Wen et al. 2010)
8	Ca/Fe3O4@SiO2	Sunflower Oil	97.0	(Feyzi and Norouzi 2016)
9	ZrO2 loaded with C4H4O6HK	Soybean oil	98.03	(Qiu et al. 2011)
10	Cs/Al/Fe3O4	Sunflower oil	94.8	(Feyzi et al. 2013)
11	ZnO nanorods	Olive oil	94.8	(Molina 2013)
12	Iron/cadmium and iron/tin oxide nanoparticles	Soybean oil	84.0	(Alves et al. 2014)
13	CaO	*Jatropha* oil	98.54	(Reddy et al. 2015)
14	Hydrotalcite-derived particles with Mg/Al oxides	*Jatropha* oil	95.2	(Deng et al. 2011)
15	TiO2-ZnO	Palm oil	92.2	(Madhuvilakku and Piraman 2013)

is much beneficial than conventional fuel as it is produced from renewable resources and has biodegradable and predominant emollient characteristics, without liberation of injurious chemicals (Feyzi and Norouzi 2016). Table 11.2 abridges various nanocatalysts and feedstock utilized for the biodiesel production and the yield belonged to each process as well.

As revealed that in this field, there are numerous different feasibilities of implementations of nanocatalysts. Several modified nanomaterials are being successfully employed in the biodiesel production. For instance, Wang et al. (2015) synthesized acid-modified magnetic nanoparticles and described their utilization as heterogeneous nanocatalysts for synthesis of biodiesel. In this investigation, acid-modified, i.e., H_3NSO_4 and $R - S(=O)_2 - OH$ silica-covered crystalline iron/iron (II, III) oxide shell/core, magnetic nanoparticles were produced and utilized for the synthesis of biodiesel from transesterification of glyceryl trioleate. Researchers observed that both the acid-modified nanocatalysts manifested the noticeable enzymatic activities. But the sulfamic acid-modified nanocatalysts presented greater activity as compared to the sulfonic acid-modified nanocatalysts. Tahvildari et al. (2015) conducted a research on the biodiesel production using cooking oils, employing calcium oxide and magnesium oxide nanoparticles produced by sol-gel and sol-gel self-combustion methods, respectively. Calcium oxide nanoparticles displayed principal improvement in the yield of biodiesel as compared to magnesium oxide nanoparticles.

A new approach known as magnetic nanomaterials is also getting importance as these are easy to recover and reuse and also favorable for economic application of the process. In this context, to produce biodiesel from soybean oil, Alves et al. (2014) suggested a quick and simple nanotechnological application. They employed a combination of iron/tin and iron/cadmium oxide nanoparticles having magnetic characteristics, formed by co-precipitation method as nanocatalysts for the production of biodiesel. Iron/tin oxide nanoparticles displayed the highest efficiency by forming approximately 84% biodiesel. These nanocatalysts proposed remarkable ability toward hydrolysis, transesterification, and esterification of soybean oil and their fatty acids. Likewise, Qiu et al. (2011) suggested the production by zirconium oxide nanocatalyst laden with potassium bitartrate, in size ranging from 10 to 40 nm. Moreover, authors explored the biodiesel production by utilizing the modified nanocatalysts for the transesterification of soybean oil and methanol in various molar ratios and other factors, e. g., concentration of nanocatalyst, time for reaction, and reaction temperature. The observations suggested that reaction mixture with oil and methanol in the ratio 1:16 having 6% nanocatalyst at 60 °C for 2 h ended in maximum production of biodiesel of around 98.03%. Use of nanomaterials for production of biodiesel is also reported in several investigations. For example, in 2010, Wen et al. (2010) produced potassium fluoride/calcium oxide nanocatalyst by saturation method with size ranging from 30 to 100 nm and further utilized it for biodiesel production from Chinese tallow seed. This study revealed about 96.8% yield of biodiesel depicting ability of such nanocatalyst for biodiesel industry. In another study, Deng et al. (2011) achieved about 95.2% production of biodiesel from pretreated *Jatropha* oil using nanocatalysts, hydrotalcite-obtained particles with magnesium/aluminum oxides created by using urea as precipitating agent through a co-precipitation method.

Feyzi et al. (2013) in a novel research found a new nanocatalyst, cesium/aluminum/iron (II, III) oxide, that was produced and assessed with respect to its catalytic efficiency for production of biodiesel. This research investigated the influence of various cesium/aluminum and cesium/iron molar ratio and calcination situations on the catalytic implementation. These nanocatalysts, in the molar ratio of 2.5:1 and 4:1, respectively, depicted prospective for biodiesel production, ending in 94.8% of process yield at 58 °C with continuous stirring for 120 min. Verziu et al. (2008) in their investigation about nanocatalysts demonstrated the establishment of magnesium oxide nanocatalyst in the form of nanosheet by aerogel method and revealed its application in the production of biodiesel from rapeseed and sunflower oil, substantiating process yield up to 98%. In a recent study, Feyzi and Norouzi (2016) developed a nanocatalyst $Ca/Fe_3O_4@SiO_2$ with powerful magnetic characteristics applying mixture of two different applications, namely, sol-gel and incipient wetness impregnation method for the formation of biodiesel. Developed nanocatalysts were suggested to be very efficient at optimum conditions and ends in maximum production of around 97%. The magnetic character of nanocatalyst favored its reuse for various times without principal is appearing in its catalytic action. In another recent study, Reddy et al. (2015) revealed the formation of new calcium oxide nanocatalyst using *Polymesoda erosa*, a seashell, through various steps like calcination, hydration, and dehydration techniques. They also found the feasibility of the formed nanocatalyst regarding production of biodiesel from the non-eatable raw oils, e.g., *Jatropha* oil. The highest production of 98.54% biodiesel was attained at conditions of oil to methanol ratio in 1:5.15 molar ratio, reaction time of 133.1 min at 0.02:1 (w/w) nanocatalyst ratio.

Bioethanol (ethyl alcohol) is usually formed by carbon means of grains, sugarcane juice, and others. However, the synthesis of this alcohol is also possible from the sugars obtained from the hydrolysis of vegetal residues and lignocellulosic biomass. The lignocellulosic biomass is generally made up of polymeric structures of carbohydrates, i.e., cellulose and hemicellulose, and lignin which is a complicated organic polymer comprised of chiefly phenolic compounds (Antunes et al. 2014). Different pretreatments of biomass are required to remove the stubborn barrier of lignin from the cellulose and hemicellulose, so that these complex polymeric frameworks can be converted into simple monomeric sugars. Generally the complex structures of cellulosic contents of biomass are hydrolyzed enzymatically into simple fermentable sugars, after an initial pretreatment. Using this procedure, simple sugars can be formed in moderate process parameters such as lower temperature, without requirement of pressure and the non-production of unwanted fermentation inhibitor compounds as compared to chemical process.

Around 18% of total expenses involved in the process of the production of bioethanol belong to the cellulases used which are responsible for the breakdown of lignocellulosic biomass. Therefore the development of advance approaches is required which could help in easy recovery and recycling of cellulases and could decrease production cost as well. Keeping this fact in mind, nanotechnology proposed immobilization of different enzymes like cellulases and hemicellulases, which are engaged in the production of bioethanol. For instance, enzyme immobilization on magnetic nanoparticles is a reliable procedure that delivers easy recovery of enzyme by assigning magnetic field that permits enzyme reuse and recovery for many runs (Rai et al. 2016). Enzymes involved in the production of bioethanol are reported to be immobilized on magnetic nanoparticles in many studies. Usually enzymes are immobilized on nanoparticles by physical adsorption or covalent binding. However covalent binding method is considered more satisfactory because this method decreases desorption of protein as a result of the formation of covalent bonds between nanoparticles and enzyme (Abraham et al. 2014). Nanomaterials need to be amended or loaded with chemically active polymer for stable immobilization of enzyme on them, to provide the functional group for binding of enzyme. Lee et al. (2010) investigated the entrapment method for the immobilization of β-glucosidase enzyme on polymer magnetic nanofibers for the production of ethanol from cellulosic matter. β-Glucosidase is literally the enzyme which catalyzes the transformation of cellobiose into glucose, and this glucose is converted into bioethanol by microorganisms. Actually, the stability of enzyme is achieved by entrapment of β-glucosidase on magnetic nanofibers, and this entrapment method also offers chances of repeated use, recycling and isolating them by implementing magnetic field. Likewise, Verma et al. (2013b) assessed immobilization of β-glucosidase isolated from fungus on magnetic nanoparticles, used as nanobiocatalyst for production of bioethanol. They reported 93% of enzyme-binding efficacy, demonstrating around 50% of its initial activity at 16th cycle. Jordan et al. (2011) also investigated the hydrolysis of microcrystalline cellulose by recycling of enzyme, employing carbodiimide as binding polymer for immobilization of enzyme on Fe_3O_4 nanoparticles. The enzyme could be restored easily and recycled for six times because of magnetic characteristics of nanoparticles. Goh et al. (2012) studied that enzyme used in the production of bioethanol was immobilized in single-walled carbon nanotubes, which was already subsumed, to give magnetic properties, by magnetic iron oxide nanoparticles. They concluded that by altering the concentration of iron oxide nanoparticles in the nanotubes, the performance of immobilized enzyme could be controlled. Thus, immobilized enzyme can be stored in acetate buffer at 4 °C for its longer storage. For immobilization of enzymes, various nanomaterials have been investigated. For instance, Xie and Wang (2012) conducted a research on the lipase immobilization on magnetic chitosan microspheres produced by chemical co-precipitation method. In this study, as linking molecules, glutaraldehyde was used for the covalent binding between magnetic chitosan microspheres and lipase enzyme. Furthermore, aiming for the use of production of bioethanol, through adsorption methods, enzyme immobilization on titanium oxide nanoparticles was also done successfully and employed for the breakdown of lignocellulosic biomasses (Ahmad and Sardar 2014). Cherian et al. (2015) in some other study immobilized cellulase enzyme isolated from *Aspergillus fumigatus* on nanoparticles of manganese dioxide by covalent binding. Researchers proved that as relative to free enzymes, enzyme immobilized on nanomaterials have improved thermostability characteristic, depicting stability up to 70 °C. Hydrolysis of agricultural waste was done in the presence of immobilized cellulase enzyme pursued by the employment of yeast for the production of bioethanol (21.96 g/L). After again and again, the use of immobilized enzyme for approximately five cycles, activity of enzyme remained 60%. Other several nanomaterials can be used in nanotechnology process in addition to magnetic nanoparticles, such as titanium oxide and silica, polymeric nanoparticles, and carbon materials like carbon nanotubes, graphene, fullerene, and others. All these materials have been successfully accounted for immobilization of various enzymes with respect to the processes of production of bioethanol (Cho et al. 2012; Pavlidis et al. 2012; Verma et al. 2013a). For example, to get enhanced catalytic activity for production of ethanol, carbon nanotubes entrapped with Rh particles can be used (Pan et al. 2007). In fact, the incorporation of materials of different interests is facilitated by free cores available on the carbon nanotubes. In some other research, Lupoi and Smith (Lupoi and Smith 2011) compared the efficiency of free cellulase enzyme with that of the cellulase enzyme immobilized on the silica nanoparticles in the hydrolysis of cellulose into glucose. They concluded that cellulase enzyme that is immobilized demonstrated high production of glucose relative to free enzyme, proving that in simultaneous saccharification and fermentation, immobilized enzymes can be employed. To employ for fermentation process of production of ethanol, microbial cells can also be immobilized on the nanoparticles. For example, a strategy was developed for immobilization of *Saccharomyces cerevisiae* cells onto the magnetic nanoparticles. Furthermore, continuous fermentation operation for the bioethanol production was performed, and reported immobilized *S. cerevisiae* cells depicted maximum ethanol production efficacy (Ivanova et al. 2011). Thus, the studies conducted either on the enzyme immobilization or whole microbial cells on the various nanomaterials proves that such approaches and technologies will be feasible for the secure and cost-effective synthesis of bioethanol from inexpensive lignocellulosic biomass.

11.5 Iron Oxide Nanomaterial Synthesis and Their Applications in Biofuel Production

Chemically or biologically nanoparticles could be synthesized. Many severe effects are linked with chemical synthesis procedures because of the existence of some toxicants which are absorbable on the surface. Biological method replaces physical and chemical methods that is an environment-friendly way of nanoparticles production using plants, microorganisms, enzymes, fungus, and plant extracts. Biological method being an environmental favorable replacement of physical and chemical methods is a way of nanoparticles production using plants, microorganisms, enzymes, fungus, and plant extracts. Development of these environmental favorable methods for production of nanoparticles is progressing into a significant domain of nanotechnology particularly silver nanocarriers, which have numerous implementations. Biosynthesis of nanoparticles by microorganisms is a green and eco-friendly technology. Green and environment favorable technology for the production of nanoparticles is biosynthesis. Variety of microorganisms both eukaryotes and prokaryotes are used for the production of metallic nanocarriers specifically palladium, gold, silver, zirconium, cadmium, iron platinum, and metal oxides such as zinc oxide, titanium oxide, etc. These microorganisms are actinomycetes, bacteria, algae, and fungi. According to the placement of nanoparticles, production of nanoparticles may be of two methods, i.e., intracellular or extracellular. Inorganic materials are formed by multicellular as well as unicellular organisms either extracellularly or intracellularly. Microorganisms such as bacteria and fungi are capable to tackle the production of metallic nanoparticles, and their ability is employed in the hunt for novel materials. Fungi have become the center of studies on biological production of metallic nanoparticles because of their resistance and metal bio-magnifications potential (Hasan 2015).

Currently iron oxide nanoparticles are getting much interest in atmospheric remediation because of their little size, magnetic attribute, and greater surface area. Among the most ubiquitous elements on earth is iron. Iron and oxygen atoms combine to form iron oxides. However, there are 16 identified iron oxides and oxyhydroxides. Feasible isolation of adsorbents from the process and its reutilization for further implementation become possible due to magnetic character of iron oxide nanoparticles. Iron oxide nanocarriers lead to a decline in the economic burden because of reusability. Feasibility of resource and facility in production presents nano-sized ferric oxides as cost-effective adsorbents for toxic metal absorption and adsorption. Elemental iron is therefore eco-favorable. Numerous encouraging applications involve utilization of iron oxide (Rasheed and Meera 2016). Several procedures have been investigated up till now for the production of iron oxide-based nanomaterials. Commonly these procedures can be categorized as chemical, biological, and physical methods. Literature exhibited that iron oxide nanoparticles are generally synthesized by chemical methods. Co-precipitation, hydrothermal synthesis, reverse micelles and micro-emulsion technology, thermal decomposition, sonochemical reactions, sol-gel production, hydrolysis and thermolysis of precursors, flow injection synthesis, electrospray production, and colloidal chemistry method are usually applied chemical methods. Chemical methods have some disadvantages, namely, the broad particle size scattering, the less dispensation in solvents, and the consistency of the size of particle of these nanomaterials that was somewhat of low quality (Nidhin et al. 2008). These procedures cause accumulation of nanoparticles. Therefore the interest in the employment of green resources for the production of nanomaterials is increasing. The following are the distinctive benefits of the biological synthesis procedures that make them superior over traditional employed chemical and physical procedures (Herlekar et al. 2014):

- Environment-friendly and clean approach because there is no use of toxicants.
- Production process is cost-effective as the active biological constituents such as enzyme itself perform as a capping and reducing agent.
- During pilot-scale synthesis, small nanocarriers can be formed.
- External investigational factors such as high pressure and high energy are not desired, thus leading to principal reduction in energy consumption.

In the area of magnetic nanomaterials in the last few years, explorations are increased on different types of iron oxide (usually involves the Fe_3O_4 magnetite, $FeIIFeIII_2O_4$, ferrimagnetic, superparamagnetic when the size is smaller than 15 nanometer), a-Fe2O3 (hematite, fragile, ferromagnetic, or antiferromagnetic), c-Fe_2O_3 (ferromagnetic, maghemite), FeO (wüstite, antiferromagnetic), e-Fe_2O_3 and b-Fe_2O_3 (wüstite, antiferromagnetic), b-Fe_2O_3, and e-Fe_2O_0 (Cornell and Schwertmann 2003). However, tackling stability, shape, size, and dispersibility of nanoparticles in interested solvents is a technological challenge. Due to greater surface to volume ratio, magnetic iron oxide nanocarriers have greater surface energies. Therefore, they try to accumulate so as to reduce the surface energies. Furthermore, the bare iron oxide nanocarriers can feasibly oxidized in atmosphere (particularly magnetite), owing to their enhanced chemical action, and usually lose dispersibility and magnetism. Therefore, it is very essential to provide appropriate surface layering and establish few effectual protection approaches to retain the stability of magnetic iron oxide nanomaterials. These approaches consist layering with or grafting of organic molecules such as little organic molecules or surfactants, biomolecules, and polymers or layering with an inorganic layer, like metal, metal oxide, silica, metal or nonmetal elementary substance, and metal sulfide. Certainly, it is notable that in several instances, the shelter shells can be employed for further functionalization too in addition to stabilizing the magnetic iron oxide nanoparticles (Wu et al. 2008).

Heterogeneous and homogeneous catalysts are used for the mass manufacture of biodiesel. The homogeneous ones have several disadvantages such as corrosiveness, catalyst recycle, influence on soap formation, and catalyst separation. The heterogeneous ones show a lot of demerits like less yields of biodiesel, costly in nature, and complex formation. Hence, it is essential to establish the cheap and potential catalyst to manufacture increased yield of biodiesel. Nowadays, metal nanoparticles perform a major role in the transesterification process as catalyst to manufacture increased productivity of biodiesel, having its large active surface area, enhanced resistance to saponification reaction, better rigidity, and increased catalytic activity. Conversely, the metal nanoparticles are unsteady in nature, and they have a propensity to make collections (Rengasamy et al. 2016). Xie and Ma (2009) described that the biodiesel were gained from soybean oil using immobilized lipase scattered on magnetic Fe_3O_4 nanoparticles. Anyhow, immobilization of enzymes needs extra cost and time. Regarding previous studies, the phytogenic metal nanoparticles are stable for long-time duration (Iravani 2011).

The most common procedures including co-precipitation, thermal decomposition, sonochemical production, hydrothermal production, microemulsion, and sonochemical synthetic route can all be directed to the manufacture of better quality of iron oxide nanoparticles. Moreover, these nanoparticles can also be manufactured by the other techniques like electrochemical formation, laser pyrolysis procedures, bacterial or microorganism formation (especially iron-reducing bacteria and the magnetotactic bacteria), etc. The most traditional method for obtaining Fe_3O_4 or c-Fe_2O_3 is by co-precipitation. In this procedure, we mix ferric and ferrous ions in a 1:2 molar ratio in very alkaline solutions at increased temperature or at room temperature. The shape and size of the iron oxide nanoparticles depend on the kind of salt used (like perchlorates, nitrates, sulfates, chlorides, etc.), the ferrous and ferric ion ratio, the ionic strength of the media, the pH value, the reaction temperature, and the other reaction factors (e.g., mixing rate, falling speed of alkaline solution). Currently it is investigated that the co-precipitation manufacture of Fe_3O_4 nanoparticles and their respective structure, apparent form, and magnetic characteristics at various temperature of reaction were studied (Wu et al. 2008). An organic solution-phase decomposition route has been largely used in iron oxide nanoparticles production, and decomposition of $Fe(cup)_3$ (cup = N-nitrosophenylhy droxylamine), $Fe(acac)_3$ (acac = acetylacetonate), or $Fe(CO)_5$ followed by oxidation can lead to excellent-quality monodispersed iron oxide nanoparticles, which mostly needs comparatively increased temperatures and a complex operation. Sun and Zeng (2002) described a common breakdown strategy for the manufacture of size-sensitive monodispersed magnetite nanoparticles established on increased temperature (265 °C) reaction of $Fe(acac)_3$ in phenyl ether in the availability of oleylamine, alcohol, and oleic acid. With the little magnetite nanoparticles as seeds, larger monodispersed magnetite nanoparticles of up to 20 nm in diameter can be manufactured and scattered into nonpolar solvent by seed-mediated growth procedure. The method does not need a size selection method and is quickly scaled up for synthesis on large scale. The as-synthesized Fe_3O_4 nanoparticle assemblies can be changed feasibly into c-Fe_2O_3 nanoparticles by annealing at increased temperature (250 °C) and oxygen for 2 h.

A thermodynamically sensible isotropic dispersion of two immiscible phases (oil and water) is microemulsion in the presence of surfactant; with the dissolved hydrophobic tails of the surfactant molecules in the oil phase, the surfactant molecules can produce a single layer at the coherence of oil and water, and in the water phase, the hydrophilic head groups are present. Self-fabricated assemblies of various kinds can be prepared, ranging, for instance, from (inverted) globular and columnar micelles to lamellar stage and bicontinuous microemulsions, which can co-occur with principally water or oil phases in the binary systems, e.g., oil/surfactant or water/surfactant (Solans et al. 2005). In that scenario, microemulsion and inverse micelles path can be applied for getting the size- and shape-checked iron oxide nanoparticles.

Iron oxide nanoparticles with authorized shape and size are significant technologically because of powerful association among these factors and magnetic characteristics. The thermal decomposition and microemulsion techniques generally take to complex method or demand comparatively increased temperatures. As a substitute, hydrothermal formation has several wet chemical strategies of crystallizing material in a closed vessel from the increased temperature aqueous solution (generally in the range from 130 to 250 °C) at increased vapor pressure (commonly in the range from 0.3 to 4 MPa). This procedure has also been used to grow dislocation-free single crystal carriers, and grains synthesized in this method could have a fitter crystallinity than that from other methods, so hydrothermal formation is susceptible to acquire the greatly crystalline iron oxide nanocarriers. Many investigators have described the production of iron oxide nanocarriers through hydrothermal technique (Hu et al. 2007).

According to whether or not use the particular surfactants, there are two main procedures. For instance, a one-step hydrothermal process to make very crystalline Fe_3O_4 nanopowders without using the surfactants has been reported by Wang et al. (2003). The nanoscale Fe_3O_4 powder (40 nm) acquired at 140 °C for 6 h possessed a saturation magnetization of 85.8 emu_g-1, a little less than that of the correspondent bulk Fe_3O_4(92 emu_g-1). It is recommended that the well-crystallized Fe_3O_4 grains made under suitable hydrothermal environment should be responsible for the enhanced saturation magnetization in nano-sized Fe_3O_4. Conversely, Zheng et al. (2006) described a hydrothermal route for synthesizing Fe_3O_4 nanoparticles with diameter of ca. 27 nm in the existence of a surfactant, sodium bis(2-ethylhexyl) sulfosuccinate (AOT). At room temperature, the magnetic characteristics of the nanoparticles showed a superparamagnetic performance. Furthermore, hydrothermal treatment is one of the well-to-do means of growing crystals for iron oxide nanoparticles.

As a ruthless substitute, the sonochemical procedure is broadly utilized to make novel materials with uncommon characteristics. The chemical effects of ultrasound emerge from acoustic cavitation, that is, the synthesis, development, and implosive bursting of bubbles in liquid. The implosive bursting of the bubble makes a confined hotspot via adiabatic compression or shock wave formation within the gas phase of the bursting bubble. The conditions formed in these hotspots have been experimentally found, with 1800 atm pressure, 5000 K transitory temperature, and cooling rates in excess of 1010 K/s (Suslick 1990). These utmost conditions were useful to make the new phase and have a shear effect for agglomeration, which is susceptible to make the very monodispersive nanoparticles. This procedure has been applied for the formation of several nanocomposites, and its versatility has been efficiently displayed in iron oxide nanoparticles synthesis (Bang and Suslick 2007). For example, magnetite nanoparticles can be effortlessly made by sonication of iron (II) acetate in water under an argon environment. Vijayakumar et al. (2000) described a sonochemical manufacturing course for synthesizing the pure nanometer-size Fe_3O_4 powder with particle size of ca. 10 nm. The synthesized Fe_3O_4 nanoparticles are superparamagnetic, and its magnetization at room temperature is very less (<1.25 emu g-1). The five aforementioned preparing ways have various merits and demerits for making iron oxide nanoparticles, respectively. In terms of size and morphology control of the iron oxide nanoparticles, thermal decomposition and hydrothermal preparing route seems the excellent procedure. For acquiring the water-soluble and biocompatible iron oxide nanoparticles, co-precipitation usually was engaged, but this procedure shows less control of the particle shape, wide dispersion of sizes, and gathering of particles. As a time ruthless substitute, sonochemical avenue can also be utilized to prepare iron oxide nanoparticles with rare magnetic characteristics. Additionally, it is notable that several green chemical preparation paths and biological construction paths are described for environment shelter causes.

11.6 Conclusion

The worldwide climatic issues like greenhouse effect caused by different chemicals such as the utilization of coal and petroleum have urged searching for substitute inexhaustible energy and biofuel means. The high requirement of substitute means that is also because of the quick consumption of occurring fuel withholds. Among the advance approaches, nanotechnology is acquiring much attention to deal with the problems of bioenergy and biofuel production by employing various implementations like alteration in raw materials and use of operative catalysts. Because of their phenomenal properties, iron oxide nanomaterials have become marvelous in biofuel industry in present years. Iron oxide nanoparticle technologies have great potentials to enhance production of biofuel.

References

Abdelsalam E, Samer M, Abdel-Hadi MA, Hassan HE, Badr Y (2015) Effect of $CoCl_2$, $NiCl_2$ and $FeCl_3$ additives on biogas and methane production. Misr J Agric Eng 32(2):843–862

Abdelsalam E, Samer M, Attia YA, Abdel-Hadi MA, Hassan HE, Badr Y (2016) Comparison of nanoparticles effects on biogas and methane production from anaerobic digestion of cattle dung slurry. Renew Energy 87:592–598

Abraham RE, Verma ML, Barrow CJ, Puri M (2014) Suitability of magnetic nanoparticle immobilised cellulases in enhancing enzymatic saccharification of pretreated hemp biomass. Biotechnol Biofuels 7(1):90

Ahmad R, Sardar M (2014) Immobilization of cellulase on TiO2 nanoparticles by physical and covalent methods: a comparative study. Indian J Biochem Biophys 51(4):314–320

Ahmed M, Douek M (2013) The role of magnetic nanoparticles in the localization and treatment of breast cancer. Biomed Res Int 2013:1

Ahrens CD (2011) Essentials of meteorology: an invitation to the atmosphere. Cengage Learning, Boston

Ali A, Hira Zafar MZ, ulHaq I, Phull AR, Ali JS, Hussain A (2016) Synthesis, characterization, applications, and challenges of iron oxide nanoparticles. Nanotechnol Sci Appl 9:49

Alves MB, Medeiros F, Sousa MH, Rubim JC, Suarez PA (2014) Cadmium and tin magnetic nanocatalysts useful for biodiesel production. J Braz Chem Soc 25(12):2304–2313

Alvira P, Tomás-Pejó E, Ballesteros MJ, Negro MJ (2010) Pretreatment technologies for an efficient bioethanol production process based on enzymatic hydrolysis: a review. Bioresour Technol 101(13):4851–4861

Ansari SA, Husain Q (2012) Potential applications of enzymes immobilized on/in nano materials: a review. Biotechnol Adv 30(3):512–523

Antunes FAF, Chandel AK, Milessi TSS, Santos JC, Rosa CA, Da Silva SS (2014) Bioethanol production from sugarcane bagasse by a novel Brazilian pentose fermenting yeast Scheffersomycesshehatae UFMG-HM 52.2: evaluation of fermentation medium. Int J Chem Eng 2014:1

Antunes FAF, Gaikwad S, Ingle AP, Pandit R, dos Santos JC, Rai M, da Silva SS (2017) Bioenergy and biofuels: nanotechnological solutions for sustainable production. In: Nanotechnology for bioenergy and biofuel production. Springer, Cham, pp 3–18

Azhar SHM, Abdulla R, Jambo SA, Marbawi H, Gansau JA, Faik AAM, Rodrigues KF (2017) Yeasts in sustainable bioethanol production: a review. Biochem Biophys Rep 10:52–61

Balat M, Balat H (2009) Recent trends in global production and utilization of bio-ethanol fuel. Appl Energy 86(11):2273–2282

Balat M, Balat H (2010) Progress in biodiesel processing. Appl Energy 87(6):1815–1835

Bang JH, Suslick KS (2007) Sonochemical synthesis of nanosized hollow hematite. J Am Chem Soc 129(8):2242–2243

Barakat A, Monlau F, Steyer JP, Carrere H (2012) Effect of lignin-derived and furan compounds found in lignocellulosic hydrolysates on biomethane production. Bioresour Technol 104:90–99

Barnard D, Casanueva A, Tuffin M, Cowan D (2010) Extremophiles in biofuel synthesis. Environ Technol 31(8–9):871–888

Beckers L, Hiligsmann S, Lambert SD, Heinrichs B, Thonart P (2013) Improving effect of metal and oxide nanoparticles encapsulated in porous silica on fermentative biohydrogen production by Clostridium butyricum. Bioresour Technol 133:109–117

Bhalla A, Bansal N, Kumar S, Bischoff KM, Sani RK (2013) Improved lignocellulose conversion to biofuels with thermophilic bacteria and thermostable enzymes. Bioresour Technol 128:751–759

Biswas A, Bayer IS, Biris AS, Wang T, Dervishi E, Faupel F (2012) Advances in top–down and bottom–up surface nanofabrication: techniques, applications & future prospects. Adv Colloid Interf Sci 170(1–2):2–27

Bolivar-Telleria M, Turbay C, Favarato L, Carneiro T, de Biasi RS, Fernandes AAR et al (2018) Second-Generation Bioethanol from Coconut Husk. Biomed Res Int:2018

Bornscheuer UT (2003) Immobilizing enzymes: how to create more suitable biocatalysts. Angew Chem Int Ed 42(29):3336–3337

Brazilian Energy Balance (2015). Final report 2015. https://ben.epe.gov.br/downloads/Relatorio_Final_BEN_2015.pdf. Accessed June 2016

Cai Y, Lai C, Li S, Liang Z, Zhu M, Liang S, Wang J (2011) Disruption of lactate dehydrogenase through homologous recombination to improve bioethanol production in Thermoanaerobacterium aotearoense. Enzym Microb Technol 48(2):155–161

Casals E, Barrena R, García A, González E, Delgado L, Busquets-Fité M et al (2014) Programmed iron oxide nanoparticles disintegration in anaerobic digesters boosts biogas production. Small 10(14):2801–2808

Cavas M, Gupta RK, Al-Ghamdi AA, Gafer ZH, El-Tantawy F, Yakuphanoglu F (2013) Preparation and characterization of dye sensitized solar cell based on nanostructured Fe_2O_3. Mater Lett 105:106–109

Chaker Ncibi M, Sillanpaa M (2013) Recent research and developments in biodiesel production from renewable bioresources. Recent Pat Chem Eng 6(3):184–193

Chang T, Yao S (2011) Thermophilic, lignocellulolytic bacteria for ethanol production: current state and perspectives. Appl Microbiol Biotechnol 92(1):13

Cherian E, Dharmendirakumar M, Baskar G (2015) Immobilization of cellulase onto MnO2 nanoparticles for bioethanol production by enhanced hydrolysis of agricultural waste. Chin J Catal 36(8):1223–1229

Cho EJ, Jung S, Kim HJ, Lee YG, Nam KC, Lee HJ, Bae HJ (2012) Co-immobilization of three cellulases on Au-doped magnetic silica nanoparticles for the degradation of cellulose. Chem Commun 48(6):886–888

Choedkiatsakul I, Ngaosuwan K, Kiatkittipong W, Laosiripojana N, Assabumrungrat S (2011) Patent review on "biodiesel production process". Recent Pat Chem Eng 4(3):265–279

Chum HL, Overend RP (2001) Biomass and renewable fuels. Fuel Process Technol 71(1–3):187–195

Cornell RM, Schwertmann U (2003) The iron oxides: structure, properties, reactions, occurrences and uses. John Wiley & Sons, New York

Davidian T, Guilhaume N, Iojoiu E, Provendier H, Mirodatos C (2007) Hydrogen production from crude pyrolysis oil by a sequential catalytic process. Appl Catal B Environ 73(1–2):116–127

Demetzos C (2016) Pharmaceutical nanotechnology: fundamentals and practical applications. Springer, Puchong

Demirbas A (2010) Use of algae as biofuel sources. Energy Convers Manag 51(12):2738–2749

Demirbas MF, Balat M (2006) Recent advances on the production and utilization trends of biofuels: a global perspective. Energy Convers Manag 47(15–16):2371–2381

Deng X, Fang Z, Liu YH, Yu CL (2011) Production of biodiesel from Jatropha oil catalyzed by nanosized solid basic catalyst. Energy 36(2):777–784

Dreizin EL (2000) Phase changes in metal combustion. Prog Energy Combust Sci 26(1):57–78

Dutta N, Mukhopadhyay A, Dasgupta AK, Chakrabarti K (2014) Improved production of reducing sugars from rice husk and rice straw using bacterial cellulase and xylanase activated with hydroxyapatite nanoparticles. Bioresour Technol 153:269–277

Eggert H, Greaker M (2014) Promoting second generation biofuels: does the first generation pave the road? Energies 7(7):4430–4445

Enweremadu CC, Rutto HL, Oladeji JT (2011) Investigation of the relationship between some basic flow properties of shea butter biodiesel and their blends with diesel fuel. Int J Phys Sci 6(4):758–767

Farfaletti A, Astorga C, Martini G, Manfredi U, Mueller A, Rey M et al (2005) Effect of water/fuel emulsions and a cerium-based combustion improver additive on HD and LD diesel exhaust emissions. Environ Sci Technol 39(17):6792–6799

Farrell AE, Plevin RJ, Turner BT, Jones AD, O'hare M, Kammen DM (2006) Ethanol can contribute to energy and environmental goals. Science 311(5760):506–508

Feng XM, Karlsson A, Svensson BH, Bertilsson S (2010) Impact of trace element addition on biogas production from food industrial waste–linking process to microbial communities. FEMS Microbiol Ecol 74(1):226–240
Feng Y, Zhang Y, Quan X, Chen S (2014) Enhanced anaerobic digestion of waste activated sludge digestion by the addition of zero valent iron. Water Res 52:242–250
Feyzi M, Norouzi L (2016) Preparation and kinetic study of magnetic Ca/Fe3O4@ SiO2 nanocatalysts for biodiesel production. Renew Energy 94:579–586
Feyzi M, Hassankhani A, Rafiee HR (2013) Preparation and characterization of Cs/Al/Fe3O4 nanocatalysts for biodiesel production. Energy Convers Manag 71:62–68
Gao L, Wang S, Xu W, Xiao G (2015) Biodiesel production from palm oil over monolithic KF/γ-- Al2O3/honeycomb ceramic catalyst. Appl Energy 146:196–201
Ghazanfar M, Irfan M, Nadeem M (2018) Statistical modeling and optimization of pretreatment of Bombax ceiba with KOH through Box–Behnken design of response surface methodology. Energy Sources, Part A:1–11
Goh WJ, Makam VS, Hu J, Kang L, Zheng M, Yoong SL et al (2012) Iron oxide filled magnetic carbon nanotube–enzyme conjugates for recycling of amyloglucosidase: toward useful applications in biofuel production process. Langmuir 28(49):16864–16873
Goldemberg J (2006) The ethanol program in Brazil. Environ Res Lett 1(1):014008
Guo T, Nikolaev P, Thess A, Colbert DT, Smalley RE (1995) Catalytic growth of single-walled manotubes by laser vaporization. Chem Phys Lett 243(1–2):49–54
Haas MJ, Scott KM, Alleman TL, McCormick RL (2001) Engine performance of biodiesel fuel prepared from soybean soapstock: a high quality renewable fuel produced from a waste feedstock. Energy Fuel 15(5):1207–1212
Haddad PS, Britos TN, Li LM, Li LDS (2015) Preparation, characterization and tests of incorporation in stem cells of superparamagnetic iron oxide. J Phys Conf Ser 617(1):012002
Hamelinck CN, Van Hooijdonk G, Faaij AP (2005) Ethanol from lignocellulosic biomass: techno-economic performance in short-, middle-and long-term. Biomass Bioenergy 28(4):384–410
Han H, Guan Y (2009) Synthesis of biodiesel from rapeseed oil using K 2 O/γ-Al 2 O 3 as nano-solid-base catalyst. Wuhan Univ J Nat Sci 14(1):75–79
Han H, Cui M, Wei L, Yang H, Shen J (2011) Enhancement effect of hematite nanoparticles on fermentative hydrogen production. Bioresour Technol 102(17):7903–7909
Hasan S (2015) A review on nanoparticles: their synthesis and types. Res J Recent Sci: ISSN 2277:2502
Herlekar M, Barve S, Kumar R (2014) Plant-mediated green synthesis of iron nanoparticles. J Nanopart 2014:1
Ho KM, Mao X, Gu L, Li P (2008) Facile route to enzyme immobilization: core−shell nanoenzyme particles consisting of well-defined poly (methyl methacrylate) cores and cellulase shells. Langmuir 24(19):11036–11042
Hu X, Yu JC, Gong J (2007) Fast production of self-assembled hierarchical α-Fe2O3 nanoarchitectures. J Phys Chem C 111(30):11180–11185
Huber DL (2005) Synthesis, properties, and applications of iron nanoparticles. Small 1(5):482–501
Hussain ST, Naheed R, Badshah A, Mehmood T (2009) Design and synthesis of nano heterogeneous supported catalysts for olefin polymerization. Afr J Pure Appl Chem 3(12):247–261
Hussein AK (2015) Applications of nanotechnology in renewable energies—A comprehensive overview and understanding. Renew Sust Energ Rev 42:460–476
Idriss H (2004) Ethanol reactions over the surfaces of noble metal/cerium oxide catalysts. Platin Met Rev 48(3):105–115
Iravani S (2011) Green synthesis of metal nanoparticles using plants. Green Chem 13(10):2638–2650
Ivanova G, Rákhely G, Kovács KL (2009) Thermophilic biohydrogen production from energy plants by *Caldicellulosiruptor saccharolyticus* and comparison with related studies. Int J Hydrog Energy 34(9):3659–3670
Ivanova V, Petrova P, Hristov J (2011) Application in the ethanol fermentation of immobilized yeast cells in matrix of alginate/magnetic nanoparticles, on chitosan-magnetite microparticles and cellulose-coated magnetic nanoparticles. arXiv preprint arXiv:1105.0619

Jia CJ, Sun LD, Yan ZG, You LP, Luo F, Han XD et al (2005) Single-crystalline iron oxide nanotubes. Angew Chem 117(28):4402–4407
John RP, Anisha GS, Nampoothiri KM, Pandey A (2011) Micro and macroalgal biomass: a renewable source for bioethanol. Bioresour Technol 102(1):186–193
Jordan J, Kumar CS, Theegala C (2011) Preparation and characterization of cellulase-bound magnetite nanoparticles. J Mol Catal B Enzym 68(2):139–146
Kaur M, Ali A (2011) Lithium ion impregnated calcium oxide as nano catalyst for the biodiesel production from karanja and jatropha oils. Renew Energy 36(11):2866–2871
Kesavan V, Dhar D, Koltypin Y, Perkas N, Palchik O, Gedanken A, Chandrasekaran S (2001) Nanostructured amorphous metals, alloys, and metal oxides as new catalysts for oxidation. Pure Appl Chem 73(1):85–91
Kidwai M, Bansal V, Saxena A, Shankar R, Mozumdar S (2006) Ni-nanoparticles: an efficient green catalyst for chemoselective reduction of aldehydes. Tetrahedron Lett 47(25):4161–4165
Kiran B, Kumar R, Deshmukh D (2014) Perspectives of microalgal biofuels as a renewable source of energy. Energy Convers Manag 88:1228–1244
Kumar P, Barrett DM, Delwiche MJ, Stroeve P (2009) Methods for pretreatment of lignocellulosic biomass for efficient hydrolysis and biofuel production. Ind Eng Chem Res 48(8):3713–3729
Kuo KK, Risha GA, Evans BJ, Boyer E (2003) Potential usage of energetic nano-sized powders for combustion and rocket propulsion. MRS Online Proc Libr 800
Laurent S, Forge D, Port M, Roch A, Robic C, Vander Elst L, Muller RN (2008) Magnetic iron oxide nanoparticles: synthesis, stabilization, vectorization, physicochemical characterizations, and biological applications. Chem Rev 108(6):2064–2110
Lee SM, Jin LH, Kim JH, Han SO, Na HB, Hyeon T et al (2010) β-Glucosidase coating on polymer nanofibers for improved cellulosic ethanol production. Bioprocess Biosyst Eng 33(1):141
Lu Y, Zhang YHP, Lynd LR (2006) Enzyme–microbe synergy during cellulose hydrolysis by Clostridium thermocellum. Proc Natl Acad Sci 103(44):16165–16169
Lupoi JS, Smith EA (2011) Evaluation of nanoparticle-immobilized cellulase for improved ethanol yield in simultaneous saccharification and fermentation reactions. Biotechnol Bioeng 108(12):2835–2843
Madhuvilakku R, Piraman S (2013) Biodiesel synthesis by TiO2–ZnO mixed oxide nanocatalyst catalyzed palm oil transesterification process. Bioresour Technol 150:55–59
Mahmood T, Hussain ST (2010) Nanobiotechnology for the production of biofuels from spent tea. Afr J Biotechnol 9(6):858–868
Mahmood T, Hussain ST, Malik SA (2010) New nanomaterial and process for the production of biofuel from metal hyper accumulator water hyacinth. Afr J Biotechnol 9(16):2381–2391
Mguni LL, Meijboom R, Jalama K (2012) Biodiesel production over nano-MgO supported on titania
Michalska K, Bizukojć M, Ledakowicz S (2015) Pretreatment of energy crops with sodium hydroxide and cellulolytic enzymes to increase biogas production. Biomass Bioenergy 80:213–221
Molina CMM (2013) ZnO nanorods as catalyst for biodiesel production from olive oil. University of Louisville, Louisville
Naik SN, Goud VV, Rout PK, Dalai AK (2010) Production of first and second generation biofuels: a comprehensive review. Renew Sust Energ Rev 14(2):578–597
Nidhin M, Indumathy R, Sreeram KJ, Nair BU (2008) Synthesis of iron oxide nanoparticles of narrow size distribution on polysaccharide templates. Bull Mater Sci 31(1):93–96
Nigam PS, Singh A (2011) Production of liquid biofuels from renewable resources. Prog Energy Combust Sci 37(1):52–68
Nizami AS, Rehan M, Waqas M, Naqvi M, Ouda OKM, Shahzad K et al (2017) Waste biorefineries: enabling circular economies in developing countries. Bioresour Technol 241:1101–1117
Osaka T, Matsunaga T, Nakanishi T, Arakaki A, Niwa D, Iida H (2006) Synthesis of magnetic nanoparticles and their application to bioassays. Anal Bioanal Chem 384:593–600
Palaniappan K (2017) An Overview of Applications of Nanotechnology in Biofuel Production. World Appl Sci J 35(8):1305–1311

Pan X, Fan Z, Chen W, Ding Y, Luo H, Bao X (2007) Enhanced ethanol production inside carbon-nanotube reactors containing catalytic particles. Nat Mater 6(7):507
Pavlidis IV, Vorhaben T, Gournis D, Papadopoulos GK, Bornscheuer UT, Stamatis H (2012) Regulation of catalytic behaviour of hydrolases through interactions with functionalized carbon-based nanomaterials. J Nanopart Res 14(5):842
Pejin D, Mojović LJ, Vučurović V, Pejin J, Denčić S, Rakin M (2009) Fermentation of wheat and triticale hydrolysates: A comparative study. Fuel 88(9):1625–1628
Prabu A (2017) Nanoparticles as additive in biodiesel on the working characteristics of a DI diesel engine. Ain Shams Eng J 9(4):2343–2349
Qiu F, Li Y, Yang D, Li X, Sun P (2011) Heterogeneous solid base nanocatalyst: preparation, characterization and application in biodiesel production. Bioresour Technol 102(5):4150–4156
Rai M, Da Silva SS (eds) (2017) Nanotechnology for bioenergy and biofuel production. Springer, Cham
Rai M, dos Santos JC, Soler MF, Marcelino PRF, Brumano LP, Ingle AP et al (2016) Strategic role of nanotechnology for production of bioethanol and biodiesel. Nanotechnol Rev 5(2):231–250
Rasheed R, Meera V (2016) Synthesis of Iron Oxide Nanoparticles Coated Sand by Biological Method and Chemical Method. Procedia Technol 24:210–216
Rawat R, Srivastava N, Chadha BS, Oberoi HS (2014) Generating fermentable sugars from rice straw using functionally active cellulolytic enzymes from Aspergillus niger HO. Energy Fuel 28(8):5067–5075
Reddy ANR, Saleh AA, Islam MS, Hamdan S, Maleque MA (2015) Biodiesel production from crude Jatropha oil using a highly active heterogeneous nanocatalyst by optimizing transesterification reaction parameters. Energy Fuel 30(1):334–343
Renewable Fuels Association (2015) World fuel ethanol in 2015. http://www.ethanolrfa.org/resources/industry/statistics/#1454098996479-8715d404-e546. Accessed June 2016
Rengasamy M, Anbalagan K, Kodhaiyolii S, Pugalenthi V (2016) Castor leaf mediated synthesis of iron nanoparticles for evaluating catalytic effects in transesterification of castor oil. RSC Adv 6(11):9261–9269
Salata OV (2004) Applications of nanoparticles in biology and medicine. J Nanobiotechnol 2(1):3
Salavati-Niasari M, Davar F, Mazaheri M, Shaterian M (2008) Preparation of cobalt nanoparticles from [bis(salicylidene)cobalt(II)]–oleylamine complex by thermal decomposition. J Magn Magn Mater 320(3–4):575–578
Satyanarayana KG, Mariano AB, Vargas JVC (2011) A review on microalgae, a versatile source for sustainable energy and materials. Int J Energy Res 35(4):291–311
Saxena RC, Adhikari DK, Goyal HB (2009) Biomass-based energy fuel through biochemical routes: a review. Renew Sust Energ Rev 13(1):167–178
Scenihr (2007) Scientific committee on emerging and newly identified health risks: the existing and proposed definitions relating to products of nanotechnologies
Shafiee S, Topal E (2009) When will fossil fuel reserves be diminished? Energy Policy 37(1):181–189
Shi W, Ma Z (2010) Amperometric glucose biosensor based on a triangular silver nanoprisms/chitosan composite film as immobilization matrix. Biosens Bioelectron 26(3):1098–1103
Shi W, Ma Z (2011) A novel label-free amperometric immunosensor for carcinoembryonic antigen based on redox membrane. Biosens Bioelectron 26(6):3068–3071
Singh P, Singh R, Borthakur A, Srivastava P, Srivastava N, Tiwary D, Mishra PK (2016) Effect of nanoscale TiO2-activated carbon composite on Solanum lycopersicum (L.) and Vigna radiata (L.) seeds germination. Energy Ecol Environ 1(3):131–140
Solans C, Izquierdo P, Nolla J, Azemar N, Garcia-Celma MJ (2005) Nano-emulsions. Curr Opin Colloid Interface Sci 10(3–4):102–110
Son SU, Lee SI, Chung YK, Kim SW, Hyeon T (2002) The First Intramolecular Pauson−Khand Reaction in Water Using Aqueous Colloidal Cobalt Nanoparticles as Catalysts. Org Lett 4(2):277–279
Srivastava N, Rawat R, Sharma R, Oberoi HS, Srivastava M, Singh J (2014a) Effect of nickel–cobaltite nanoparticles on production and thermostability of cellulases from newly isolated

thermotolerant Aspergillus fumigatus NS (class: Eurotiomycetes). Appl Biochem Biotechnol 174(3):1092–1103

Srivastava N, Rawat R, Oberoi HS (2014b) Application of thermostable cellulases in biofuels production. Recent Adv Bioenergy Res III:121–129. ISBN 978–81–927097-2-7

Srivastava N, Srivastava M, Mishra PK, Singh P, Ramteke PW (2015a) Application of cellulases in biofuels industries: an overview. J Biofuels Bioenergy 1(1):55–63

Srivastava N, Rawat R, Singh Oberoi H, Ramteke PW (2015b) A review on fuel ethanol production from lignocellulosic biomass. Int J Green Energy 12(9):949–960

Srivastava N, Singh J, Ramteke PW, Mishra PK, Srivastava M (2015c) Improved production of reducing sugars from rice straw using crude cellulase activated with Fe3O4/Alginate nanocomposite. Bioresour Technol 183:262–266

Srivastava N, Srivastava M, Mishra PK, Ramteke PW (2016) Application of ZnO nanoparticles for improving the thermal and pH stability of crude cellulase obtained from *Aspergillus fumigatus* AA001. Front Microbiol 7:514

Srivastava N, Srivastava M, Mishra PK, Singh P, Pandey H, Ramteke PW (2017a) Nanoparticles for biofuels production from lignocellulosic waste. In: Nanoscience in food and agriculture, vol 4. Springer, Cham, pp 263–278

Srivastava N, Srivastava M, Manikanta A, Singh P, Ramteke PW, Mishra PK (2017b) Nanomaterials for biofuel production using lignocellulosic waste. Environ Chem Lett 15(2):179–184

Staniszewski M, Kujawski W, Lewandowska M (2007) Ethanol production from whey in bioreactor with co-immobilized enzyme and yeast cells followed by pervaporative recovery of product–Kinetic model predictions. J Food Eng 82(4):618–625

Statista, The statistic portal (2016a) Biodiesel production in the European Union from 2010 to 2025 (in million gallons). http://www.statista.com/statistics/202236/eu-biodiesel-productionfrom-2010/. Accessed June 2016

Statista, The statistic portal (2016b) The world's biggest biodiesel producers in 2014, by country (in billion liters). http://www.statista.com/statistics/271472/biodiesel-production-in-selectedcountries/. Accessed June 2016

Sun Y, Cheng J (2002) Hydrolysis of lignocellulosic materials for ethanol production: a review. Bioresour Technol 83(1):1–11

Sun S, Zeng H (2002) Size-controlled synthesis of magnetite nanoparticles. J Am Chem Soc 124(28):8204–8205

Suslick KS (1990) Sonochemistry. Science 247(4949):1439–1445

Svoboda K, Slowinski G, Rogut J, Baxter D (2007) Thermodynamic possibilities and constraints for pure hydrogen production by iron based chemical looping process at lower temperatures. Energy Convers Manag 48(12):3063–3073

Tahvildari K, Anaraki YN, Fazaeli R, Mirpanji S, Delrish E (2015) The study of CaO and MgO heterogenic nano-catalyst coupling on transesterification reaction efficacy in the production of biodiesel from recycled cooking oil. J Environ Health Sci Eng 13(1):73

Tan KT, Lee KT, Mohamed AR (2008) Role of energy policy in renewable energy accomplishment: the case of second-generation bioethanol. Energy Policy 36(9):3360–3365

Teja AS, Koh PY (2009) Synthesis, properties, and applications of magnetic iron oxide nanoparticles. Prog Cryst Growth Charact Mater 55(1–2):22–45

Tischer W, Wedekind F (1999) Immobilized enzymes: methods and applications. In: Biocatalysis-from discovery to application. Springer, Berlin/Heidelberg, pp 95–126

Ubale AU, Belkhedkar MR (2015) Size dependent physical properties of nanostructured α-Fe2O3 thin films grown by successive ionic layer adsorption and reaction method for antibacterial application. J Mater Sci Technol 31(1):1–9

Verma ML, Barrow CJ (2015) Recent advances in feedstocks and enzyme-immobilised technology for effective transesterification of lipids into biodiesel. In: Microbial Factories. Springer, New Delhi, pp 87–103

Verma M, Godbout S, Brar SK, Solomatnikova O, Lemay SP, Larouche JP (2012) Biofuels production from biomass by thermochemical conversion technologies. Int J Chem Eng 2012:1

Verma ML, Barrow CJ, Puri M (2013a) Nanobiotechnology as a novel paradigm for enzyme immobilisation and stabilisation with potential applications in biodiesel production. Appl Microbiol Biotechnol 97(1):23–39

Verma ML, Chaudhary R, Tsuzuki T, Barrow CJ, Puri M (2013b) Immobilization of β-glucosidase on a magnetic nanoparticle improves thermostability: application in cellobiose hydrolysis. Bioresour Technol 135:2–6

Verziu M, Cojocaru B, Hu J, Richards R, Ciuculescu C, Filip P, Parvulescu VI (2008) Sunflower and rapeseed oil transesterification to biodiesel over different nanocrystalline MgO catalysts. Green Chem 10(4):373–381

Viikari L, Alapuranen M, Puranen T, Vehmaanperä J, Siika-Aho M (2007) Thermostable enzymes in lignocellulose hydrolysis. In: Biofuels. Springer, Berlin/Heidelberg, pp 121–145

Vijayakumar R, Koltypin Y, Felner I, Gedanken A (2000) Sonochemical synthesis and characterization of pure nanometer-sized Fe3O4 particles. Mater Sci Eng A 286(1):101–105

Wang J, Wan W (2008) Effect of Fe2+ concentration on fermentative hydrogen production by mixed cultures. Int J Hydrog Energy 33(4):1215–1220

Wang J, Sun J, Sun Q, Chen Q (2003) One-step hydrothermal process to prepare highly crystalline Fe3O4 nanoparticles with improved magnetic properties. Mater Res Bull 38(7):1113–1118

Wang Y, Wang X, Tang R, Yu S, Zheng B, Feng Y (2010) A novel thermostable cellulase from Fervidobacterium nodosum. J Mol Catal B Enzym 66(3–4):294–301

Wang H, Covarrubias J, Prock H, Wu X, Wang D, Bossmann SH (2015) Acid-functionalized magnetic nanoparticle as heterogeneous catalysts for biodiesel synthesis. J Phys Chem C 119(46):26020–26028

Waqas M, Aburiazaiza AS, Miandad R, Rehan M, Barakat MA, Nizami AS (2018) Development of biochar as fuel and catalyst in energy recovery technologies. J Clean Prod 188:477–488

Wei Y, Li X, Yu L, Zou D, Yuan H (2015) Mesophilic anaerobic co-digestion of cattle manure and corn stover with biological and chemical pretreatment. Bioresour Technol 198:431–436

Wen L, Wang Y, Lu D, Hu S, Han H (2010) Preparation of KF/CaO nanocatalyst and its application in biodiesel production from Chinese tallow seed oil. Fuel 89(9):2267–2271

Wolf EL, Medikonda M (2012) Understanding the nanotechnology revolution. Wiley, Weinheim

Wu N, Fu L, Su M, Aslam M, Wong KC, Dravid VP (2004) Interaction of fatty acid monolayers with cobalt nanoparticles. Nano Lett 4(2):383–386

Wu W, He Q, Jiang C (2008) Magnetic iron oxide nanoparticles: synthesis and surface functionalization strategies. Nanoscale Res Lett 3(11):397

Xie W, Ma N (2009) Immobilized lipase on Fe3O4 nanoparticles as biocatalyst for biodiesel production. Energy Fuel 23(3):1347–1353

Xie W, Wang J (2012) Immobilized lipase on magnetic chitosan microspheres for transesterification of soybean oil. Biomass Bioenergy 36:373–380

Yang WS, Gao HY, Xiang HW, Yin DH, Yang Y, Yang JL, Li YY (2001) Cobalt supported mesoporous silica catalyst for Fischer-Tropsch synthesis. Acta Chim Sin 59(11):1870–1877

Yang Z, Huang R, Qi W, Tong L, Su R, He Z (2015) Hydrolysis of cellulose by sulfonated magnetic reduced graphene oxide. Chem Eng J 280:90–98

Yearbook Enerdata (2015a) Global energy statistical yearbook 2015. https://yearbook.enerdata.net/renewable-in-electricity-production-share-by-region.html#energy-primary-production.Html

Yearbook Enerdata (2015b) Global energy statistical yearbook 2015. https://yearbook.enerdata.net/renewable-in-electricity-production-share-by-region.html#renewable-in-electricity-productionshare-by-region.html. Accessed June 2016

Yearbook Enerdata (2015c) Global energy statistical yearbook 2015. https://yearbook.enerdata.net/renewable-in-electricity-production-share-by-region.html#wind-solar-share-electricityproduction.html. Accessed June 2016

Yeoman CJ, Han Y, Dodd D, Schroeder CM, Mackie RI, Cann IK (2010) Thermostable enzymes as biocatalysts in the biofuel industry. In: Advances in applied microbiology, vol 70. Academic Press, New York, pp 1–55

Zhang Y, Shen J (2007) Enhancement effect of gold nanoparticles on biohydrogen production from artificial wastewater. Int J Hydrog Energy 32(1):17–23

Zhang C, Zhang R, Li X, Li Y, Shi W, Ren X, Xu X (2011) Bench-scale fluidized-bed fast pyrolysis of peanut shell for bio-oil production. Environ Prog Sustain Energy 30(1):11–18

Zhao X, Zhang L, Liu D (2012) Biomass recalcitrance. Part I: the chemical compositions and physical structures affecting the enzymatic hydrolysis of lignocellulose. Biofuels Bioprod Biorefin 6(4):465–482

Zhao W, Zhang Y, Du B, Wei D, Wei Q, Zhao Y (2013) Enhancement effect of silver nanoparticles on fermentative biohydrogen production using mixed bacteria. Bioresour Technol 142:240–245

Zheng YH, Cheng Y, Bao F, Wang YS (2006) Synthesis and magnetic properties of Fe3O4 nanoparticles. Mater Res Bull 41(3):525–529

Correction to: Application of Plant-Based Natural Product to Synthesize Nanomaterial

Mohammad Khajeh Mehrizi and Zahra Shahi

Correction to:
Chapter 3 in: M. Srivastava et al. (eds.),
Nanomaterials in Biofuels Research,
Clean Energy Production Technologies,
https://doi.org/10.1007/978-981-13-9333-4_3

The book was inadvertently published without including the name of the second author (Zahra Shahi) in Chapter 3. This has now been corrected.

The updated online version of this chapter can be found at
https://doi.org/10.1007/978-981-13-9333-4_3

M. Srivastava et al. (eds.), *Nanomaterials in Biofuels Research*,
Clean Energy Production Technologies, https://doi.org/10.1007/978-981-13-9333-4_12